EMC in Railways

Electromagnetic Field Measurement

EMC in Railways

Electromagnetic Field Measurement

First Edition

Andrea Mariscotti

ASTM

Image used for cover: Anthony Ivanoff, 2008.

ASTM Analysis, Simulation, Test and Measurement Sagl
Via Comacini, 7 - 6830 Chiasso (Switzerland)

Printed by CreateSpace

ISBN 978-88-941091-1-5

All change is not growth, as all movement is not forward.

Ellen Glasgow

Preface

A modern railway system may be a very complex system, with a large extension in space and a deep impact on the territory and existing installations. Many subsystems interact with more or less evident EMC interfaces spanning various coupling means and configurations for sources and victims. Although analysis and simulation are nowadays extensively used since system design, for several reasons experimental assessment and verification still play a major role: complexity of phenomena does not lend easily to modeling and simulation, provisions and countermeasures need to be verified and compared, or, as it is commonly the case, applicable standards and regulations require so. The aim of this book is the definition of the correct procedures for the preparation and execution of tests and measurements on-site for the assessment of EMC interfaces or phenomena in railway transportation systems, addressing problems of theoretical and practical nature. The attention is focused on methods of measurement and characterization of electromagnetic field emissions, that involve subsystems and third parties at various extent: emissions to outside world for railway system certification, disturbance to radio communication systems and sensitive scientific and medical equipment, human exposure.

Many elements characterize a test or measurement: adequate instrumentation must be selected, not only for performance, but also for suitability to real conditions; the test setup must be clearly described and as simple as possible; operating conditions shall be defined from different viewpoints of driver, test operator and site manager; the test procedure must be unambiguous and detailed enough for the different purposes; the means of verification of operations and conditions need to be identified and understood, as well as methods for processing and evaluation of results. A correct, unambiguous, exhaustive and complete test procedure requires the knowledge of many elements, requisites and exigencies: materials and equipment, measurement theory and uncertainty, personnel, impact on circulation and safety. It will sound strange, but even mature standards focusing on a well defined EMC interface or phenomenon may fail to give a complete description and characterization of the problem, leaving out important details.

This book is for engineers and technicians in charge of on-site EMC tests and is conceived to address the preparation and the execution of tests, as well as the interpretation of results, including the quantification of uncertainty.

Feedback on errors, questions, as well as suggestions for new problems and references are warmly welcome.

Chiasso, April 2016

Do, or do not. There is no try.

–Star Wars, The Empire strikes again

Contents

II Electromagnetic field emissions   175

Part I

Theory and background

1

Electromagnetic theory fundamentals

Electromagnetic theory plays a central role to the field of electromagnetic compatibility, it is therefore important that a basic understanding of the electromagnetic theory and its application in railway electromagnetic compatibility be treated in context. It is not the intention of this chapter to cover in detail electromagnetic as this is covered in various classic textbooks; however in this chapter an overview of electromagnetic theory is presented with the view of enhancing the understanding of analytical and theoretical aspects of electromagnetic compatibility in railway applications.

1.1 Transmission lines

There are many textbooks dealing with the subject in a more or less detailed and complete way: this section contains the useful relationships and quantities, starting from operating principle and physical explanation. Telegrapher's equations are the starting point, with the intuitive approach that may be found in [35, 158] or the more formal one, for example appearing in [123]. A detailed, yet concise, treatment may be found in [99], where also multi-port transmission lines and mixed-mode scattering parameters are discussed and applied.

1.1.1 Telegrapher's equations

Let's consider an infinitesimal section of line of length $\mathrm{d}z$ at the longitudinal position z. Starting from the p.u.l. parameters r, l, c and g, assumed constant over the entire transmission line, we may say that this section has a series resistance $r\,\mathrm{d}z$, series inductance $l\,\mathrm{d}z$, shunt capacitance $c\,\mathrm{d}z$ and shunt conductance $g\,\mathrm{d}z$.

For the most general approach, transmission line equations must be in time domain, to account for transients and arbitrary waveforms. Using instantaneous quantities for voltage and current at the left and right ports of the section, Kirchhoff equations give

$$v(z+\mathrm{d}z,t)-v(z,t)=-r\,\mathrm{d}z\,i(z,t)-l\,\mathrm{d}z\,\frac{\partial i(z,t)}{\partial t} \tag{1.1.1}$$

$$i(z+\mathrm{d}z,t)-i(z,t)=-g\,\mathrm{d}z\,v(z,t)-c\,\mathrm{d}z\,\frac{\partial v(z,t)}{\partial t} \tag{1.1.2}$$

Since the infinitesimal section has the shunt elements on the right, one may argue that the voltage and its derivative that produce the current in the second equation is in reality $v(z+\mathrm{d}z,\,t)$. The objection is, first, that we couldn't do otherwise to derive a symmetric set of equations, and, second, that the term $v(z+\mathrm{d}z,\,t)$ may be developed in Taylor's series and all the higher order terms would be anyway negligible.

Dividing by $\mathrm{d}z$ and letting $\mathrm{d}z$ approach zero gives the well known set of linear partial differential equations

$$\frac{\partial v(z,t)}{\partial t}=-r\,i(z,t)-l\,\frac{\partial i(z,t)}{\partial t} \tag{1.1.3}$$

$$\frac{\partial i(z,t)}{\partial t}=-g\,v(z,t)-c\,\frac{\partial v(z,t)}{\partial t} \tag{1.1.4}$$

whose solution will explicit v and i in terms of z and t, once the boundary conditions are known, that is the characteristics of source at $z=0$ and load at $z=L$.

If the line section is analyzed in the frequency domain, then phasors $V(z)$ and $I(z)$ take the place of the instantaneous quantities $v(z,t)$ and $i(z,t)$, and similarly

$$\frac{\mathrm{d}V(z)}{\mathrm{d}z}=-(r+j\omega l)\,I(z) \qquad \frac{\mathrm{d}I(z)}{\mathrm{d}z}=-(g+j\omega c)\,V(z) \tag{1.1.5}$$

The solution of equations (1.1.5) leads to second order differential equations that admit a time-harmonic integral of the form

$$V(z)=V_{\mathrm{f}}\exp(-\gamma z)+V_{\mathrm{b}}\exp(+\gamma z) \tag{1.1.6}$$

$$I(z)=I_{\mathrm{f}}\exp(-\gamma z)+I_{\mathrm{b}}\exp(+\gamma z) \tag{1.1.7}$$

The solution indicates that there exist two voltage waves (and two current waves) traveling along the transmission line, the leftmost one in the direction of increasing z and the rightmost one in that of decreasing z: they are usually named *progressive wave* and *regressive wave*, or *forward wave* and *backward wave*.

1.1.1.1 Propagation, attenuation and phase constants

The quantity γ, named *propagation constant*, is derived from the p.u.l. parameters as

$$\gamma = \sqrt{(r+j\omega l)(g+j\omega c)} \tag{1.1.8}$$

and is in general a complex number. Its real and imaginary parts

$$\gamma = \alpha + j\beta \tag{1.1.9}$$

are named *attenuation constant* and *phase constant* respectively (or better, especially for α, attenuation coefficient and phase propagation coefficient, since the former is in reality far from being constant because of skin effect and other power loss effects), and their function is clarified if (1.1.9) is substituted into (1.1.6):

$$V(z) = V_f \exp(-\alpha z)\exp(-j\beta z) + V_b \exp(+\alpha z)\exp(+j\beta z) \tag{1.1.10}$$

Going to the instantaneous voltage applying the $\operatorname{Re}\{\cdot\}$ operator

$$\begin{aligned} v(z,t) = \ & |V_f|\exp(-\alpha z)\operatorname{Re}\{\exp(j(\omega t - \beta z + \varphi_1))\} + \\ & |V_b|\exp(+\alpha z)\operatorname{Re}\{\exp(j(\omega t + \beta z + \varphi_2))\} \end{aligned} \tag{1.1.11}$$

the amplitude (given by the peak values $|V_f|$ and $|V_b|$ multiplied by the attenuation terms $\exp(-\alpha z)$ and $\exp(+\alpha z)$) and the orientation (given by the combination of angular pulsation ωt, initial phase φ_1 and φ_2, and phase rotation $\pm\beta z$) can be separated.

The explicit expression of α and β with respect to line per-unit-length parameters is not simple and is not given in textbooks. A low frequency approximation exists

$$\alpha = \frac{1}{2}\sqrt{lc}\left(\frac{r}{l} + \frac{g}{c}\right) = \frac{1}{2}\left(rY_c + gZ_c\right) \tag{1.1.12}$$

$$\beta = \omega\sqrt{lc} \tag{1.1.13}$$

For α we see the usual distinction of conduction loss (the first term depends on conductor series resistance r) and dielectric loss (the second term contains the shunt dielectric conductance g).

For β a more complete form exists that includes terms up to the second order in ω and that is valid for almost all cases:

$$\beta = \omega\sqrt{lc}\left[1+\frac{r^2}{8\omega^2 l^2}-\frac{rg}{4\omega^2 lc}+\frac{g^2}{8\omega^2 C^2}\right]=\omega\sqrt{lc}\left[1+\frac{1}{2}\left(\frac{r}{2\omega l}-\frac{g}{2\omega c}\right)^2\right] \qquad (1.1.14)$$

In this expression terms r and g appear and make β slightly dependent on losses: by grouping and isolating line losses, a compact expression is obtained where the term $\beta^0 = \omega\sqrt{lc}$ stand for the value valid for a lossless line,

$$\beta = \beta^0\left[1+\frac{1}{2}\left(\frac{\alpha_c}{\beta^0}-\frac{\alpha_d}{\beta^0}\right)^2\right] \qquad (1.1.15)$$

and analogously the wavelength

$$\lambda = \lambda^0\left[1-\frac{1}{2}\left(\frac{\alpha_c}{\lambda^0}-\frac{\alpha_d}{\lambda^0}\right)^2\right] \qquad (1.1.16)$$

The line is slightly "slower" in general if losses are considered, but there exists one case in which losses have no impact on the propagation, that is when

$$\frac{r}{\omega l}=\frac{g}{\omega c} \qquad (1.1.17)$$

also known as Heaviside's relationship.

The propagation constant may also be expressed as amplitude $|\gamma|$ and phase $\angle\gamma$, that have an obvious relationship with α and β, and may also be related back to the per-unit-length parameters of the line:

$$|\gamma| = \sqrt{\sqrt{(r^2+\omega^2 l^2)}+\sqrt{(g^2+\omega^2 c^2)}} \qquad (1.1.18)$$

$$\angle\gamma = \frac{1}{2}\left[\arctan\left(\frac{\omega l}{r}\right)+\arctan\left(\frac{\omega c}{g}\right)\right] \qquad (1.1.19)$$

The measuring unit of α and β is 1/m, since both appears multiplied by z in the exponential argument, that must be dimensionless. Yet, a distinction is needed. It was felt to be convenient to find a unit for the dimensionless argument αz, that expresses the attenuation, and this unit for α is the neper/m: so, a voltage or current is said to undergo an attenuation of X nepers[1] when its magnitude changes by a factor $\exp(-X)$. The phase constant β analogously is measured in radians/m; by definition of wavelength λ, as the length in space traveled by a sinusoidal wave after a phase rotation of 2π, the known relationship $\beta = 2\pi/\lambda$ derives.

---

[1] The name "neper" derives from the latinization of Napier, the Scottish mathematician who invented the logarithms.

1.1.1.2 Line velocity and propagation time

Let's consider a "point" at the longitudinal position z_0 and time t_0 and the voltage of the first forward wave; at a slightly later instant of time $t_0 + dt$ the point has moved from z_0 to $z_0 + dz$ and the voltage value in z_0 is no longer the same. By neglecting attenuation of amplitude, and thus equating the angular positions to have (z_0, t_0) and $(z_0 + dz, t_0 + dt)$ points correspond with a simple horizontal shift, we may write $\omega t_0 - \beta z_0 + \varphi_1 = \omega(t_0 + dt) - \beta(z_0 + dz) + \varphi_1$, that leads to $\omega dt - \beta dz = 0$: the velocity, represented by the ratio dz/dt, with dz and dt approaching zero, is called *phase velocity*.

$$v_{ph} = \frac{\omega}{\beta} \tag{1.1.20}$$

The same result is obtained for the backward wave, except for the sign of the direction of propagation. So, the wave looks still to an observer moving at its velocity; if the observer is fixed in space observing a longitudinal position, the time variation of the instantaneous voltage is sinusoidal.

There is another type of velocity, named *group velocity*, that refers to several frequency components (e.g. part of a radio transmission, such as the modulated signal transmitted over a physical radio channel) that have not the same phase velocity. This case is however not of interest here and for non-dispersive lines the two velocity quantities are equal [99].

For a transmission line of length L the time for the wave to travel is called propagation delay, propagation time, or traveling time, and is given by length divided by velocity:

$$\tau = L/v_{ph} \tag{1.1.21}$$

1.1.2 The Characteristic Impedance

There is a beautiful demonstration of what the characteristic impedance is in [77], p. 142. Starting from the definition of the propagation delay, above in (1.1.21), they imagine to charge the line capacitance with a step voltage of amplitude V and evaluate the amount of current flowing in the line. The step voltage is applied and propagates along the line; let's consider 1 m of line, so that the capacitance C is exactly the p.u.l. line capacitance c, the charge across it equals the capacitance C multiplied by the voltage difference, that is equal to V (since the step hasn't "arrived" yet ahead of 1 m). The time during which that line section is charged is the propagation delay multiplied by the length, in our case 1 m; the total charge divided by that time gives the current

$$I = \frac{cV}{\sqrt{lc}} = V/\sqrt{\frac{l}{c}} \tag{1.1.22}$$

and the last term is right the characteristic impedance!

Thus, the characteristic impedance Z_c is ideally calculated from a uniform line with no reflected waves ($V_2 = 0$ and $I_2 = 0$), by taking the ratio of the voltage and current waves, and it may be demonstrated that this ratio is constant at any point along the transmission line.

By taking (1.1.10) with only the forward wave for simplicity and by differentiating with respect to z

$$\frac{\mathrm{d}V}{\mathrm{d}z} = -(\alpha + j\beta)V_1 \exp(-\alpha z)\exp(-j\beta z) \qquad (1.1.23)$$

The direct comparison with (1.1.5) gives

$$Z_c = \frac{V}{I} = \frac{V_1}{I_1} = \frac{r + j\omega l}{\alpha + j\beta} = \sqrt{\frac{r + j\omega l}{g + j\omega c}} = \sqrt{\frac{Z_{\text{series}}}{Y_{\text{shunt}}}} \qquad (1.1.24)$$

As seen Z_c may be expressed as the square root of the ratio of the series impedance and the shunt admittance terms, and that will be used soon in sec. 1.1.4.

The quantity Z_c has the dimension of an impedance and it is a characteristic of the line behavior (neglecting reflections), so that it was named *characteristic impedance*. Its dependency on frequency is twofold: direct, since numerator and denominator have reactive terms multiplied by $j\omega$, and indirect, since the p.u.l. parameters may depend on frequency. If the resistance r and the conductance g are considered negligible (as it is, for example, for high quality coaxial cables), then the characteristic impedance becomes a pure resistance $Z_c = \sqrt{l/c}$. At low frequency, when the resistive part at the numerator is relevant if compared with the inductive reactance term on the right, we say that the line operates with a RC behavior; at higher frequency the characteristic impedance becomes constant. The "critical" frequency that distinguishes these two intervals is given by $f_c = r/2\pi l$.

If one is interested in keeping line losses low, of course, the longitudinal resistance r shall be low (including skin effect) and the conductance term g shall be low (that is high-quality dielectrics shall be used, like for transmission lines in air). For the vast majority of cables the attenuation will follow a curve that is dictated by the skin effect, so with a square-root dependency on frequency.

At low frequency the characteristic impedance is thus not real valued as we may expect, but has a non negligible imaginary component, that depends on loss parameters, resistance and conductance. The variability of the characteristic impedance for traction lines was investigated in [93] and the result of the scattering of characteristic impedance decomposed in real and imaginary parts for randomly varying line parameters is shown in Figure 1.1.1. A traction line has several non-idealities, related to the heavy skin effect in the running rails and shunt conductance and capacitance influenced by the rails supports. The large value of the characteristic impedance at low frequency due to the preponderant imaginary part is of no consequence, being no exigency of impedance matching at low frequency. The characteristic impedance has an asymptotic behavior with the imaginary part vanishing above the critical frequency f_c (this is the macroscopic variation of the characteristic impedance).

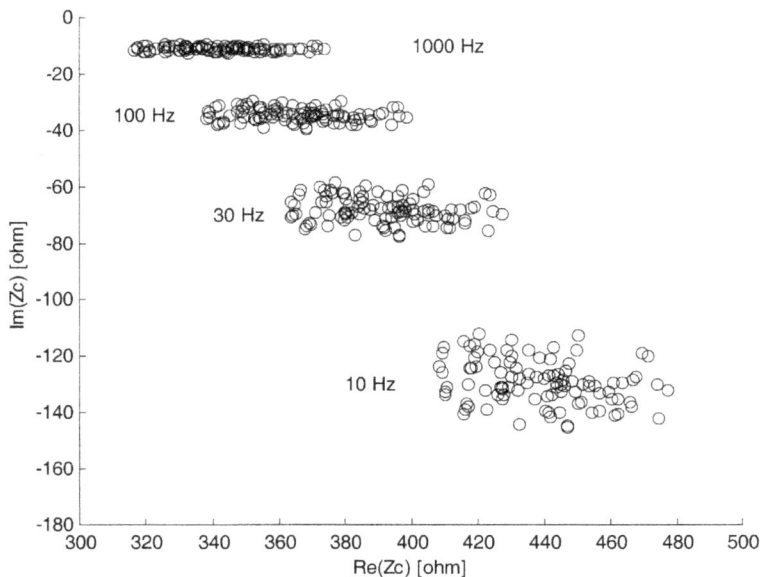

Figure 1.1.1 – Scatter plot of $\mathrm{Re}\{Z_c\}$ and $\mathrm{Im}\{Z_c\}$ at 10, 30, 100, and 1000 Hz for traction line p.u.l. parameters r, l and c uniformly distributed.

Additionally, there is frequency dependency also for the inductance (due to skin effect and to the variation of internal inductance[2]) and capacitance (slight reduction of the dielectric permittivity and increase of dielectric losses). The reduction of the inductance with frequency makes the characteristic impedance slightly increase; for high permeability materials magnetic losses also increase with frequency and this adds an imaginary component to the already stabilized characteristic impedance value; analogously and more commonly, the same effect occurs for the increase of dielectric losses, that never ceases nor saturates with frequency.

It is finally commented that the characteristic impedance is a somewhat intangible entity, in that it cannot be measured directly with a RCL bridge and its resistive part is not dissipative. It may be determined indirectly by some impedance measurements (at least two), as it will be seen shortly (see sec. 1.1.4).

The line velocity v_0 (that corresponds to the *phase velocity* $v_{ph.}$) also can be expressed in p.u.l. parameters, and, again under the assumption of negligible r and g parameters, the known formula is derived:

$$v_{ph.} = 1/\sqrt{lc} \tag{1.1.25}$$

---

[2]   See A. Mariscotti and P. Pozzobon, "Resistance and Internal Inductance of traction rails: a survey", *IEEE Transactions on Vehicular Technology*, Vol. 53, no. 4, July 2004, pp. 1069-1075, for a discussion of skin effect and internal parameters for heavy ferromagnetic conductors, i.e. running rails.

1.1.3   Input impedance and reflection coefficient

The *input impedance* of a transmission line is the ratio of voltage and current at the input port, that is $Z_{in} = V_1/I_1$, so that for a non-reflective transmission line the input impedance and the characteristic impedance coincide. Since it was said that the ratio V/I is constant over the entire line, at the far end $z = L$, the same relationship holds for the voltage and current phasors across the terminating load Z_t, so that $Z_t = Z_0$ becomes the condition for a non-reflecting transmission line termination (this condition is often referred to as *matched load condition*).

In general, the input impedance and the matching conditions at either end (source and/or load side) may be evaluated in the following way.

By differentiation of (1.1.10) with respect to z, and writing the equality with (1.1.5), the following expression for the current I is derived:

$$I(z) = \frac{\gamma}{r+j\omega l} V_f \exp(-\gamma z) - \frac{\gamma}{r+j\omega l} V_b \exp(+\gamma z) \qquad (1.1.26)$$

where the term $\gamma/(r+j\omega l)$ is equal to $1/Z_0$. The phasor current coefficients for the forward and backward waves are thus related to the corresponding coefficients of the phasor voltage by

$$I_f = \frac{V_f}{Z_0} \qquad I_b = -\frac{V_b}{Z_0} \qquad (1.1.27)$$

In the general case of possibly mismatched source and load impedances, the impedance at any point along the line is given by

$$Z(z) = Z_0 \frac{V_f \exp(-\gamma z) + V_b \exp(+\gamma z)}{V_f \exp(-\gamma z) - V_b \exp(+\gamma z)} \qquad (1.1.28)$$

Let's assume now that at the end of the line a terminating impedance Z_t is connected; when the traveling wave arrives at $z = L$ where the terminating impedance is connected, $V(L)$ and $I(L)$ shall be such that their ratio is equal to Z_t and (1.1.28) shall be satisfied for all positions and in particular for $z = L$. The same holds for the wave traveling in the opposite direction that meets the terminating impedance at $z = 0$: in this case calculations are much easier because the exponents are all equal to one and we adopt this configuration to conclude the reasoning.

$$Z_t = \frac{V(0)}{I(0)} = Z_0 \frac{V_f + V_b}{V_f - V_b} \qquad (1.1.29)$$

By solving this equation for V_f and V_b it is possible to obtain a similar relationship that gives the ratio of the forward and backward waves as a function of Z_t and Z_0:

$$\frac{V_b}{V_f} = \frac{Z_t - Z_0}{Z_t + Z_0} \qquad (1.1.30)$$

This quickly demonstrates where the reflection comes from: a traveling wave, called *incident wave*, with its amplitude set by the characteristic impedance of the line approaching a longitudinal position where a terminating impedance is connected, give rise to a forward and a backward wave, whose amplitude is determined by the expression above.

The *reflection coefficient*, either indicated by ρ or Γ, is defined as the ratio of the values of the reflected wave, going in the direction of decreasing z and associated to the term $\exp(+\gamma z)$, and of the incident wave, going in the direction of increasing z and associated to the term $\exp(-\gamma z)$. At the load port $z = L$, the reflection coefficient Γ_t is

$$\Gamma_t = (V_b/V_f)\exp(2\gamma L) \tag{1.1.31}$$

and is complex. Taking (1.1.28) at $z = L$ and dividing all the terms by the same quantity $(V_f \exp(-\gamma z))$, it is possible to isolate the reflection coefficient at the load as

$$\frac{Z_t}{Z_0} = \frac{1+\Gamma_t}{1-\Gamma_t} \tag{1.1.32}$$

that solved for Γ_t gives

$$\Gamma_t = \frac{Z_t - Z_0}{Z_t + Z_0} \tag{1.1.33}$$

For some extreme terminating conditions physical reasoning also helps in finding the value of the reflection coefficient. If $Z_t = 0$, that is a short circuit, then the incident and reflected voltage wave must be identically equal at $z = L$, since their sum must be zero; the reflection coefficient is thus $\Gamma_t = -1$. On the contrary, if the line is terminated on an open circuit, $Z_t = \infty$, the same reasoning should be applied to the phasor current waves; the equation above confirms that the reflection coefficient is $\Gamma_t = +1$. If Z_t is a generic reactive impedance the modulus of the reflection coefficient is still unity, but the phase is variable; this is reasonable observing that a reactive impedance is not dissipative, cannot absorb power from the incident wave and is thus totally reflecting.

Correspondingly, it is sometimes defined a *transmission coefficient* T_t that complements the reflection coefficient: $T_t = 1 + \Gamma_t$. The transmission coefficient gives the amount of the incident wave transmitted beyond an interface (e.g. between two line sections). Just to summarize, the reflection coefficient is always calculated as difference divided by sum of the two characteristic impedances, taking as first the one for the line section ahead where the wave is directed; the same applies if one of the line sections is replaced by a lumped circuit. Reversing the direction changes the sign of the reflection coefficient. Following the example above with a terminating Z_t impedance

$$T_t = 1 + \Gamma_t = \frac{2Z_t}{Z_t + Z_0} \tag{1.1.34}$$

Dividing (1.1.28) by V_1 and substituting the ratio V_2/V_1 rewriting (1.1.31), the line impedance $Z(z)$ is

$$Z(z) = Z_0 \frac{\exp(-\gamma z) + \Gamma_t \exp(-2\gamma L)\exp(+\gamma z)}{\exp(-\gamma z) - \Gamma_t \exp(-2\gamma L)\exp(+\gamma z)} \tag{1.1.35}$$

Multiplying all the terms at numerator and denominator by $\exp(\gamma L)$, the expression is put in a form where the characteristic impedance, the reflection coefficient at the terminating load and the distance $x = L - z$ between the point z and the load at $z = L$ are the sole parameters used.

$$Z(x) = Z_0 \frac{\frac{Z_t}{Z_0}[\exp(+\gamma x) + \exp(-\gamma x)] + [\exp(+\gamma x) - \exp(-\gamma x)]}{[\exp(+\gamma x) + \exp(-\gamma x)] + \frac{Z_t}{Z_0}[\exp(+\gamma x) - \exp(-\gamma x)]} \tag{1.1.36}$$

The input impedance at $z = 0$, that is $x = L$, may be readily obtained, where the complex exponential is replaced by the hyperbolic tangent:

$$Z_{in} = Z_0 \frac{\frac{Z_t}{Z_0} + \tanh(\gamma L)}{1 + \frac{Z_t}{Z_0}\tanh(\gamma L)} = Z_0 \frac{Z_t + Z_0\tanh(\gamma L)}{Z_0 + Z_t\tanh(\gamma L)} \tag{1.1.37}$$

When the line is terminated with two extreme cases of short circuit ($\Gamma_t = -1$) and open circuit ($\Gamma_t = +1$), the input impedance expressions are as follows:

$$Z_{in,sc} = jZ_0\tan(\gamma L) \tag{1.1.38}$$

$$Z_{in,oc} = -jZ_0\cot(\gamma L) \tag{1.1.39}$$

When an integer multiple n of half the wavelength $\lambda/2$ fits line length $L = n\lambda/2$, then the terminating impedance Z_t is seen directly at the line input port, as if there is no line in between, because the term $\tanh(\cdot) = 0$ and vanishes.

When an odd integer multiple $(2n+1)$ of a quarter wavelength $\lambda/4$ fits line length $L = (2n+1)\lambda/4$, then it is said that the line is resonating in quarter-wave conditions and the input impedance becomes

$$Z_{in(\lambda/4)} = \frac{Z_0^2}{Z_t} \tag{1.1.40}$$

The line behaves like a transformer (quarter-wave transformer), so that a large (e.g. larger than Z_0) Z_t value at the far end is transformed into a small impedance value, proportionally smaller than Z_0. This of course works for a narrow frequency interval around the $\lambda/4$ condition. Quarter-wave matching is achieved when the line characteristic impedance is equal to geometric mean (square root of product) of the terminating impedance Z_t at the far end and the desired impedance value at the input Z_{in}^* (e.g.

equal to that of the connected source, or to the complex conjugate of it, depending on the desired match).

It is now possible to consider the problem of the determination of the line characteristics on the basis of impedance measurements.

1.1.4 Line parameters from measured input impedance

Under the two extreme conditions of short circuit and open-circuit termination, the measured input impedance is expressed as

$$Z_{in,sc} = Z_c \tanh(\gamma L) \qquad Z_{in,oc} = \frac{Z_c}{\tanh(\gamma L)} \tag{1.1.41}$$

and the multiplication of the two expressions gives

$$Z_c = \sqrt{Z_{in,sc}\, Z_{in,oc}} \tag{1.1.42}$$

This was already evident in 1.1.24, observing that a measurement in short-circuit conditions rules out the shunt terms leaving only those of the series impedance, and conversely an open-circuit measurement reduces to nearly zero the flowing current and thus the voltage drop on the series terms, leaving only the shunt admittance terms.

This method for estimating the characteristic impedance is really adopted in practice, especially for low and medium frequency applications, when the measurements of the short-circuit and open-circuit input impedance values may be performed at a single or few frequency values using ordinary instrumentation, such as oscilloscopes and current probes.

Besides reasoning on the accuracy of the measurement method and the propagation of uncertainty, it is evident that both $Z_{in,sc}$ and $Z_{in,oc}$ must be in the scale of the used instrument, so neither too small nor too large; we may state that the appropriate length L is close to an odd number of eighths of the test wavelength, or, put in another more practical way, that the measurements are to be taken at somewhat different frequency values to avoid singularities, based on the assumption that the line p.u.l. parameters are not varying appreciably for changes of the test frequency within a few %.

Also the attenuation factor α and the phase factor β may be estimated from these measurements. Taking the square root of the ratio of $Z_{in,sc}$ and $Z_{in,oc}$ expressions (obtained by setting $Z_t = 0$ and $Z_t = \infty$ in (1.1.37), respectively) and replacing it with the $\tanh(\gamma L)$ term

$$\sqrt{Z_{in,sc}/Z_{in,oc}} = \tanh(\gamma L) = \frac{1 - \exp(-2\gamma L)}{1 + \exp(-2\gamma L)} \tag{1.1.43}$$

it is possible to isolate the $\exp(-2\gamma L)$ term; changing the sign in the exponent and correspondingly reversing the fraction gives:

$$\exp(2\gamma L) = \frac{1 + \sqrt{Z_{in,sc}/Z_{in,oc}}}{1 - \sqrt{Z_{in,sc}/Z_{in,oc}}} \tag{1.1.44}$$

Then, taking the log and separating the real and imaginary part of γ,

$$\alpha = \frac{1}{2L} \ln \left| \frac{1 + \sqrt{Z_{in,sc}/Z_{in,oc}}}{1 - \sqrt{Z_{in,sc}/Z_{in,oc}}} \right| \tag{1.1.45}$$

$$\beta = \frac{1}{2L} \left\{ \arg \left[\frac{1 + \sqrt{Z_{in,sc}/Z_{in,oc}}}{1 - \sqrt{Z_{in,sc}/Z_{in,oc}}} \right] + 2n\pi \right\} \tag{1.1.46}$$

where the $2n\pi$, n integer, impedes to determine β uniquely, but rather a series of values that differ by π/L.

When the phase factor β has been determined, also the phase velocity is known by applying (1.1.20).

It appears that for the correct determination of α, a piece of line is needed, with a length that causes a loss of some dB as a minimum; on the contrary, the unambiguous determination of β requires a piece of line so short that the discrimination of π/L is possible (put in another way, since an estimate of β may be independently determined by means of the simple relationship $\beta = 2\pi/\lambda$, the value of λ in the transmission line needs to be known with enough accuracy to discriminate which value of n above is the most appropriate).

1.1.5 Standing wave pattern and VSWR

Let's take again the phasor voltage equation (1.1.10), that may be reformulated as

$$
\begin{aligned}
V(z) &= V_1 \left[\exp(-\gamma z) + \frac{V_2}{V_1} \exp(+\gamma z) \right] = \\
&= V_1 \left[\exp(-\gamma z) + \Gamma_t \exp(-2\gamma L) \exp(+\gamma z) \right] = \\
&= V_1 \exp(-\gamma L) \left[\exp(+\gamma x) + \Gamma_t \exp(-\gamma x) \right]
\end{aligned}
\tag{1.1.47}
$$

where $x = L - z$. The phasor current equation may be soon determined dividing by Z_0.

When two waves of identical frequency travel in opposite directions, the interference phenomenon called *standing waves* occurs. In the extreme case of a lossless ($\alpha = 0$) line with maximum reflection at the load termination (open circuited, thus $\Gamma_t = 1$), the magnitude of the voltage produced by the composition of the two waves is

$$|V(x)| = |2V_1 \exp(-\gamma L)| \, |\cos(\beta x)| \tag{1.1.48}$$

The magnitude oscillates thus between minima (zero valued) and maxima; the separation of consecutive minima (or maxima) is $\beta(x_{i+1} - x_i) = \pi$ or $x_{i+1} - x_i = \lambda/2$; the minima are located at $\beta x_i = \pi/2 + n\pi$, n integer, and the maxima $\pi/2$ apart.

The instantaneous voltage at the points of minimum is always zero, while at the points of maximum oscillates in time at the same frequency of the signal and spans to the maximum level.

The most common characterization of the standing wave voltage pattern is the VSWR (Voltage Standing Wave Ratio), defined as

$$\text{VSWR} = \frac{|V(x)|_{\max}}{|V(x)|_{\min}} = \frac{1 + |\Gamma_t|}{1 - |\Gamma_t|} \tag{1.1.49}$$

If the line has losses ($\alpha > 0$) the expression above is modified as

$$\text{VSWR} = \frac{|V(x)|_{\max}}{|V(x)|_{\min}} = \frac{1 + |\Gamma_t| \exp(-2\alpha x)}{1 - |\Gamma_t| \exp(-2\alpha x)} \tag{1.1.50}$$

and implies that the VSWR is not a constant, since minima and maxima are distributed along the line and they have thus a different value of the position x. Practically speaking, if the attenuation is not so relevant ($\alpha/\beta \ll 1$) with terminating impedance values at the source and load side that do not produce too large reflection coefficients (let's say less than 0.5), the determination of the VSWR as the ratio of adjacent maximum and minimum is accurate enough.

The reflection coefficient Γ_t may be calculated back from the VSWR as

$$\Gamma_t = \frac{\text{VSWR} - 1}{\text{VSWR} + 1} \tag{1.1.51}$$

Several problems and adverse effects may be put in relationship with the observation of a standing wave pattern and large VSWR values:

- the voltage at the points of maximum may exceed or be too close to the voltage rating of some of the connected equipment or the cable itself, thus causing voltage breakdown and contributing to losses;

- similarly the current at the points of maximum may exceed or be too close to the cable rating, thus causing local heating;

- because of the presence of an impedance mismatch (at the output and then proceeding backward causing a variation of the input impedance with frequency), the power-transfer efficiency from the source will vary with frequency.

Referring to the last two points, it is underlined that thus in real cases cable attenuation itself is not the only factor that affects and determines power loss (and cable heating in extreme cases), but also the impedance mismatch at the ends, and the resulting increase in voltage and current waveforms due to standing waves, shall be accounted for. Such increase of voltage and current intensity raises dielectric losses and ohmic losses, roughly proportional to the square of the rms values of voltage and current, respectively.

The beneficial effect of attenuators for impedance matching is considered later on in sec. 1.5.3: in many cases, when the load is expect to vary, such as for the input impedance of an antenna used over a wide frequency range, the use of moderate attenuation (e.g. 6 or 10 dB) ensures a high reliability of the results, in terms of controlled standing waves and reflection coefficients. For power applications this solution cannot clearly be implemented easily due to large dissipated power and over-heating.

1.2 Electromagnetic Field

This section provides a review of the electromagnetic theory and of the electromagnetic field fundamentals. The concepts of static electric and magnetic fields are first considered, as a preliminary basis for wave equation and electromagnetic wave propagation.

1.2.1 Electric field

Electric field is generated by the electric charge, that is measured in coulomb; it is indicated by $\mathbf{E}$ and defined as the vector force exerted on a unitary test charge. The measuring unit of the electric field is V/m from the definition

$$N/C = J/m/(A\,s) = W\,s/m/(A\,s) = V/m$$

The electric field is in relationship with the flux density $\mathbf{D}$ and the permittivity ε, that for an isotropic medium is a scalar, so that the flux density is a vector with the same direction as $\mathbf{E}$ and with strength proportional to the quantity of charge. The $\mathbf{E}$ field originates at positive charges and terminates on negative charges or at infinity.

Gauss's Law states that the integral of the flux density over a closed surface is equal to the total charge Q therein. The permittivity is assumed for our purposes to be a constant scalar and for a given material is expressed as the relative permittivity ε_r (a real positive number) times the permittivity of free space $\varepsilon_0 = 8.854\,\mathrm{pF/m}$.

When the electric field is applied to two regions made of different materials with two permittivity values ε_1 and ε_2, it is possible to calculate the direction and intensity of the two vectors $\mathbf{E}$ and $\mathbf{D}$ in the two regions (see Figure 1.2.1):

1) by conservation of the normal component of the flux density vector $\mathbf{D}$ at the interface ($D_{N1} = D_{N2}$), the relationship between the normal components of the electric field is $E_{1n} = \dfrac{\varepsilon_2}{\varepsilon_1}E_{2n}$; more generally the relationship between the two normal components of the flux density vector at the interface is dictated by the possible presence of an electric surface charge density ρ_S, such that $D_{1n} - D_{2n} = \rho_S$;

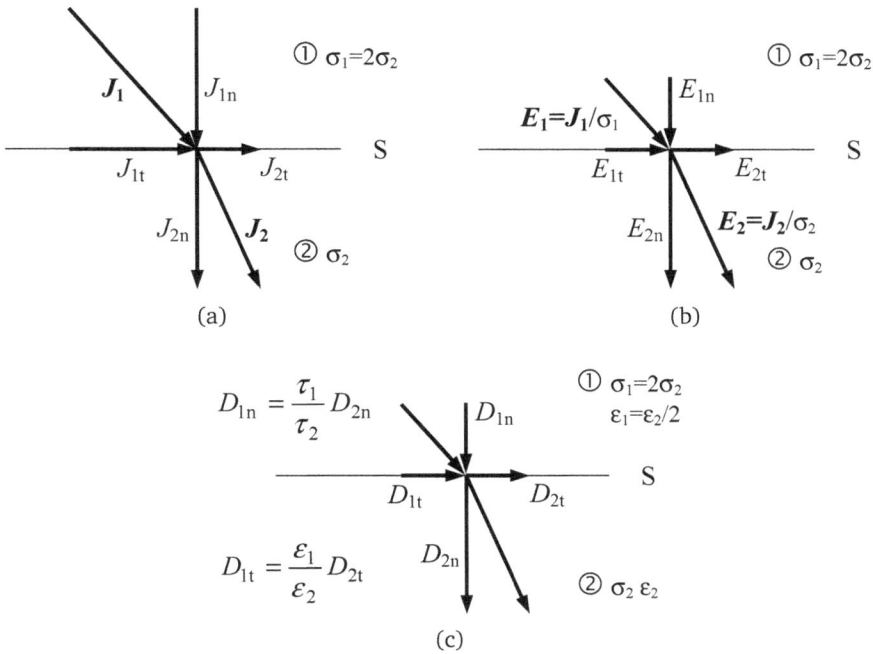

Figure 1.2.1 – **E** and **D** for two regions with different conductivity σ and permittivity ε.

2) by conservation of the of the tangential component of the electric field **E** at the interface ($E_{1t} = E_{2t}$, since the rotor of the electric field **E** is zero) , the angle of the direction of the electric field in either of the two regions may be expressed in terms of the angle of the direction in the other region and the permittivity values: $\phi_2 = \arctan(E_{2n}/E_{2t}) = \arctan(\frac{\varepsilon_2\,E_{1n}}{\varepsilon_1\,E_{1t}}) = \arctan(\frac{\varepsilon_2}{\varepsilon_1}\tan(\phi_1))$; so the amount of diffraction at the interface depends on the grazing angle ϕ_1 and the permittivities ε_1 and ε_2.

1.2.2 Magnetic field

Magnetic field is generated by steady current flow or by magnetic materials; as it was done for the electric field there is a magnetic field strength (or simply magnetic field) **H** and a magnetic flux density (or induction field vector) **B**. For linear materials (those with no hysteresis behavior) they are related by the magnetic permeability, a scalar, so that $\mathbf{B} = \mu\mathbf{H}$, so that the two vectors have the same direction. The Biot-Savart law (a more complete formulation of the Ampère's Law) in its integral form

$$\mathbf{H} = \frac{1}{4\pi}\int \frac{I\mathbf{dl}\times\mathbf{r}}{r^3} \qquad (1.2.1)$$

says that the magnetic field vector **H** at a point is given by the integral of the current elements $I\mathbf{dl}$ in vector product with the vector between each current element and the point itself, normalized by the same distance amplitude to the third power. From this it is evident that the unity of measure of **H** is A/m. The unity of measure of

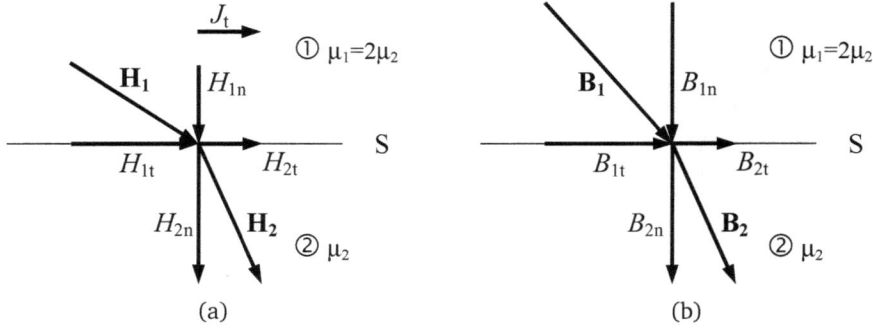

Figure 1.2.2 – **H** and **B** for two regions with different permeability μ.

the magnetic permeability is H/m, so that the unity of measure of the induction field vector **B** is $\mathrm{Wb/m^2 = T}$. Again the magnetic permeability μ may be expressed as a relative magnetic permeability μ_r times the magnetic permeability of free space $\mu_0 = 4\pi\,10^{-7}\mathrm{H/m}$.

When the magnetic field is applied to two regions made of different materials with two permeability values μ_1 and μ_2, it is possible to calculate the direction and intensity of the two vectors **H** and **B** in the two regions:

1) by conservation of the of the normal component of the induction magnetic field vector **B** at the interface ($B_{1\mathrm{n}} = B_{2\mathrm{n}}$, since div **B** $= 0$), the relationship between the normal components of the magnetic field is $H_{1\mathrm{n}} = \dfrac{\mu_2}{\mu_1}H_{2\mathrm{n}}$; more generally the relationship between the two normal components of the induction field vector at the interface is dictated by the possible presence of a magnetic surface charge density τ_S, such that $B_{1\mathrm{n}} - B_{2\mathrm{n}} = \tau_S$;

2) by the relationship between the tangential component of the magnetic field **H** at the interface ($H_{1\mathrm{t}} - H_{2\mathrm{t}} = J_\mathrm{t}$, since rot **H** $= J_\mathrm{t}$, that is the tangential current density, i.e. along the surface), the angle of the direction.

1.2.3 Electromagnetic wave

A time-varying electric field produces a magnetic field, possibly related to a flowing current and a time-varying magnetic field produces an electric field, related to the derivative of the linked magnetic flux. The two rotor equations lead to the harmonic expression of the wave equation, that under the assumption of orthogonality between the electric field, the magnetic field and the direction of propagation, takes the form

$$\frac{\partial^2 E_x}{\partial t^2} = \mu\varepsilon\frac{\partial^2 E_x}{\partial z^2} \tag{1.2.2}$$

The solution is the wave equation of the electric field component along the x axis, that under the assumption of orthogonality is the only non-zero electric field component, that propagates along the z axis. The solution is a wave equation, function of z and t, composed of a progressive (indicated by "f", forward) and a regressive (indicated by "b", backward) component, perfectly analogous to that derived using the telegrapher's equations for transmission lines.

$$E_x(z,t) = E_f \, \exp(\omega t - \gamma z) + E_b \, \exp(\omega t + \gamma z) \tag{1.2.3}$$

The arguments of the exponential $(\omega t - \gamma z)$ and $(\omega t + \gamma z)$ describe the propagation of the wave in time and space.

1.2.3.1 Wave impedance and medium characteristic impedance

The impedance Z_w of a wave impinging normally to a surface is $Z_w = E_t/H_t$, where the tangential fields E_t and H_t are mutually perpendicular to each other. If we consider a conducting surface of thickness t, with a uniform linear current density $\hat{J}$ equal to J_t above, the conduction equation based on the conductivity σ is $J = \sigma E$. If t approaches zero and σ increases so that the product $G = \sigma t$ remains constant,

$$\begin{aligned} J' &= GE \\ E &= RJ' \end{aligned} \tag{1.2.4}$$

where G and R are called the surface conductance and surface resistance of the sheet, respectively. More generally, the surface admittance Y_s and the surface impedance Z_s (both complex quantities) are defined by the equations:

$$\begin{aligned} J' &= Y_s E \\ E &= Z_s J' \end{aligned} \tag{1.2.5}$$

The primary electromagnetic properties of a medium are conductivity σ, dielectric constant and permeability μ. In transmission line theory, two secondary constants were introduced (see sec. 1.4.1.3), the propagation constant Γ and the characteristics impedance Z_c. In three dimensional theory, the important constants are the intrinsic propagation constant γ and the intrinsic impedance η, defined as:

$$\gamma = \sqrt{j\omega\mu\,(\sigma + j\omega\varepsilon)} \qquad \eta = \sqrt{\frac{j\omega\mu}{\sigma + j\omega\varepsilon}} \tag{1.2.6}$$

These constants are independent of the geometry of the wave, and in general both are complex quantities. For a perfect dielectric:

$$\gamma = j\beta \qquad \beta = \frac{2\pi}{\lambda} \qquad \eta = \sqrt{\frac{\mu}{\varepsilon}} \qquad \nu = \frac{1}{\sqrt{\mu\varepsilon}} \tag{1.2.7}$$

where the terms have their usual meanings.

1.2.3.2 Polarization

Polarization of an antenna in a given direction is defined as "the polarization of the wave transmitted (radiated) by the antenna. Note: When the direction is not stated, the polarization is taken to be the polarization in the direction of maximum gain." In practice, polarization of the radiated energy varies with the direction from the center of the antenna, so that different parts of the pattern may have different polarizations. Polarization of a radiated wave is defined as "that property of an electromagnetic wave describing the time-varying direction and relative magnitude of the electric-field vector; specifically, the figure traced as a function of time by the extremity of the vector at a fixed location in space, and the sense in which it is traced, as observed along the direction of propagation." Polarization then is the curve traced by the end point of the arrow (vector) representing the instantaneous electric field.

The polarization of a wave can be defined in terms of a wave radiated or received by an antenna in a given direction. The polarization of a wave radiated by an antenna in a specified direction at a point in the far field is defined as "the polarization of the (locally) plane wave which is used to represent the radiated wave at that point. At any point in the far field of an antenna the radiated wave can be represented by a plane wave whose electric-field strength is the same as that of the wave and whose direction of propagation is in the radial direction from the antenna. As the radial distance approaches infinity, the radius of curvature of the radiated wave's phase front also approaches infinity and thus in any specified direction the wave appears locally as a plane wave." This is a far-field characteristic of waves radiated by all practical antennas. The polarization of a wave received by an antenna is defined as the "polarization of a plane wave, incident from a given direction and having a given power flux density, which results in maximum available power at the antenna terminals." Polarization may be classified as linear, circular, or elliptical.

1.3   Antennas and electromagnetic field radiation

An antenna may be defined as a structure associated with the region of transition between a guided wave and a space wave (transmission), or vice-versa (reception); the term "space wave" is used, and not "free space wave" in the sense that in many cases there is always a certain degree of interaction of the antenna with the surroundings and the propagation is not that of free space. For a broad class of linear antennas the Reciprocity Theorem [7, 8] says that the radiation pattern (describing the distribution of electromagnetic field around the antenna itself) is the same for a transmitting or receiving antenna.

Since an antenna is a source of electromagnetic field, composed in turn in the far field region by the E-field component and by the orthogonal H-field component, there are two main planes at least, each containing the E- and H-field vectors and used usually as reference planes for the antenna placement and orientation in drawings.

To distinguish between portions of space with intense radiation and portions of space with weak radiation, the term lobe is often used, since these regions of relatively higher radiation have all the shape of a balloon. In a radiation pattern it is possible to distinguish between the main lobe, minor lobes, side lobes and back lobes (or back radiation lobes).

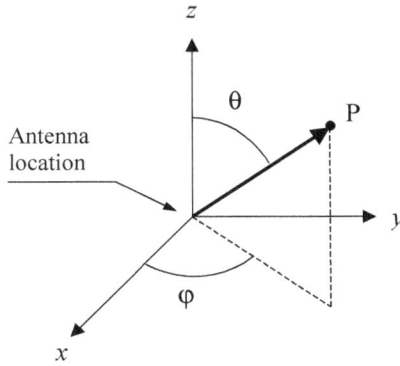

Figure 1.3.1 – Coordinate system for antenna characterization.

The main variables that are used in the following to describe antenna parameters are, for an antenna in transmission mode: P_g, the power generated by an external source (e.g. radio frequency amplifier or generator), P_a the "accepted" power, entering the antenna connector, and P_r, the power effectively radiated outside the antenna into the surrounding space. We define the antenna efficiency as $\eta = P_r / P_a$.

1.3.1 Antenna parameters

1.3.1.1 Radiated power and radiation intensity

The radiated power may be defined as the integral over the entire volume of the radiation intensity $\Phi(\theta, \varphi)$, that under far-field assumption depends only on coordinates θ and φ. The radiation intensity has the unity of measure of a power density $\mathrm{W/m^2}$.

$$P_r = \int \int \Phi(\theta, \varphi) \sin \theta \, \mathrm{d}\theta \mathrm{d}\varphi \qquad (1.3.1)$$

The average radiation intensity is then defined as

$$\Phi_{\mathrm{avg}} = \frac{P_r}{4\pi} \qquad (1.3.2)$$

The radiation intensity $\Phi(\theta, \varphi)$ is derived from the radiation density and then from the application of the Poynting Theorem to the E-field and H-field components. The radiation density $W_{\mathrm{rad}}(\theta, \varphi)$ is given by

$$W_{\mathrm{rad}}(\theta, \varphi) = \frac{1}{2} \mathrm{Re}[\mathbf{E} \times \mathbf{H}] \qquad (1.3.3)$$

and the product of it and the square of the distance gives the radiation intensity, that is the radiation density per unit solid angle (1 srad).

$$\Phi(\theta, \varphi) = r^2 W_{\mathrm{rad}}(\theta, \varphi) \tag{1.3.4}$$

1.3.1.2 Radiation pattern

The *radiation pattern* is defined as a mathematical function or representation of the radiation properties of the antenna in the space coordinates. The radiation properties may include the radiation intensity (see function $\Phi(\theta, \varphi)$ above), the field strength or field polarization. Based on this it is possible to define:

- *isotropic radiator*, as a lossless antenna having equal radiation in all directions; although not really existing, it is taken as a reference radiator to quantify the distribution of the radiation intensity around a particular antenna;

- *directional antenna*, with the property of receiving or radiating the electromagnetic field in some directions than others; in particular, this definition, even if of general value, is applied to those antennas that show a significant directivity in a particular direction;

- *omni-directional antenna*, that is a particular type of directional antenna with a non-directional pattern in a given plane, as it will be a dipole (see sec. 1.3.3) or a loop antenna (see sec. 1.3.6); it is anticipated here that the radiation pattern of these antennas (see sec. 1.3.3) is often approximated by a power of $\sin(\theta)$, that is $|\sin^n(\theta)|$, and independent on φ, so omni-directional in the $0 \le \varphi \le 2\pi$ plane.

1.3.1.3 Gain and directivity

The directivity is then defined as the ability of an antenna to concentrate the emitted power in a particular portion of space; the directivity function is the ratio of the radiation intensity with the average radiation intensity

$$D(\theta, \varphi) = \frac{\Phi(\theta, \varphi)}{\Phi_{\mathrm{avg}}} \tag{1.3.5}$$

so that its maximum is called simply directivity using the symbol D or D_0.

There is a practical formula for antennas with only one narrow major lobe and negligible minor lobes, that gives an estimation of the maximum directivity with respect to the half-power beam-width (HPBW) in two orthogonal planes (usually the horizontal and vertical plane)

$$D_0 = \frac{32400}{\Theta_h \Theta_v} \tag{1.3.6}$$

The HPBW angles Θ_h and Θ_v are measured in degrees and correspond to the angle in a plane containing the direction of maximum that intercepts the half-power points (-3 dB) of the power radiation pattern or, more generally, the points where the value is half the maximum. This formula is very useful for directive antennas, such as radar, radio link and access point antennas, for which, given the radiation patterns, it is possible to estimate directivity, and compare it to the reported gain.

Directivity is often confused with gain: the difference is that the latter takes into account also antenna efficiency; we have a gain function

$$G(\theta, \varphi) = \eta D(\theta, \varphi) \tag{1.3.7}$$

and the maximum (that corresponds to the same direction in (θ, φ)) is called gain G.

Balanis proposes a practical formula related to the HPBW angles [8], similar to the one for directivity, where the numerator is only 30000; from this an implicit estimation of the antenna efficiency of 92.5 %, that is a power loss of 0.33 dB.

By putting in relationship the input power P_a and the radiation intensity in one direction $\Phi(\theta, \varphi)$, it is possible to obtain an operative relationship that involves the gain function

$$P(\theta, \varphi) = \frac{\Phi(\theta, \varphi)}{r^2} = G(\theta, \varphi) \frac{P_a}{4\pi r^2} \tag{1.3.8}$$

Practically, the gain function may be found immediately by taking the radiation pattern used for the directivity and by assigning the gain G to the point of maximum of the plot. The factor $P_a/4\pi r^2$ represents the power density that would result if the power accepted by the antenna were radiated by a lossless isotropic antenna.

1.3.1.4 Antenna factor

This parameter is a transducer factor that relates the intensity of the incident field strength and the voltage across a matched load at antenna input. The antenna factor K can be related to antenna gain and wavelength by the following formula

$$K = \frac{9.73}{\lambda \sqrt{G}} \tag{1.3.9}$$

where the parameters G and K are in linear scale; the same expression in dB takes the form

$$K = -29.8 \, \text{dB} + 20 \log(f/1 \, \text{MHz}) - G \tag{1.3.10}$$

1.3.1.5 Bandwidth

Until now antenna characteristics have been considered independent on frequency. Yet, they variously depend on frequency and for this reason the concept of antenna bandwidth is ambiguous, if not related to a specific characteristic, such as gain, input impedance, half-power beam-width, etc.: it may be said simply that the bandwidth is the range of frequency values within which the performance of the antenna for one characteristic meets a given requirement; the requirement normally is to stay within a given percentage of the nominal or maximum value, as it happens for gain and directivity; the bandwidth is thus centered around a given frequency which the nominal or maximum value of the characteristic corresponds to.

Depending on the parameter more or less tight percentages may be specified: for the gain of a measuring antenna the tolerance is quite narrow, almost never larger than $-3\,\mathrm{dB}$; for input impedance a wider variation is tolerated (see sec. 1.3.5), being the impact on VSWR halved and in any case effectively controlled by the use of attenuators (see sec. 1.5.3).

From a more general point of view a frequency response may be defined: an approximate $-6\,\mathrm{dB/octave}$ slope is usually expected at the extremes of the bandwidth, possibly with overlapped resonances. For arrays (e.g. log-periodic antenna) each element has its own resonance frequency and combines with the adjacent ones, so that periodic fluctuations of the frequency response between local maxima and minima are commonplace, with the antenna bandwidth extended to cover the combination of the elementary responses.

1.3.1.6   Input impedance

The input impedance of an antenna Z_A is simply given by the ratio of voltage and current at the input connector of the antenna itself. The input impedance may be thought composed of a real resistive part R_A and an imaginary reactive part jX_A; the resistive part is in turn composed of the radiation resistance R_r and the antenna loss resistance R_l.

A simple circuit is considered, where the antenna is supplied by a generator with amplitude E_g and series impedance $Z_g = R_g + jX_g$; the antenna is thus operated in transmitting mode. The calculations that follow aim at estimating the power losses and the conditions for the best transmission of power from the generator to the antenna and then to the space around; the antenna is a simple antenna for which R_r and R_l may be considered in series.

The current supplied by the generator to the antenna input port is

$$I_g = \frac{E_g}{Z_g + Z_A} = \frac{E_g}{(R_g + R_r + R_l) + j(X_g + X_A)} \qquad (1.3.11)$$

The power delivered for radiation P_r and dissipated as losses P_l are

$$P_r = R_r \left|I_g\right|^2 \qquad P_l = R_l \left|I_g\right|^2 \qquad (1.3.12)$$

The remaining power is dissipated on the generator resistance R_g or exchanged by the reactive components.

The maximum power transfer occurs when the generator and the antenna have conjugate matching of their impedances:

$$R_g = R_r + R_l \qquad X_g = -X_A \qquad (1.3.13)$$

and the power terms become:

$$
\begin{aligned}
P_r &= |V_g|^2 \frac{R_r}{4(R_r+R_l)^2} \qquad P_l = |V_g|^2 \frac{R_l}{4(R_r+R_l)^2} \\
P_g &= |V_g|^2 \frac{1}{4R_g} = |V_g|^2 \frac{R_r+R_l}{4(R_r+R_l)^2}
\end{aligned}
\qquad (1.3.14)
$$

If the antenna is lossless ($R_l = 0$), then half of the power supplied by the generator reaches the antenna and is radiated, the other half being dissipated in the generator resistance; in all other cases the fraction of radiated power is even less.

Analogously, if the antenna is operated in receiving mode, the terminating load (represented by the measuring or receiving apparatus) is characterized by an impedance $Z_t = R_t + jX_t$, that replaces the generator impedance in the expressions above. The same expressions apply, where Z_t, P_t and V_t replace Z_g, P_g and V_g respectively. In particular, since the objective in receiving mode is to deliver power to the terminating load, and we know that under conjugate it reaches half of the received power, the power delivered to R_r and R_l sum to the other half of the received power; the first term across R_r is called scattered (or re-radiated) power.

1.3.1.7 Effective aperture and directivity

When considering for simplicity a receiving antenna, the effective aperture (or effective area) is that antenna parameter that puts in relationship the power density W_{inc} of the incident wave and the power P_t appearing at antenna terminals on the terminating load and has thus the unity of measure of a surface:

$$
A_e = \frac{P_t}{W_{inc}}
\qquad (1.3.15)
$$

The maximum effective aperture A_{em} is obtained for the conjugate matching conditions seen above. For a transmitting antenna the effective aperture puts in relationship the radiation density W_{rad} and the accepted power P_a.

Other areas may be defined that correspond to other two terms of the power balance above, that is scattering area and loss area. The aperture efficiency of an antenna is the ratio between the maximum effective aperture and the physical area. There are types of antennas for which the effective area can be even equal to the physical area (so reaching a 100 % efficiency), but this doesn't imply that the received power is larger than one half, since also the scattering area (neglecting any power loss for simplicity) can be as large, so reaching a scattering efficiency of 100 %, thus keeping the balance between the two area terms at 50 % (as it should be for conjugate matching).

The relationship between the (maximum) effective aperture A_e and the (maximum) directivity D is (without demonstration)

$$
A_e = \frac{\lambda^2}{4\pi} D
\qquad (1.3.16)
$$

1.3.2   Elementary antenna - the Hertzian dipole

An infinitesimal current element of length dz is considered at the centre of the co-ordinate system, normally referred to as "Hertzian Dipole"; with reference to Figure 1.3.1 the current element is positioned along the z axis. The three components of the electric field E and magnetic field H along r, θ and φ are determined as follows:

$$E_r = \frac{\eta}{2\pi} k^2 I\,dz \left[\frac{1}{(kr)^2} - j\frac{1}{(kr)^3}\right] \cos\theta\, e^{-jkr} \tag{1.3.17}$$

$$E_\theta = j\frac{\eta}{4\pi} k^2 I\,dz \left[\frac{1}{(kr)} - j\frac{1}{(kr)^2} - \frac{1}{(kr)^3}\right] \sin\theta\, e^{-j\beta r} \tag{1.3.18}$$

$$H_\varphi = j\frac{k^2}{4\pi} I\,dz \left[\frac{1}{(kr)} - j\frac{1}{(kr)^2}\right] \sin\theta\, e^{-j\beta r} \tag{1.3.19}$$

$$E_\varphi = H_r = H_\theta = 0 \tag{1.3.20}$$

As a note, it is underlined that the parameter k is called the "wave number", and is equivalent to the phase constant $\beta = 2\pi/\lambda$. It is interesting to observe that the ratio of the first order components of the electric and magnetic field, that are projected along θ and φ axes (with propagation occurring along the r axis), give the free space impedance, that is $Z_0 = E_\theta^{(1)}/H_\varphi^{(1)} = 120\pi = 376.7$, the unity of measure being ohm.

The other components of higher order form the reactive field. A general definition of the three regions function of distance from the antenna is: that region in which the reactive component predominates is said the *reactive near-field region* (some call it simply "near-field region"); the outer region where the radiating component predominates, is further subdivided into two regions, *radiating near-field region* (where the relative phases and amplitudes of contributions from various elements of the antenna are functions of the distance from the antenna), and *radiating far-field region* (where, at larger distance, straight lines from the antenna elements to the observation point are essentially parallel and the angular distribution of radiated energy is independent on distance). The radiating near field-region is sometimes called intermediate region, since it separates the near-field region *tout-court* from the far-field region.

Following the development in [8], the Hertzian dipole equations above may be computed for the three regions, where $kr \ll 1$, $kr > 1$ and $kr \gg 1$ respectively.

<u>Reactive near-field region, $kr \ll 1$:</u>

$$E_r = -j\frac{\eta}{2\pi} I\,dz \frac{1}{kr^3} \cos\theta\, e^{-jkr} \tag{1.3.21}$$

$$E_\theta = j\frac{\eta}{4\pi} kI\,dz \frac{1}{r} \sin\theta\, e^{-jkr} \tag{1.3.22}$$

$$H_\varphi = \frac{1}{4\pi} I\,dz \frac{1}{r^2} \sin\theta\, e^{-jkr} \tag{1.3.23}$$

The two E-field components have both a $-j$ coefficient with respect to the H-field component, so that they are in time-quadrature; this implies that their time-averaged product is zero and so it is the associated power flow. Furthermore, the two E-field components have similar expressions and are in an in-phase condition.

Radiating near-field region, $kr > 1$:

$$E_r = \frac{\eta}{2\pi} I\,dz \frac{1}{r^2} \cos\theta\, e^{-jkr} \tag{1.3.24}$$

$$E_\theta = j\frac{\eta}{4\pi} kI\,dz \frac{1}{r} \sin\theta\, e^{-jkr} \tag{1.3.25}$$

$$H_\varphi = j\frac{1}{4\pi} kI\,dz \frac{1}{r} \sin\theta\, e^{-jkr} \tag{1.3.26}$$

The two E-field components lose their in-phase condition, so that the resulting E-field vector is a rotating vector with the extremity tracing an ellipse. However, the two components E_θ and H_φ are in time-phase condition and this indicates the formation of an average power flow directed radially in the outward direction, thus starting the formation of the radiation phenomenon, typical of the far-field region; from this the correctness of naming this region *radiating* near-field region.

Far-field region, $kr \gg 1$:

$$E_r = 0 \tag{1.3.27}$$

$$E_\theta = j\frac{1}{4\pi} kI\,dz \frac{1}{r} \sin\theta\, e^{-jkr} \tag{1.3.28}$$

$$H_\varphi = j\frac{1}{4\pi} kI\,dz \frac{1}{r} \sin\theta\, e^{-jkr} \tag{1.3.29}$$

The condition of orthogonal propagation is reached for E- and H-field components projected only on θ and φ axes, with a 120π ratio (free space impedance) and with a linear dependence on distance; the associated power density, given by the time-averaged product of the E- and H-field, shows a dependence on distance of the $1/r^2$ type.

The distance r^* at which the transition from the reactive field to the radiating field region occurs is conventionally taken as $r^* = \lambda/2\pi$, that corresponds to the condition $\beta r = 1$ above. It is important to observe that at this distance the amplitude of the far field and reactive field components is only equal, so the meaning of this distance is that above it the far field components start prevailing, but in no way it indicates a distance at which the reactive field components are negligible.

1.3.3   Dipole antenna

A dipole antenna is a very common antenna, since several sources may be approximated by this antenna with a satisfactory and accurate enough treatment: the bi-conical and log-periodic measuring antennas stem from a basic design built around a dipole; a slot in an enclosure may be modeled by an orthogonal dipole; the overhead contact wire excited by the electric arc at pantograph may be seen as a dipole with variable length, depending on frequency, as well as a long wire antenna; another very useful antenna, the monopole antenna, may be and is calculated by halving a dipole antenna.

By assuming a finite length dipole with negligible wire diameter, the current distribution may be assumed sinusoidal, going to zero at the two extremes, with the dipole center-fed. Even if it can be shown that a closed form solution for this current distribution is available, that is valid in all the regions (and all points in space), the followed approach is that of limiting the validity to the far-field region, so that it is applicable in general to a wide variety of distributions and leads to simpler formulations.

The dipole is subdivided in several infinitesimal dipoles of length dz, whose far-field E-field and H-field components can be computed by using (1.3.28) and (1.3.29). The sum of all the contributions from each infinitesimal element reduces, in the limit $dz \to 0$, to an integral. The integral gives an expression for the E-field (and in turn for the H-field, by simply dividing by the free space impedance) that is composed of a first term that is very similar to the one obtained for the infinitesimal dipole (1.3.28) multiplied by the integral of the current distribution. By assuming the sinusoidal current distribution above, the mathematical expression of the E-field component E_θ is

$$E_\theta = j60kI_0\frac{1}{r}e^{-jkr}\left[\frac{\cos(kl/2\cos\theta) - \cos(kl/2)}{\sin\theta}\right] \quad (1.3.30)$$

The power distribution around the dipole antenna is given by the real part of the Poynting vector, that corresponds to the multiplication of E_θ and H_φ; the radiation intensity is readily obtained by multiplying it by the unit solid angle, that is r^2.

$$U = \frac{30}{2\pi}kI_0^2\left[\frac{\cos(kl/2\cos\theta) - \cos(kl/2)}{\sin\theta}\right]^2 \quad (1.3.31)$$

It is instructive to look at the radiation intensity plot as a function of θ for different length/wavelength ratios; the plots shown in Figure 1.3.2 are normalized plots (with the maximum value corresponding to 0 dB) for $l = \lambda/4$, $l = \lambda/2$, $l = 3\lambda/4$ and $l = \lambda$ cases, together with the reference case of the infinitesimal dipole, for which $l \ll \lambda$.

All the patterns have one lobe with different width: as the length of the dipole increases the lobe becomes narrower and thus the dipole directivity increases; the half-power beam-width of the examined cases is reported in Table 1.3.1.

If the l/λ ratio is further increased, the number of lobes begins to increase. By using the expression (1.3.31) of the radiation intensity (that defines the radiation pattern of

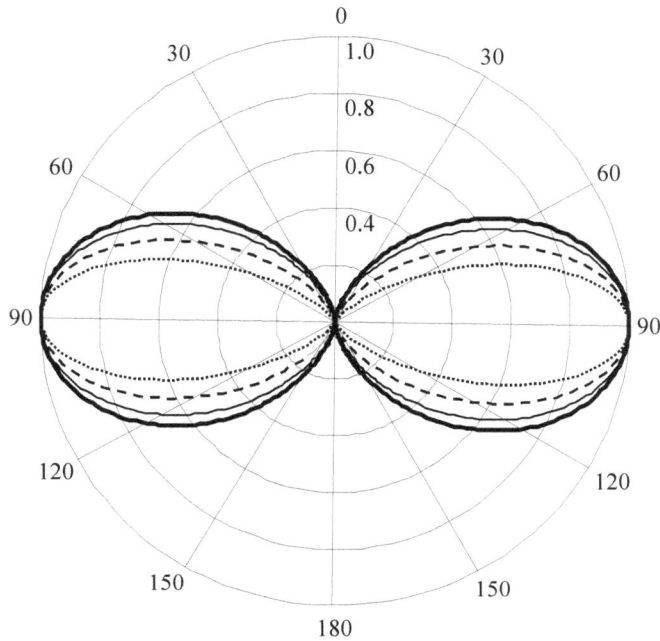

Figure 1.3.2 – Radiation intensity patterns in the elevation plane for sinusoidal current distribution and $l = \lambda/4$, $\lambda/2$, $3\lambda/4$, λ (going from the black dotted inner smaller lobes to the solid black larger external lobes).

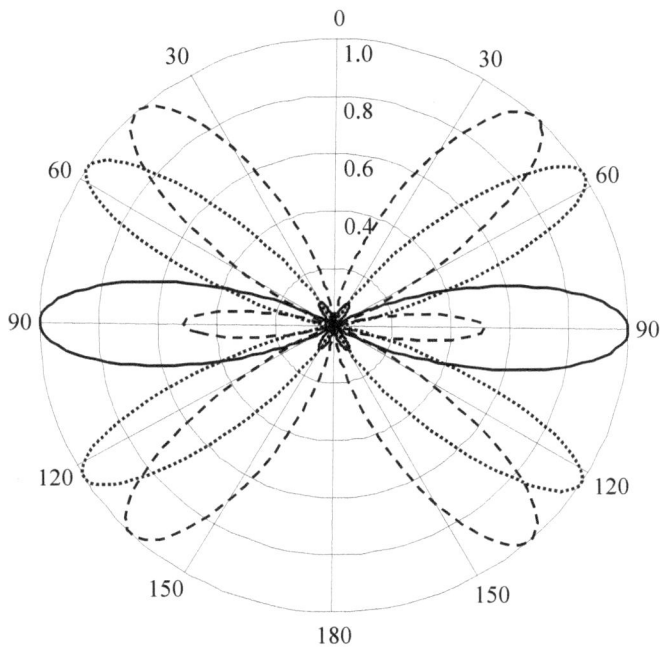

Figure 1.3.3 – Radiation intensity patterns in the elevation plane for a sinusoidal current distribution and $l = 5\lambda/4$, $3\lambda/2$, 2λ (again going from the dotted to the solid black curves).

Case	HPBW
$l \ll \lambda/4$	$87°$
$l \ll \lambda/2$	$78°$
$l \ll 3\lambda/4$	$64°$
$l \ll \lambda$	$47.8°$

Table 1.3.1 – Half-power beam-width of the dipole for $l \leq \lambda$.

the dipole for any l/λ ratio), the behavior for l/λ ratio larger than unity is investigated in Figure 1.3.3.

The general concepts of radiation resistance and input resistance may be clarified for the dipole; some of the calculations and simplifications are omitted, but the complete development may be found in [8].

The calculation and integration over a sphere of radius r of the Poynting vector give the radiated power

$$
P_{rad} = \eta \frac{|I_0^2|}{4\pi} \left\{ C + \ln(kl) - C_i(kl) + \frac{1}{2}\sin(kl)\left[S_i(2kl) - 2S_i(kl)\right] + \right.
$$
$$
\left. + \frac{1}{2}\sin(kl)\left[C + \ln(kl/2) + C_i(2kl) - 2C_i(kl)\right] + \right\}
\tag{1.3.32}
$$

where

$C = 0.5572$ is the Euler's constant;

$C_i = -\int_x^\infty \dfrac{\cos y}{y}\mathrm{d}y$ is the cosine integral function;

$S_i = -\int_0^x \dfrac{\sin y}{y}\mathrm{d}y$ is the sine integral function;

I_0 is the peak value of the current distribution.

From the radiated power the radiation resistance R_r is calculated by simply dividing by the rms value of the current $\dfrac{1}{2}|I_0|^2$:

$$
R_{rad} = \frac{2P_{rad}}{|I_0|^2}
\tag{1.3.33}
$$

This expressions will be used again in sec. 1.3.3.2 below.

1.3.3.1   Half-wavelength dipole

This is a special case of the formulas above with length $l = \lambda/2$ and it is very useful in practice, since several transmitting and receiving antennas are or may be modeled with the help of the half-wavelength dipoles: measuring antennas such as calibrated dipoles and biconical antennas; amateur radio antennas made of a Marconi semi-dipole in the

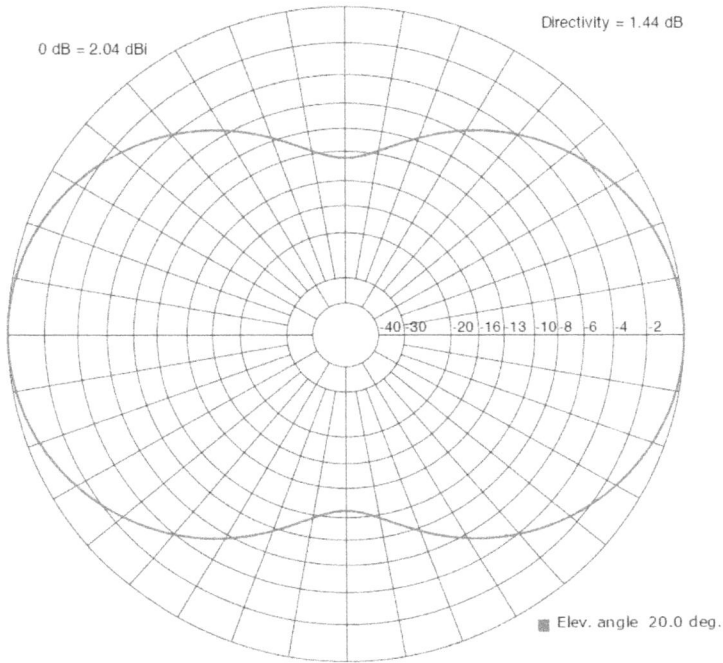

0 dB = 2.04 dBi

Directivity = 1.44 dB

-40 -30 -20 -16 -13 -10 -8 -6 -4 -2

Elev. angle 20.0 deg.

Figure 1.3.4 – Two-dimensional radiation pattern of the half-wave dipole.

27 and 144 MHz bands; FM radio and TV broadcasting towers, equipped with arrays of tuned vertical dipoles.

The electric and magnetic field equations are taken from (1.3.30) and repeated here for $l = \lambda/2$:

$$E_\theta = j\eta I_0 \frac{1}{2\pi r} e^{-jkr} \left[\frac{\cos(\pi/2 \cos\theta)}{\sin\theta} \right] \tag{1.3.34}$$

$$H_\varphi = j I_0 \frac{1}{2\pi r} e^{-jkr} \left[\frac{\cos(\pi/2 \cos\theta)}{\sin\theta} \right] \tag{1.3.35}$$

The radiation intensity is again defined starting from the time-average power density, given by the multiplication of E_θ and H_φ

$$U = \eta \frac{I_0^2}{8\pi^2} \sin^3\theta \tag{1.3.36}$$

and its two-dimensional pattern is plotted in Figure 1.3.4.

The total radiated power is calculated in [8] and is found to be

$$P_{rad} = \eta \frac{I_0^2}{8\pi} C_{in}(2\pi) \simeq \eta \frac{I_0^2}{8\pi} 2.435 \qquad (1.3.37)$$

(where $C_{in}(\cdot)$ indicates the cos-integral function). The directivity, by definition, is the ratio of the maximum radiation intensity over the average power density, that is $P_{rad}/4\pi$: $D_0 \simeq 4/2.435 = 1.643$; the maximum effective area is in force of (1.3.16) $A_{em} \simeq 0.13\lambda^2$.

The radiation resistance R_r may be calculated from the radiated power P_{rad}, by equating (1.3.37) to the simple definition (1.3.12):

$$R_r = \frac{\eta}{4\pi} C_{in}(2\pi) \simeq 73\,\Omega \qquad (1.3.38)$$

1.3.3.2   Effect of finite diameter

If the assumption of infinitesimal wire radius is relaxed and the dipole radius is a, with a length over radius ratio l/a that is not infinite, then the calculations become more complex. If two antennas are close to each other or the antenna is close to ground, then the current distribution over the antenna surface is no longer uniform. For the derivation of the self impedance and radiation properties the current density is assumed, however, uniform; central feeding is also assumed, so that the profile of the current is symmetrical.

There are several formulations available for the input impedance of cylindrical antennas, depending on the used series expansion and the approximations. The Moment Method with either Pocklington's or Hallén's integral equations is the most accurate for the calculation of the input impedance, but as a numerical method it doesn't give closed form expressions. Balanis proposes closed form expressions using the Induced EMF Method [8], p. 405, where the radiation impedance and the input impedance are computed referred to "at the current maximum" and "at the current at the input terminals", respectively; depending on the length/wavelength ratio the current maximum I_m may not occur at the antenna input terminals, where I_{in} is applied.

The relationships for the real and imaginary part of the radiation impedance are

$$
\begin{aligned}
R_r = {} & \frac{\eta}{2\pi} \left\{ C + \ln(kl) - C_i(kl) + \frac{1}{2}\sin(kl)\left[S_i(2kl) - 2S_i(kl)\right] + \right. \\
& \left. \frac{1}{2}\cos(kl)\left[C + \ln\left(\frac{kl}{2}\right) + C_i(2kl) - 2C_i(kl)\right] \right\}
\end{aligned}
\qquad (1.3.39)
$$

$$
\begin{aligned}
X_r = {} & \frac{\eta}{4\pi} \left\{ 2S_i(kl) - \cos(kl)\left[S_i(2kl) - 2S_i(kl)\right] + \right. \\
& \left. + \sin(kl)\left[C_i(2kl) - 2C_i(kl) - C_i(2ka^2/l)\right] \right\}
\end{aligned}
\qquad (1.3.40)
$$

and

$$R_{in} = \left(\frac{I_m}{I_{in}}\right)^2 R_r = \frac{R_r}{\sin^2(kl/2)} \qquad (1.3.41)$$

$$X_{in} = \left(\frac{I_m}{I_{in}}\right)^2 X_r = \frac{X_r}{\sin^2(kl/2)} \qquad (1.3.42)$$

where

$C = 0.5572$ is the Euler's constant;

$C_i = -\int_x^\infty \frac{\cos y}{y} dy$ is the cosine integral function;

$S_i = -\int_0^x \frac{\sin y}{y} dy$ is the sine integral function;

$C_{in}(x) = 0.5572 + \ln(x) - C_i(x)$

These expressions are very similar to those also reported by [84], sec. 13-5, for the thin dipole.

As a proof R_r and X_r are calculated for the $\lambda/2$ dipole by using (1.3.39) and (1.3.40): by replacing $kl = \pi$ the known $R_r = 73.13\,\Omega$ is obtained, together with $X_r = 42.545\,\Omega$ that indicates that for a perfectly resonant dipole (with $X_r = 0$) the length needs to be shorten by a few %; by successive approximation using the same expressions X_r has been brought to nearly zero for a 0.4883 length, so shorter by 2.34%, and the corresponding resistance becomes $R_r = 68.16\,\Omega$.

The above formulas (1.3.39) and (1.3.40) are plotted and shown in Figure 1.3.5, using the l/λ as the normalized length. For those who wish to recalculate the expressions above, it is underlined that, since $C_i(\cdot)$, $S_i(\cdot)$ and $C_{in}(\cdot)$ are computed by series expansions, the number of terms required to reach convergence of the series and get a meaningful and stable result may be large, depending on the value of the argument.

1.3.4 The monopole

When a monopole is mounted on an ideal infinite ground plane, its impedance and radiation characteristics can be deduced from that of a dipole of twice its length in free space. Also known as $\lambda/4$ stub antenna, a base-driven monopole antenna can be modeled starting from an electric dipole and using a transverse ideal plane of symmetry represented by a perfect conductor sheet. In this ideal configuration the radiation pattern is that of the $\lambda/2$ dipole taken in the upper half space, being zero the electric field at and below the reference ground plane. The input impedance is half of that of the $\lambda/2$ dipole. When the ground plane is of finite size, the image theorem does not fully apply.

This type of antenna may be used to model some real antennas, such as whip antennas (helical antennas in the broadside radiation mode), used for example in mobile telephones and portable transceivers, or receiving/transmitting monopoles for HF and VHF ranges. The effect of the real ground is that of changing slightly the elevation

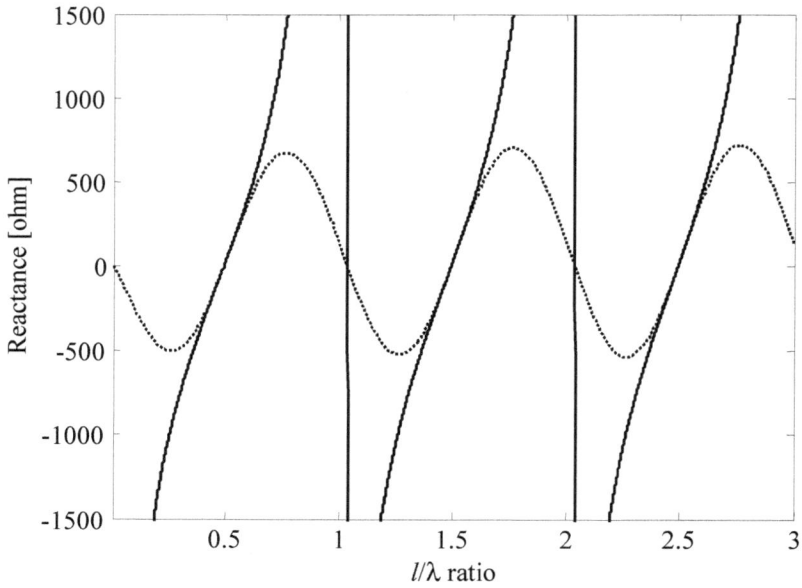

(a)

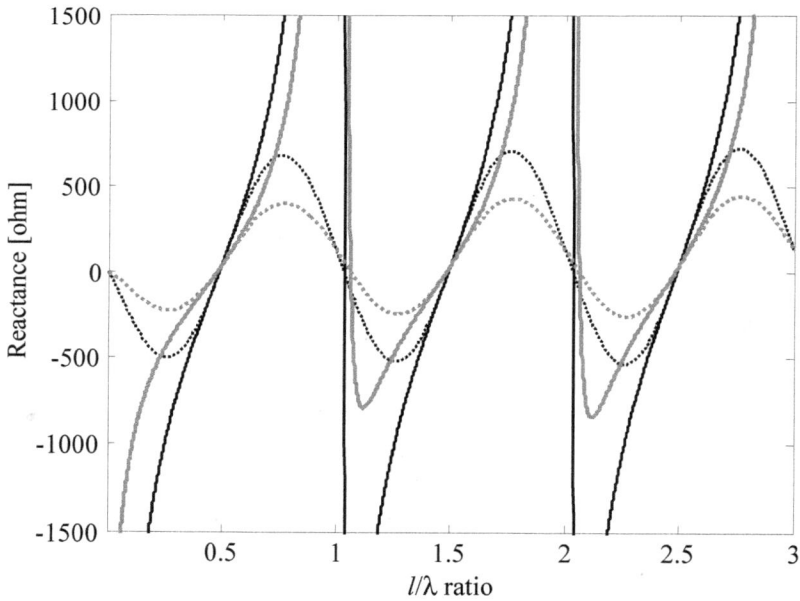

(b)

Figure 1.3.5 – (a) Self-resistance and (b) self-reactance for a dipole antenna with wire radius of $10^{-5}\lambda$ (black) and $10^{-3}\lambda$ (gray); solid lines represent the input resistance and reactance, dotted lines the radiation resistance and reactance.

of the radiation pattern lobe with field intensity drastically smaller very close to the ground (the zero elevation is no longer a maximum, as it was for perfect ground, but the radiation patter goes to zero).

The results for various ground conductance values are shown reported in [119].

1.3.5   Elementary antenna - small loop antenna

Let's consider a small circular loop antenna with center in the origin of the coordinates system and axis directed along the z axis; the distribution of current flowing in the φ axis is assumed constant and equal to I_0. As it was done for the Hertzian dipole, the small loop is subdivided into infinitesimal elements of length dz; with reference to Figure 1.3.1 the current elements are positioned along the φ axis. The three components of the electric field E and magnetic field H along r, θ and φ are determined as follows:

$$E_r = \frac{\eta}{2\pi} k^2 I \, dz \left[\frac{1}{(kr)^2} - j\frac{1}{(kr)^3} \right] \cos\theta \, e^{-jkr} \tag{1.3.43}$$

$$E_\theta = j\frac{\eta}{4\pi} k^2 I \, dz \left[\frac{1}{(kr)} - j\frac{1}{(kr)^2} - \frac{1}{(kr)^3} \right] \sin\theta \, e^{-jkr} \tag{1.3.44}$$

$$H_\varphi = j\frac{k^2}{4\pi} I \, dz \left[\frac{1}{(kr)} - j\frac{1}{(kr)^2} \right] \sin\theta \, e^{-jkr} \tag{1.3.45}$$

$$E_\varphi = H_r = H_\theta = 0 \tag{1.3.46}$$

The ratio of the first order components of the electric and magnetic field, that are projected along θ and φ axes (with propagation occurring along the r axis), give the free space impedance.

Following the development in [8], the small loop equations above may be computed for two extreme regions, where $kr \ll 1$ and $kr \gg 1$, respectively.

Reactive near-field region, $kr \ll 1$:

$$H_r = I \, a^2 \frac{1}{r^3} \cos\theta \, e^{-jkr} \tag{1.3.47}$$

$$H_\theta = I \, a^2 \frac{1}{4r^3} \sin\theta \, e^{-jkr} \tag{1.3.48}$$

$$E_\varphi = -jI \, a^2 \frac{k}{r^2} \sin\theta \, e^{-jkr} \tag{1.3.49}$$

Again the E-field component has a $-j$ coefficient in time-quadrature with respect to the H-field components: their time-averaged product is zero and so the associated power flow. Furthermore, the two H-field components have similar expressions and are in an in-phase condition.

Far-field region, $kr \gg 1$:

$$H_\theta = -I\,a^2\,\frac{k}{4r}\sin\theta\,e^{-jkr} \qquad (1.3.50)$$

$$E_\varphi = \eta I\,a^2\,\frac{k}{4r}\sin\theta\,e^{-jkr} \qquad (1.3.51)$$

This is the condition of orthogonal propagation of the E-field and H-field components projected only on θ and φ axes, with a 120π ratio (free space impedance) and with a linear dependence on distance; the associated power density, given by time-average product of the E-field and H-field, shows a dependence on distance of the type $1/r^2$.

The radiation resistance (and radiation efficiency) may be improved by increasing the number of turns; the same of course cannot be done with the Hertzian dipole.

1.3.6 Loop antenna

A loop antenna is again a very common antenna, that is applied in several applications, particularly in railway signaling and communication systems, where loop antennas may form the core of a uni or bi-directional communication system; in the field of measuring antennas, the loop is used as a directive search antenna or, on a smaller scale, as a field probe (*sniffer*) to diagnose the operation of electronic circuits and equipment; many sources may be modeled by means of a loop antenna, such as large and extended circuits where the structure of the ground connection is "circular" (the so called *ground loops*) or parts of commutation circuits (for example the same *snubber circuits*) inside power static converters, where large currents may flow during switching transients; any re-closure of the absorbed traction current and the traction return current forms loops of large extension, that involve the aerial and the grounded conductors. The loop has a very low radiation resistance that makes it an inefficient radiator; so, when used for radio communication applications it is used as a receiving, and not a transmitting, antenna. It is possible to increase the radiation resistance either by increasing the number of turns or by inserting ferrite beads (with high magnetic permeability) along the loop wire. The shape of the field pattern of electrically small loop antennas resembles that of a dipole, with a null on the loop axis (perpendicular to its plane) and the maxima around the loop circumference on the plane itself. When the loop circumference is close to one free-space wavelength the loop approaches its resonance and the maximum of the field pattern moves towards the loop axis.

1.3.7 Long wire antenna

The long wire antenna is very useful to model radiation of long conductors, such as catenary or third rail, when the simple low-frequency loop approximation is not sufficient any longer and multiple wavelengths unfold along conductor length. The long wire antenna model is also mentioned by the ERTMS standard for Euroloop. A complete solution may be found in [8], but the solution given in [87] with a pure sinusoidal (standing wave) current distribution is quite handy:

$$E_\theta = \frac{60I}{r} \underbrace{\left[\frac{1}{\sin\theta}\right.}_{A} \underbrace{\vphantom{\frac{1}{\sin\theta}}\cos}_{B} \left. \underbrace{\sin\left(\frac{\pi L}{\lambda}\cos\theta\right)}_{C} \right] \tag{1.3.52}$$

The coefficient A relates the field strength to the anti-node current I in amperes and the distance r in meters from the center of the wire. The coefficient B gives the envelope for the pattern lobes, θ being the angle to the wire axis. The coefficient C oscillates from $+1$ to -1 and contains the information about the angles of the zeros in the pattern. The cos or sin function in the coefficient C is used when the number of half wavelengths is odd or even, respectively. Coefficients B and C describe the shape of the pattern.

For a perfect non-attenuated traveling wave the current distribution of any length the far-field pattern is given as:

$$E_\theta = \frac{60I}{r} \underbrace{\left[\frac{\sin\theta}{1-\cos\theta}\right.}_{A} \underbrace{\vphantom{\frac{\sin\theta}{1-\cos\theta}}}_{B} \left. \underbrace{\sin\left(\frac{\pi L}{\lambda}(1-\cos\theta)\right)}_{C} \right] \tag{1.3.53}$$

For the angles of the series of maxima of diminishing amplitudes starting with the largest at θ_{m1} nearest to the forward axis of the wire, the angle of maxima can be determined from:

$$\theta_{mm} = \mathrm{hav}^{-1}\frac{K}{2L/\lambda} \tag{1.3.54}$$

where Haversines $\equiv (1-\cos\theta)/2$ and K has values of 0.371, 1.466, 2.480, 3.486, and 4.495. For the angles θ_{0m} for the mth zero with respect to the axis, from θ_{01} to $180°$

$$\theta_{0m} = \mathrm{hav}^{-1}\frac{m}{2L/\lambda} \tag{1.3.55}$$

1.4 Circuit theory and devices

This section covers the elements of circuit theory and the characteristics of some devices that are useful for several reasons, e.g. for the quantification of antenna performance and to describe its impact on and interaction with the setup, the behavior of the spectrum analyzer and the evaluation of results and related uncertainty, to cite the most relevant. The considered topics have been limited to network port representations, showing impedance and admittance parameters, transmission parameters, scattering parameters, the relationship with transmission lines and how they are usually quantified and manipulated. To simplify only attenuators have been considered among

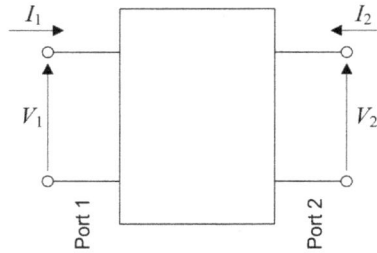

Figure 1.4.1 – Two-port network and related variables.

the devices for their usefulness for impedance matching (see sec. 1.5); the reader is invited to deal into other devices such as filters, baluns and couplers referring to more specific books [45, 129, 141].

1.4.1 Ports and matrix representation

In several disciplines it is useful and comfortable to use the concept of port, made of two terminal wires, where the input currents are bounded to sum to zero (or, in other words, are required to be equal in magnitude and of opposite polarity). Several electric circuits and equipment are suitable to be treated on a port basis (such as amplifiers, cables, transformers, etc.), which several matrix representations are related to. In the following a two-port network like in Figure 1.4.1 is considered, but n-port networks may be treated using the examples developed herein.

Depending on the choice of the input and output variables, starting from the four available variables, that is the voltages and currents of the two ports, V_1, I_1, V_2 and I_2, some matrix representations are possible.

1.4.1.1 Impedance matrix

This is the more "natural" representation of a two-port. We see immediately that there are many equivalent representations, and that impedance representation is not the most handy for transistors. The input variables are the two port currents I_1 and I_2; the output variables are the two port voltages V_1 and V_2.

$$\begin{bmatrix} V_1 \\ V_2 \end{bmatrix} = \begin{bmatrix} Z_{11} & Z_{12} \\ Z_{21} & Z_{22} \end{bmatrix} \begin{bmatrix} I_1 \\ I_2 \end{bmatrix} = [Z] \begin{bmatrix} I_1 \\ I_2 \end{bmatrix} \tag{1.4.1}$$

The relationships that define the terms inside the $[Z]$ matrix are very straightforward and identify a volt-amperometric test on the two-port network, while setting one quantity to a known condition.

$$Z_{11} = \left. \frac{V_1}{I_1} \right|_{I_2=0} \qquad Z_{12} = \left. \frac{V_1}{I_2} \right|_{I_1=0} \qquad Z_{21} = \left. \frac{V_2}{I_1} \right|_{I_2=0} \qquad Z_{22} = \left. \frac{V_2}{I_2} \right|_{I_1=0} \tag{1.4.2}$$

This representation is suitable for electric supply networks and to express the concept of input or loading impedance in various circuits (probes, transformers, etc.).

1.4.1.2 Admittance matrix

The input variables are the two voltages V_1 and V_2. The output variables are the two currents I_1 and I_2.

$$\begin{bmatrix} I_1 \\ I_2 \end{bmatrix} = \begin{bmatrix} Y_{11} & Y_{12} \\ Y_{21} & Y_{22} \end{bmatrix} \begin{bmatrix} V_1 \\ V_2 \end{bmatrix} = [Y] \begin{bmatrix} V_1 \\ V_2 \end{bmatrix} \tag{1.4.3}$$

The relationships that define the terms inside the $[Y]$ matrix are directly related to the one above for the impedance terms. The admittance matrix $[Y]$ is the inverse of the impedance matrix $[Z] = [Y]^{-1}$ and this can be easily demonstrated by substituting (1.4.1) into (1.4.3), or vice-versa.

The use of this representation is the same as that of the impedance matrix above; the choice between the two is based on the availability of either the two sets of variables when defining the problem, but for transistors admittance representation is particularly attractive, especially when including parasitic capacitance.

1.4.1.3 Transmission matrix

The input variables are the current and voltage on one port I_1 and V_1. The output variables are the current and voltage on the other port $-I_2$ and V_2.

$$\begin{bmatrix} V_2 \\ I_2 \end{bmatrix} = \begin{bmatrix} T_{11} & T_{12} \\ T_{21} & T_{22} \end{bmatrix} \begin{bmatrix} V_1 \\ I_1 \end{bmatrix} = [T] \begin{bmatrix} V_1 \\ I_1 \end{bmatrix} \tag{1.4.4}$$

The $[T]$ matrix contains non homogeneous terms, that are either the ratio of two voltages or two currents (so dimensionless), or the ratio of one current and one voltage, thus taking the dimension of an impedance or admittance. In analogy with the impedance terms they are defined as

$$T_{11} = \frac{V_2}{V_1}\bigg|_{I_1=0} \qquad T_{12} = \frac{V_2}{I_1}\bigg|_{V_1=0} \qquad T_{21} = \frac{I_2}{V_1}\bigg|_{I_1=0} \qquad T_{22} = \frac{I_2}{I_1}\bigg|_{V_1=0} \tag{1.4.5}$$

The opposite relationships between port 2 variables on the right-hand side and port 1 variables on the left-hand side are not used since they would lead to expressions that couldn't be realized, when trying to adopt the procedure for the measurement of the transmission parameters: for example, $T_{11} = \frac{V_1}{V_2}\bigg|_{I_2=0}$ requires the application of a driving signal V_2 to a port where also the condition $I_2 = 0$ applies.

This representation is suitable to assemble a series of building blocks to model more complex electrical networks, such as electric supply networks, complex circuits, etc. The choice of the input and output variables allows the direct cascade connections of different blocks, resulting simply in the chain multiplication of the respective $[T]$ matrices; to this aim one of the two currents is reversed in polarity and marked positive while exiting the port.

1.4.1.4 Wave parameters matrix

Wave representations and scattering parameters are reviewed with focus on the relationship with the other representations and on the underlying assumptions.

With a port perspective, the two sides of the matrix relationship are characterized by the pair of voltage and current at port 1 and at port 2 respectively:

$$\begin{bmatrix} V_1 \\ I_1 \end{bmatrix} = \begin{bmatrix} a_{11} & a_{12} \\ a_{21} & a_{22} \end{bmatrix} \begin{bmatrix} V_2 \\ -I_2 \end{bmatrix} = [A] \begin{bmatrix} V_2 \\ -I_2 \end{bmatrix} \tag{1.4.6}$$

$$\begin{bmatrix} V_2 \\ -I_2 \end{bmatrix} = \begin{bmatrix} b_{11} & b_{12} \\ b_{21} & b_{22} \end{bmatrix} \begin{bmatrix} V_1 \\ I_1 \end{bmatrix} = [B] \begin{bmatrix} V_1 \\ I_1 \end{bmatrix} \tag{1.4.7}$$

These two representations are the so-called *chain parameter* matrices.

For each port, once the reference impedance is established (corresponding usually to the characteristic impedance of the lines), called Z_{01} and Z_{02}, possibly different, we may write a set of expressions, relating current and voltage at each port, as governed by the reference impedance property.

$$a_1 = \frac{V_1 + Z_{01} I_1}{2\sqrt{Z_{01}}} \qquad b_1 = \frac{V_1 - Z_{01} I_1}{2\sqrt{Z_{01}}} \tag{1.4.8}$$

$$a_2 = \frac{V_2 + Z_{02} I_2}{2\sqrt{Z_{02}}} \qquad b_2 = \frac{V_2 - Z_{02} I_2}{2\sqrt{Z_{02}}} \tag{1.4.9}$$

These wave quantities, identified by letters a and b (maybe looking as parameters of matrices $[A]$ and $[B]$) are used to define the scattering parameters:

$$\begin{bmatrix} b_1 \\ b_2 \end{bmatrix} = \begin{bmatrix} s_{11} & s_{12} \\ s_{21} & s_{22} \end{bmatrix} \begin{bmatrix} a_1 \\ a_2 \end{bmatrix} = [S] \begin{bmatrix} a_1 \\ a_2 \end{bmatrix} \tag{1.4.10}$$

and going back to a port perspective, we obtain a scattering transfer matrix (or chain matrix) representations by means of the two matrices $[T]$ and $[U]$, its inverse:

$$\begin{bmatrix} b_1 \\ a_1 \end{bmatrix} = \begin{bmatrix} t_{11} & t_{12} \\ t_{21} & t_{22} \end{bmatrix} \begin{bmatrix} b_2 \\ a_2 \end{bmatrix} = [T] \begin{bmatrix} b_2 \\ a_2 \end{bmatrix} \tag{1.4.11}$$

$$\begin{bmatrix} b_2 \\ a_2 \end{bmatrix} = \begin{bmatrix} u_{11} & u_{12} \\ u_{21} & u_{22} \end{bmatrix} \begin{bmatrix} b_1 \\ a_1 \end{bmatrix} = [U] \begin{bmatrix} b_1 \\ a_1 \end{bmatrix} \tag{1.4.12}$$

Parameters appearing in (1.4.6) are indicated in [52] as ABCD parameters, where $A = a_{11}$, $B = a_{12}$, $C = a_{21}$ and $D = a_{22}$.

1.4.1.5 Scattering parameters of a 2-port network

Let's consider in more detail scattering parameters: we define V_1, I_1 and V_2, I_2 as the total voltage and current at the two ports, that may be grouped into two column vectors $\mathbf{I}$ and $\mathbf{V}$. Each port is considered connected to a test voltage source with internal impedance equal to the network reference impedance R.

It is possible to define incident quantities when a reference load R is connected to each voltage source, $\mathbf{V^i}$ and $\mathbf{I^i}$, each with two components, for port 1 and 2. The two vectors are related by a diagonal matrix of R values, since the two loads are decoupled and independent.

$$\mathbf{V^i} = \begin{bmatrix} R & 0 \\ 0 & R \end{bmatrix} \mathbf{I^i} \tag{1.4.13}$$

The incident currents and voltages are defined as the current through and the voltage across the two ports when the 2-port network is not connected and each generator is loaded by its reference impedance R. Hence

$$\mathbf{I^i} = \begin{bmatrix} \dfrac{E_1}{2R} \\ \dfrac{E_2}{2R} \end{bmatrix} = \begin{bmatrix} \dfrac{V_1 + RI_1}{2R} \\ \dfrac{V_2 + RI_2}{2R} \end{bmatrix} \qquad \mathbf{V^i} = \begin{bmatrix} \dfrac{E_1}{2} \\ \dfrac{E_2}{2} \end{bmatrix} = \begin{bmatrix} \dfrac{V_1 + RI_1}{2} \\ \dfrac{V_2 + RI_2}{2} \end{bmatrix} \tag{1.4.14}$$

Replacing the two test loads with the 2-port network, we define also the reflected variables as the difference between the defined incident current and voltage and the total current and voltage, respectively:

$$\mathbf{I^r} = \mathbf{I^i} - \mathbf{I} \qquad\qquad \mathbf{V^r} = \mathbf{V^i} - \mathbf{V} \tag{1.4.15}$$

$$\mathbf{I^r} = \begin{bmatrix} \dfrac{E_1}{2R} - I_1 \\ \dfrac{E_2}{2R} - I_2 \end{bmatrix} = \begin{bmatrix} \dfrac{V_1 - RI_1}{2R} \\ \dfrac{V_2 - RI_2}{2R} \end{bmatrix} \qquad \mathbf{V^r} = \begin{bmatrix} V_1 - \dfrac{E_1}{2} \\ V_2 - \dfrac{E_2}{2} \end{bmatrix} = \begin{bmatrix} \dfrac{V_1 - RI_1}{2} \\ \dfrac{V_2 - RI_2}{2} \end{bmatrix} \tag{1.4.16}$$

The relation between reflected and incident waves may be written as

$$\mathbf{I^r} = \mathbf{S^I} \mathbf{I^i} \qquad \mathbf{S^I} = \begin{bmatrix} S_{11}^I & S_{12}^I \\ S_{21}^I & S_{22}^I \end{bmatrix} \tag{1.4.17}$$

$$\mathbf{V^r} = \mathbf{S^V} \mathbf{V^i} \qquad \mathbf{S^V} = \begin{bmatrix} S_{11}^V & S_{12}^V \\ S_{21}^V & S_{22}^V \end{bmatrix} \tag{1.4.18}$$

The so-defined scattering parameters remain unchanged, provided that each port has its own reference impedance and its variables are defined upon that (the two reference impedances are taken equal).

Going from incident and reflected currents/voltages to a corresponding concept for wave quantities, the notation and the number of variables necessary to describe the network is simplified. Currents and voltages differ for R appearing at the denominator; splitting it into two multiplicative $\sqrt{R}$ terms split among current and voltage, the latter become equal, so that unique wave quantities may be defined: they are a for incident terms and b for reflected terms, and both are vectors, one row for each port:

$$\mathbf{a} = \begin{bmatrix} \dfrac{V_1 + RI_1}{2\sqrt{R}} \\ \dfrac{V_2 + RI_2}{2\sqrt{R}} \end{bmatrix} \qquad \mathbf{b} = \begin{bmatrix} \dfrac{V_1 - RI_1}{2\sqrt{R}} \\ \dfrac{V_2 - RI_2}{2\sqrt{R}} \end{bmatrix} \qquad (1.4.19)$$

$$a = S\,b \qquad \mathbf{S} = \begin{bmatrix} S_{11} & S_{12} \\ S_{21} & S_{22} \end{bmatrix} \qquad (1.4.20)$$

$$\mathbf{S} = \mathbf{S^I} = \mathbf{S^V} \qquad (1.4.21)$$

Taking a simple two-port network example, the most relevant parameters are:

- S_{11} identifies the reflection coefficient at the input port 1, and is commonly referred to as "return loss", for it indicates the amount of voltage that is returned reflected at the incident port and is not transmitted through to the other port;

- S_{21} identifies what is not transmitted to the output port 2, and is commonly referred to as "insertion loss", for it indicates the amount of signal that is lost in traveling from port 1 to port 2 (even if what is really indicated by S_{21} is the amount of signal that is transmitted from port 1 to port 2).

The two parameters, S_{11} and S_{21}, are not independent on each other; by invoking the conservation of energy we may write $S_{11}^2 + S_{21}^2 = 1$. The conservation of energy can be invoked if no significant coupling to adjacent traces and radiation losses occur. A coaxial cable is a type of interconnect that meet these requisites over a wide frequency range, while for PCB traces this is not absolutely true due to radiation and coupling to nearby lines.

As complex numbers, S parameters can be displayed in separate amplitude and phase plots or in polar plot, as shown below in sec. 1.4.2.

Reflection and transmission coefficients depend on the impedance matching between source (or load) and transmission line. From this the dependency of S parameters from a reference impedance value, that indicates the source/load impedance. Once S parameters are known with a given impedance reference value, any other impedance value may be selected and the S parameters changed accordingly, by using a matrix transformation. Of course if we are considering measured values, measurement errors may slightly vary due to scale change, thus affecting measurement accuracy; in other words measuring with $50\,\Omega$ or $100\,\Omega$ reference impedance will not lead to the same results in terms of uncertainty.

In case of an arbitrary loading impedance Z_L at the output port 2 of the network

$$S_{11}' = S_{11} + \frac{S_{12}S_{21}\Gamma_L}{1 - S_{22}\Gamma_L} \tag{1.4.22}$$

If the impedance mismatching is at the input port 1 with source impedance Z_S, then

$$S_{22}' = S_{22} + \frac{S_{12}S_{21}\Gamma_S}{1 - S_{11}\Gamma_S} \tag{1.4.23}$$

1.4.1.6 Relationships of scattering parameters with line parameters

It is known that for a linear network several representations of its input-output transfer functions are equivalent. The transformation relationships between the most common two-port representations are reported in [99] together with relevant references.

The S parameters are now considered in relationship to transmission line parameters: most of the elements that are part of RF and microwave setups are transmission lines or may be modeled as such; the relationship is briefly reviewed between transmission line quantities (namely characteristic impedance Z_c and propagation constant γ) and the S parameters measurable at line ports. The transmission line is symmetric and so only S_{11} and S_{22} are calculated:

$$S_{11} = S_{22} = \frac{\left[\left(\frac{Z_c}{Z_r}\right)^2 - 1\right]\sinh(\gamma L)}{2\left(\frac{Z_c}{Z_r}\right)\cosh(\gamma L) + \left[\left(\frac{Z_c}{Z_r}\right)^2 + 1\right]\sinh(\gamma L)} \tag{1.4.24}$$

$$S_{21} = S_{12} = \frac{2\left(\frac{Z_c}{Z_r}\right)}{2\left(\frac{Z_c}{Z_r}\right)\cosh(\gamma L) + \left[\left(\frac{Z_c}{Z_r}\right)^2 + 1\right]\sinh(\gamma L)} \tag{1.4.25}$$

where L is the transmission line length and Z_r is the reference impedance of measured S parameters.

These relationships can be inverted, so that the transmission line quantities may be determined experimentally using measured S parameters. In general Z_c and γ may be determined by curve fitting of measured S parameters, using e.g. a least mean square approach, provided that the variability with frequency of per-unit-length parameters was adequately modeled: the larger the frequency range, the more complex the task.

1.4.2   Interpretation of return and insertion loss

Taking a 50 Ω transmission line, like most coaxial cables, we go further with the interpretation of the behavior of S_{11} and S_{21} parameters, considering the output of a Vector Network Analyzer. If a sinusoidal wave is applied to port 1 using a 50 Ω source on a 50 Ω transmission line, it is known that theoretically the reflection coefficient is 0 and the entire wave is transmitted to the other port (except for line losses and coupling or radiation to the surrounding elements, as already said). The amplitude of S_{11} and S_{21} parameters are thus 0 and 1 respectively. From a practical standpoint, when measured with a VNA, they will look like a large negative value in dB and a nearly 0 dB curve when plotted versus frequency; including losses, the amplitude of S_{21} will begin to decrease with frequency.

1.4.2.1   Insertion loss S_{21} interpretation

The amplitude of S_{21} for increasing frequency detaches from the 0 dB value for dc and low frequency and begins decreasing in the negative dB range (so the figure gets larger, but negative) with a more or less linear shape, showing the typical periodic oscillations when the sample is not perfectly matched or line losses are not negligible.

The slope of the amplitude of insertion loss curve gives the attenuation constant α: this factor has always been considered in linear units, while an insertion loss measurement will often return a quantity expressed in dB ((1.4.27) below makes this explicit). The envelope of the real and imaginary part curves appearing in Figure 1.4.2(a) is exactly the magnitude value expressed in linear units. The magnitude is also shown in polar coordinates (Figure 1.4.2(b)), where the spiral is slightly looser at the lower frequency values and then gets denser and more packed.

What about the phase of the insertion loss curve? The interpretation of the phase curve requires that the behavior of the line under test is recalled when the sinusoidal test wave is applied (see also sec. 1.1.1.1): depending on the frequency, less than one period or many periods may be spatially present along the line at a given instant of time, depending on the ratio of wavelength and line length. With impedance mismatch at cable ends, reflections and standing wave will take place. Because perfect matching is never attained, a slight periodicity might be always visible in measured S_{21} curves (and of course in S_{11}). For a line of length L the applied sine-wave of frequency f (and wavelength $\lambda = v/f$, where $v = kc$ is the velocity of light through the line and k is the velocity factor of the cable with respect to the speed of light in vacuum c) will undergo a phase rotation φ of a complete turn when $\lambda = L$. So, at a given instant of time, when the input and the output sine waves are measured by the VNA to calculate S_{21}, their phase displacement will be increasing with frequency passing one period for every wavelength that is contained within line length. With the time delay TD of the line, i.e. $\mathrm{TD} = L/v$, the phase displacement between output and input sine waves will be increasing with frequency. If we observe that the measured phasor of the sine wave at port 2 is an old copy of the phasor applied to port 1 TD seconds ago, that has traveled through the line, and that at the same time instant the new phasor at port 1 that can be measured by the VNA has advanced in time and phase, we understand that the sign of the S_{21} phase is negative. The phase of S_{21} is thus expressed by $\varphi = -f\,\mathrm{TD}$.

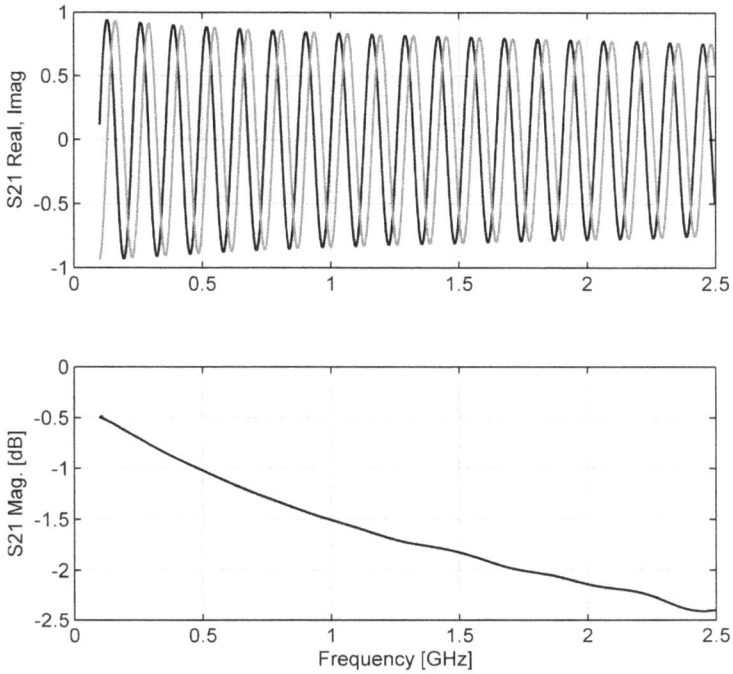

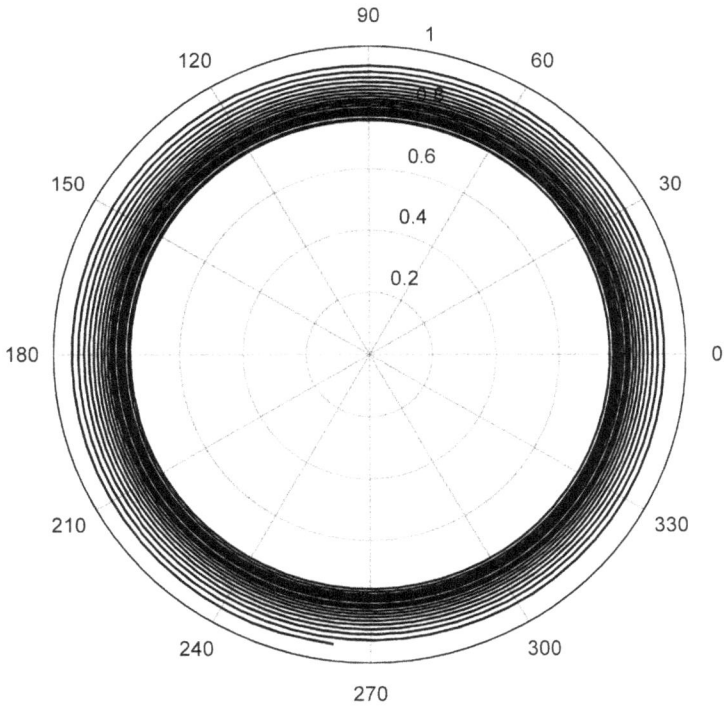

Figure 1.4.2 – Interpretation of S_{21} for a 10 m low-loss cable: (a) real and imaginary parts (above) and log mag (below); (b) polar plot of magnitude vs. unwrapped phase.

The direct quantification of attenuation of a cable sample is obtained by measuring insertion loss S_{21}: this is a two-port measurement that requires that both ends of the cable sample are accessible and available at the same location to plug into the VNA. The typical S_{21} curve shows the downward slope discussed above in sec. 1.4.2.1, due to the various loss mechanisms.

There is also an alternative using the return loss measurement alone, as seen below: $\alpha(f)$ may be estimated by fitting the S_{11} curve.

1.4.2.2 Return loss S_{11} interpretation

S_{11} interpretation follows that of S_{21}: the phase rotation is explained similarly, by observing that for the round-trip it is twice that evaluated for the S_{21} parameter. Assuming a reflective termination, to determination of S_{11} contributes not only the first reflection at the near port 1, but also the signal that propagates to the remote port 2 and that is reflected back to port 1: the phase relationship determines how these two terms compose and if they cancel out partially, totally or if they add in amplitude. At low frequency the reflection at port 1 is always negligible; the wave reflected back at port 2 has an opposite phase with respect to the one entering at port 1, so that they almost cancel out and the S_{11} curve starts from a large negative dB value. As the frequency increases the phase shift for the wave traveling from port 1 along the line increases; at a quarter of the cycle the reflected wave comes back at port 1 with half a period phase rotation: in this case the two signals entering port 1 from the generator and reflected back to port 1 almost cancel out, so that the reflection at port 1 is at its maximum and the S_{11} amplitude is equal to twice the reflection coefficient (assuming equal reflection coefficients for both ports). Correspondingly S_{21} will be at its minimum (by the conservation of energy). With the frequency further increasing S_{11} and S_{21} will alternatively cycle between their minima and maxima.

Observing (1.1.50), one may wonder how effective is a return loss measurement to determine line attenuation, as opposed to the more intuitive use of the insertion loss measurement. With some knowledge of real measurements and the shape of S_{11} curves, we know that a long enough line is needed for the satisfactory determination of α, but at the expense of many phase rotations and very frequent minima and maxima, that clutter the S_{11} curve and require complicated smoothing and interpolation. An amplitude-only measurement is considered.

$$S_{11} = 10 \log \left[\frac{Z_{in} - 50}{Z_{in} + 50} \right] \tag{1.4.26}$$

By replacing (1.1.37) into (1.4.26), a relationship between return loss S_{11}, cable length L and propagation constant γ is obtained:

$$S_{11} = 10 \log \left[\frac{\tanh(\gamma L) - 1}{\tanh(\gamma L) + 1} \right] = 10 \log \left[\exp(-\gamma L) \right] =$$

$$\tag{1.4.27}$$

$$= 10 \log \left[\exp(-\alpha L) \right] + 10 \log \left[\cos(-\beta L) + j \sin(-\beta L) \right]$$

The first term represents the attenuation and, assuming a constant value of α, it would be a horizontal line, constant with frequency; since α increases with frequency, the curve has a downward slope and $\alpha(f)$ may be estimated by curve fitting. The second term is a superimposed ripple, oscillating by the same mechanism shown for (1.1.48), where the distance between two adjacent peaks is such that $2\pi L/\lambda = \pi$ and thus $L = \lambda/2$. By observing the distance on the frequency axis between successive local minima or maxima, the resonant frequency f_0 is derived and wave velocity is $v = 2f_0 L$.

1.5 Attenuators

Attenuators are largely used, as their name says, to attenuate signal amplitude to better feed it to an instrument or circuit; attenuation, of course, may be expressed in terms of voltage attenuation, or more rarely as power attenuation. Normally they are designed for a given reference impedance value that is $50\,\Omega$, but we will see that they are useful also when there is an impedance difference (or mismatch) between two circuits, or parts of a network, and they are thus used to ease impedance matching at the expense of some attenuation.

Attenuator behavior and design with respect to its main characteristics (that is attenuation and impedance matching at input and output) may be considered and evaluated from different viewpoints. Different degrees of simplification are possible, if symmetry or a known reference impedance are assumed.

1.5.1 Circuit approach and calculation

The most common circuits used to build an attenuator are the "Pi" (or π) and "T" cell, as shown in Figure 1.5.1 (only for generality the elements of the cells are indicated by "Z" rather than by "R", but attenuators are normally built around resistive elements).

Since they are used in circuits where the nominal impedance of devices and equipment is the network reference impedance (i.e. $50\,\Omega$), the same termination at the source and at the load is assumed, even if for generality source and load impedances are kept separate; for this reason attenuators of this kind are symmetric, so that $R_1 = R_2$.

For the two attenuators in π and T configuration, there exist balanced counterparts with symmetric structure, where both terminals of the input and output ports are loaded in the same way (balanced). These configurations are obtained by splitting the longitudinal resistors in two halves and moving one of them to the lower part of the circuit, as shown in Figure 1.5.2: this brings benefits for common mode signals, but has no effect for differential signals.

The relevant quantities that characterize the attenuator when inserted in a larger circuit are the input impedance Z_{in} seen looking into either port 1 or port 2, and the attenuation A. The input impedance calculation is initially done for matched load conditions (where the opposite port is terminated on the correct reference impedance of $50\,\Omega$), but will be extended also to mismatched conditions, i.e. terminating the op-

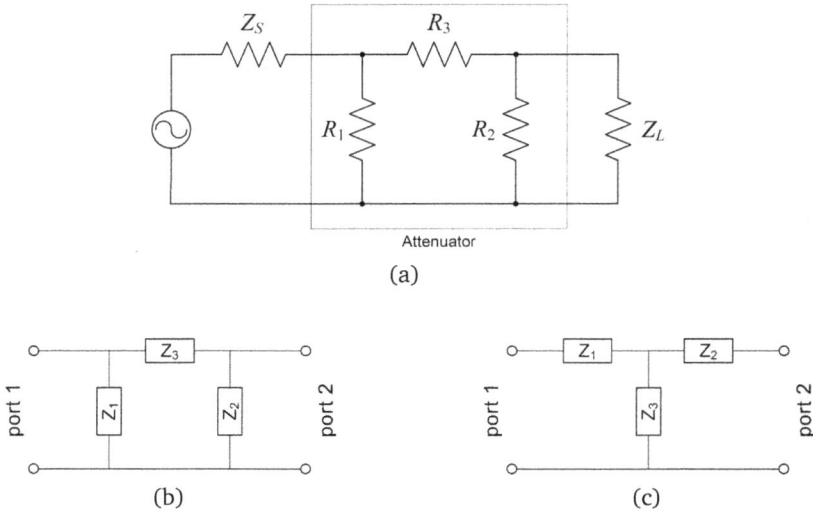

Figure **1.5.1** – (a) Attenuator π-cell and complete circuit including source and load; (b) general equivalent circuit for π-cell; (c) general equivalent circuit for T-cell.

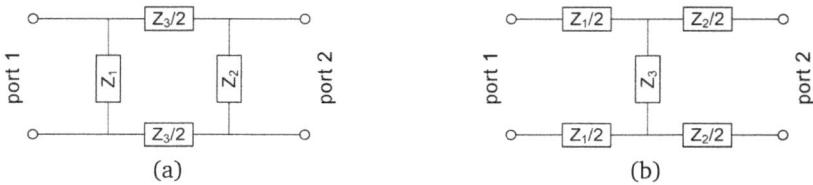

Figure **1.5.2** – Balanced versions of (a) π-cell attenuator and (b) T-cell attenuator.

posite port on a short or open circuit. Similarly, attenuation shall be always determined in matched load conditions, and then checked when the opposite port is terminated in extreme conditions.

Equations for the π attenuator

$$Z_{in} = Z_1//(Z_3+(Z_2//Z_L)) =$$

$$= \frac{Z_1 Z_2 Z_L + Z_1 Z_2 Z_3 + Z_1 Z_3 Z_L}{Z_1 Z_2 + Z_1 Z_L + Z_2 Z_L + Z_2 Z_3 + Z_3 Z_L} \tag{1.5.1}$$

$$V_{in} = E\frac{Z_{in}}{Z_S + Z_{in}} \tag{1.5.2}$$

$$V_{out} = E\frac{Z_1 Z_2 Z_L}{Z_1 Z_2 Z_S + Z_1 Z_S Z_L + Z_3(Z_1+Z_S)(Z_2+Z_L) + Z_1 Z_2 Z_L + Z_2 Z_S Z_L} \tag{1.5.3}$$

$$A = \frac{V_{out}}{V_{in}} = \frac{Z_2 Z_L}{Z_2 Z_3 + Z_2 Z_L + Z_3 Z_L} \tag{1.5.4}$$

Equations for the T attenuator

$$Z_{in} = Z_1 + \frac{Z_3(Z_2 + Z_L)}{Z_3 + Z_2 + Z_L} \qquad V_{in} = E \frac{Z_{in}}{Z_S + Z_{in}} \tag{1.5.5}$$

$$V_{out} = E \frac{Z_3 Z_L}{Z_3(Z_S + Z_1) + (Z_S + Z_1 + Z_3)(Z_2 + Z_L)} \tag{1.5.6}$$

$$A = \frac{V_{out}}{V_{in}} = \frac{Z_3 Z_L}{Z_2 Z_3 + Z_3 Z_L + Z_1(Z_2 + Z_3 + Z_L)} \tag{1.5.7}$$

Results are shown in Figure 1.5.3 and 1.5.4: scale and size of figures were chosen to fit the wide range of reference impedance and attenuation values.

1.5.2 Design formulas [156]

For the required attenuation A (expressed in linear units in dB voltage) and the reference impedance Z_{ref} (with which input and output impedance shall be matched, so assuming $Z_S = Z_L = Z_{\mathrm{ref}}$), the following expressions are available for the determination of the internal resistor values[3]. For the Pi and T attenuator respectively, we have:

$$R_1^\pi = R_2^\pi = Z_{\mathrm{ref}} \frac{1+A}{1-A} \qquad R_3^\pi = \frac{2R_1^\pi}{\left(\dfrac{R_1^\pi}{Z_{\mathrm{ref}}}\right)^2 - 1} \tag{1.5.8}$$

for which R_1 (and R_2) is larger than the reference source impedance Z_S and R_3 is always smaller than R_1;

$$R_1^T = R_2^T = Z_{\mathrm{ref}} \frac{1-A}{1+A} \qquad R_3^T = \frac{Z_{\mathrm{ref}}^2 - \left(R_1^T\right)^2}{2R_1^T} \tag{1.5.9}$$

where, conversely, R_1 (and R_2) is smaller than Z_S and R_3 is larger than Z_S. Other results and cross-checking of equations can be found in [99].

1.5.3 Improving matching with attenuators

It is a known practice placing attenuators at cable ends when connected to a mismatched load (such as an antenna with a varying VSWR), to reduce cable reflections and resonances. Used attenuation values are a compromise between good matching and limited signal loss: 3 dB or 6 dB attenuators are usually chosen. Attenuator performance is verified by terminating it with a varying Z_L, below and above the reference impedance ($50\,\Omega$) for a 5:1 range. Such a Z_L variation will cause an impedance mismatch characterized by a $(Z_L - 50)/(Z_L + 50)$ reflection coefficient, ranging between -0.667 and $+0.667$, and correspondingly a VSWR up to 5:1. Results are shown in

---

[3] See also Wikipedia: http://en.wikipedia.org/wiki/Attenuator_(electronics).

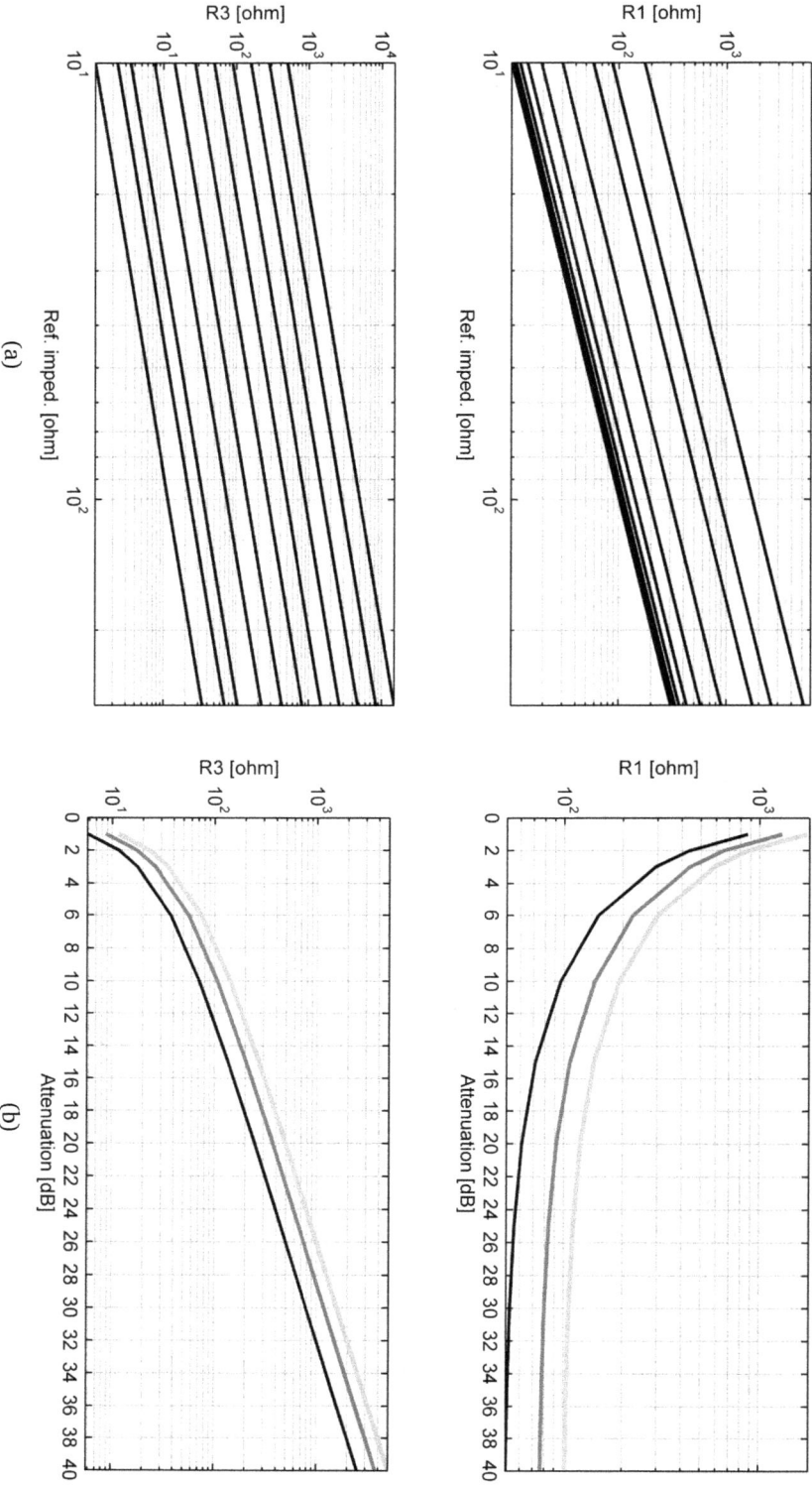

Figure 1.5.3 – π attenuator charts for (a) variable reference impedance in abscissa and attenuation (1, 2, 3, 6, 10, 15, 20, 25, 30, 35, 40 dB from bottom to top), (b) variable attenuation and $Z_S = Z_L = 50\,\Omega$ (black line), $75\,\Omega$ (dark gray line), $100\,\Omega$ (light gray line).

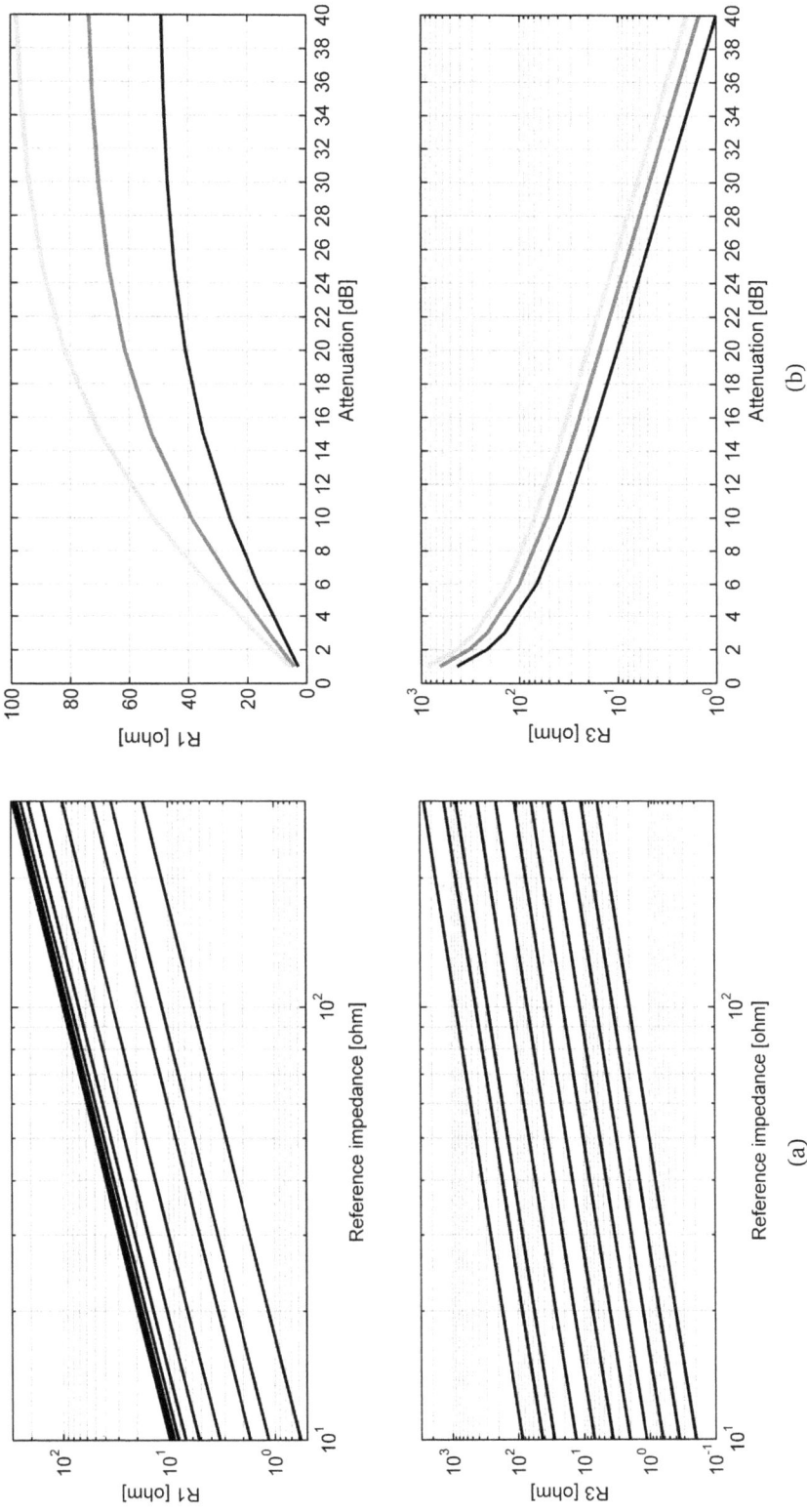

Figure 1.5.4 – T attenuator charts for (a) variable reference impedance in abscissa and attenuation (1, 2, 3, 6, 10, 15, 20, 25, 30, 35, 40 dB from bottom to top), (b) variable attenuation and $Z_S = Z_L = 50\,\Omega$ (black line), $75\,\Omega$ (dark gray line), $100\,\Omega$ (light gray line).

Figure 1.5.5: it may be seen that already a 6 dB attenuator limits the resulting VSWR to about 1.4 when Z_L spans to 10 or 250 Ω; a tighter control of the VSWR may be achieved then with a 10 dB attenuator, or a 20 dB one; going above these values it really unnecessary.

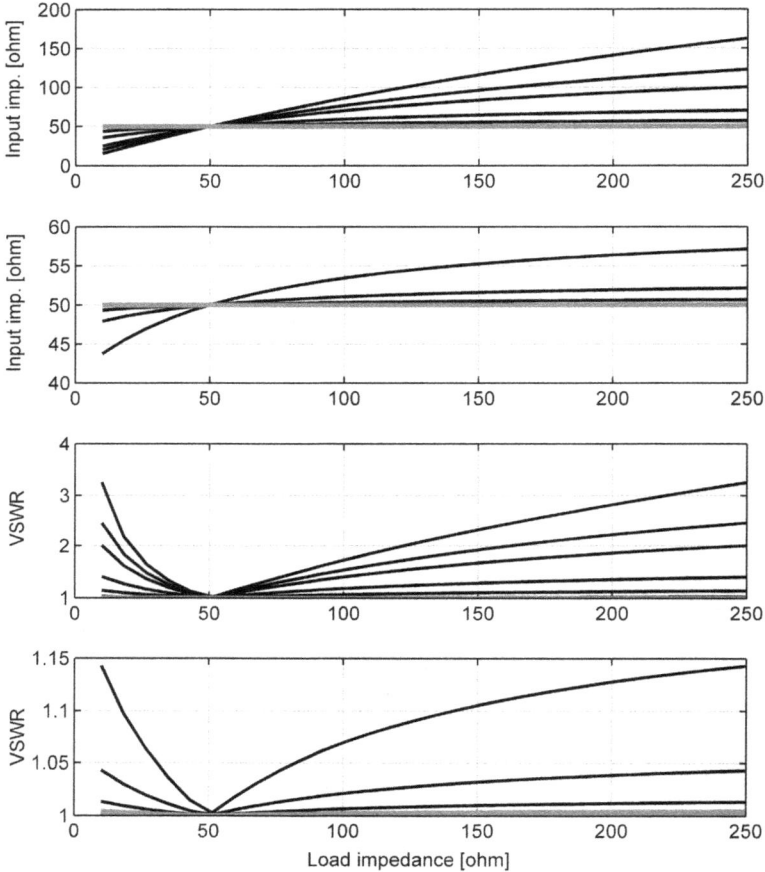

Figure 1.5.5 – Matching of input impedance enforced by T attenuators connected to a varying load: Att. = 1, 2, 3, 6, 10, 15, 20, 25, 30, 35, 40 dB, $Z_L = 10 \div 250\,\Omega$, $Z_{\mathrm{ref}} = 50\,\Omega$. First two subplots show the input impedance: (top) the five heavy black curves refer to Att. = 1, 2, 3, 6, 10 dB, (bottom) the other three heavy black curves refer to Att. = 10, 15, 20 dB; other two subplots show the resulting VSWR at the attenuator input: (top) the five heavy black curves refer to Att. = 1, 2, 3, 6, 10 dB, (bottom) the other three heavy black curves refer to Att. = 10, 15, 20 dB. The gray thick curve indicates the area occupied by the other values not explicitly taken into account.

2

Measurement fundamentals

In this Chapter there are reported and reviewed those concepts and properties for signals, signal transformation, statistics, that form a starting ground for carrying out the measurement of radiated emissions, the analysis of the results, their representation and evaluation, and the correct setting and verification of measuring equipment.

2.1   Signal quantities

The basic quantities and relationships involved in circuit and electromagnetic theory are briefly reviewed as a starting point.

Under the assumption of sinusoidal signals the peak (usually indicated by "p" or "pk") and root mean square (rms, or RMS) values are related by $X_{\mathrm{pk}} = \sqrt{2}\,X_{\mathrm{rms}}$. The general definition of rms for a signal $x(t)$ is based on power (or thermal) equivalent $X_{\mathrm{rms}} = \frac{1}{T}\int_T x(\xi)\mathrm{d}\xi$, where T is the period of the signal, but to extend also to aperiodic signals it may be generally considered as an observation time interval.

2.1.1   Decibel and distribution of measured quantities

The decibel (abbreviated dB) is a logarithmic representation of ratios, so that values, ranging several orders of magnitude, may be kept on a handy scale. The use of dB is so widespread since the application of gain or attenuation is translated directly into sum and difference operations.

The dB is defined as $10 \log_{10}(X_2/X_1)$, where X_1 and X_2 represent two homogeneous variables with a unit of power. It is evident that if X_2 is ten times X_1, then their ratio equals 10 dB. The variables X_1 and X_2 may represent two variables of a circuit or system (for example the input and the output power of a module, so that their ratio represents gain or attenuation) or, if X_1 is taken equal to a reference value that expresses the reference unit of measure, then X_2 becomes expressed in dB of this unit of measure: so, if $X_1 = 1\,\mathrm{W}$, then X_2 is expressed in dBW; if $X_1 = 1\,\mathrm{mW}$, then X_2 is expressed in dBmW or more commonly dBm; it is immediate to relate watts to milliwatts by observing that $1\,\mathrm{W} = 30\,\mathrm{mW}$.

By recalling the power law that for electric variables puts power in relationship with either voltage or current, the dB representation of the latter is obtained by a coefficient of 20 in front of the log operation: $20 \log_{10}(Y_2/Y_1)$. Again, the expression may represent a true ratio between two variables in a circuit or system, or the ratio of a variable with respect to a reference variable, thus obtaining dBV if $X = 1\,\mathrm{V}$, dBmV if $X = 1\,\mathrm{mV}$, dBA if $X = 1\,\mathrm{A}$ and so on; the relationship between different powers of the same measuring unit is obtained with the criterion of 20 dB/decade, so that $1\,\mathrm{dBV} = 60\,\mathrm{mV}$.

In this transformation we have momentarily neglected the value of the resistance R, which the voltage or current is applied to; power may be related to voltage or current when the reference resistance value is known.

Example 2.1: Relationship between power and voltage in dB

The very common transformation between X [dBm] and Y [dBV] commonly used in radio frequency measurements and calculations is developed for $R = 50\,\Omega$, the reference resistance for radio frequency equipment, such as antennas, cables, spectrum analyzers, amplifiers.

$$
\begin{aligned}
P &= 1\,\mathrm{mW}; \qquad P = V^2/R \\
X\,[\mathrm{dBm}] &= 10 \log_{10}(P) = 10 \log_{10}(V^2) - 10 \log_{10}(R) = \\
&= 20 \log_{10}(V) - 10 \log_{10}(50\,\Omega) = Y\,[\mathrm{dBV}] - 17
\end{aligned}
$$

As known, Y [dBV] may be transformed to Y_2 [dBµV] by adding 120 dB.

If the reference resistance value had been different, let's say $R_3 = 150\,\Omega$, then the new voltage Y_2 [dBµV] would be larger by $K = 10 \log_{10}(150\,\Omega) - 10 \log_{10}(50\,\Omega) = 21.8 - 17\,\mathrm{dB} = 4.8\,\mathrm{dB}$. Pay attention that the so calculated K is applied to a 20 dB/decade variable, Y_2 [dBµV], to get back another 20 dB/decade variable, Y_3 [dBµV].

Decibel is extremely useful when dealing with multiplicative composition of factors, such as cascaded blocks featuring gain or attenuation, interfaces characterized by impedance mismatching and as a consequence by a reflection-transmission coefficients pair. When expressing cable attenuation dB per unit of length (e.g. m, 100 m or 100 feet) is normally used as commonplace in cable datasheets; if using the attenuation constant α expressed in neper/m, a transformation factor is necessary. As explained at the end of sec. 1.1.1.1, a voltage or current is said to undergo an attenuation of X nepers when its magnitude changes by a factor $\exp(-X)$; Both nepers and dB may be used to express attenuation with respect to two voltages (or currents) V_1 and V_2 taken at two positions z_1 and z_2. By definition of neper the attenuation between V_1 and V_2 is X neper, if $V_1/V_2 = \exp(-X)$; then, by definition of dB $A = 20 \log_{10}(V_1/V_2) = 20 \log_{10}(\exp(-X)) = 8.686X$.

The dB scale is widely used for RF measurements and the most common attitude is to use it as a linear scale, forgetting that the value expressed in dB is obtained by means of a log operation. In particular, there are two questions that deserve an answer:

1. a confidence interval for a symmetric distribution is symmetric on the original measurement unit: is it still uniform if expressed in dB scale? if not, what is the expected skewness?

2. how the assumption on the original PDF of a variable (e.g. white Gaussian noise) preserves while going through the measurement chain and getting out finally represented on a dB scale?

It is necessary to go into the details of the log operation to answer both questions. Let's assume an input random variable x with a given probability distribution function PDF $p(x)$, that might represent the input noise signal samples or a set of lumped readings to be transformed in dB. The resulting dB value is $y = 10 \log_{10}(x)$ if x is a power, or $y = 20 \log_{10}(x)$ if x is a voltage or current. Let's proceed assuming that x is a voltage without losing generality. It is evident that if the span of x values is not large (i.e. its dispersion is small with respect to its mean value), a linear approximation may be valid: the first order Taylor's approximation of $\log_{10}(x)$ holds and y distribution does not deviate much from that of x.

Let's consider an intermediate r.v. $y' = \ln(x)$: it has a log-normal distribution and the mean and standard deviation of y' can be determined starting from those of x.

Then, y is related to y' by a linear transformation, that is a multiplying factor of 20 (or 10), together with $\log_{10}(e) \simeq 0.434$: $y = y' [20 \log_{10}(e)]$.

When considering the intensity of a Gaussian random process passing through a Fourier transform, with normally distributed in-phase and in-quadrature components, the resulting intensity (i.e. the modulus) is Rayleigh distributed (please see [99] for a graphical interpretation and quantitative evaluation). The Rayleigh PDF is not symmetric, has its maximum in a point near zero and has non-negative values. Thus the amplitude on a linear scale looks asymmetric, with a distribution far from the original normally distributed noise. Taking the log of a Rayleigh causes a long left tail in the negative dB value, thus having the resulting distribution much different from a Gaussian. It is of course not really an issue nor noticeable, when the attention is focused on the largest values around and above the mean, that should look satisfactorily Gaussian.

2.2 Probability and statistics

Probability and stochastic processes are essential to model phenomena, measurement results and the measuring instrument itself. When considering measured data the statistic perspective develops through probabilities and distributions, confidence intervals, and up to correlation and power spectral density functions.

Good references that treat exhaustively theory and examples are: Carlson [25], Drake [47], Montgomery and Runger [114], Papoulis [122].

2.2.1 Expectations and statistical moments

The expected value noted by $E\{\cdot\}$ is an average operation weighted by the likelihood of occurrence of the argument, or of the contained variables, that is to say weighted by the Probability Distribution Function (PDF) indicated by $p(x)$.

Considering a random variable x or a stationary random process $x(t)$ (for this, please, see sec. 2.5), its moments are defined as the expected values of increasing powers of x, so that the order of the moment corresponds to the exponent of x:

$$\mu_k = E\left\{x^k\right\} = \int_{-\infty}^{+\infty} x^k p(x)\,\mathrm{d}x \tag{2.2.1}$$

having used for brevity x as argument of the integral (should have been X, to distinguish it from the variable x).

Thus, the moment of order 0, μ_0, is simply equal to 1 because it is the integral of the PDF; the moment of order 1, μ_1, gives the mean value; the moment of order 2 gives the mean square value, indicated by ψ^2. For the latter and for higher order moment it is often convenient to express them as central moments, that is to calculate them about the mean value; the usual notation uses an appended "c".

$$\mu_k^c = E\left\{(x-\mu)^k\right\} = \int_{-\infty}^{+\infty} (x-\mu)^k p(x)\,\mathrm{d}x \tag{2.2.2}$$

The central moment of order 2 becomes thus the *variance*, indicated normally by σ^2. Descriptive statistics say that "variance" attempts to capture (or "model") in a single value the extent to which data vary from some basis; the chosen basis here is the mean value. The square root of variance is *dispersion*, which is the expected difference each sample has from the base value.

Another parameter normally considered a signal property rather than a statistical property is the *crest factor*: it is defined as the ratio between the peak value of a signal and its rms. Its use is for the definition of the input scale of data-acquisition systems to avoid over-range or to decide on the dynamic range of a receiver or transmitter to maintain linearity and avoid significant distortion. For a Normally distributed noise as input signal, a relationship between its statistical properties and the crest factor may be derived (see Table 2.2.1): in principle, the Normal distribution doesn't exclude the occurrence (very rare, but possible) of huge values located at the extremes of the distribution and above any threshold that may be selected.

Prob. %	Crest factor
4.6	2.0
1.0	2.6
0.37	3.0
0.1	3.3
0.01	3.9
0.006	4.0
0.001	4.4
0.0001	4.9

Table 2.2.1 – Probability of exceedance of the assumed crest factor for Normal distribution.

2.2.2 Time averages

When a random variable is characterized by a time axis being the result of the measurement of a random process, then the question may be posed of the use of time averages, noted by $\langle x \rangle$, rather than ensemble averages (in sec. 2.5 this is further considered).

$$\langle x \rangle = \frac{1}{T} \int_0^T x(t)\, \mathrm{d}t \qquad (2.2.3)$$

2.2.3 Statistical distributions

Some statistical distributions assume particular importance for our problems.

2.2.3.1 Uniform distribution

The uniform distribution is often used to indicate the maximum uncertainty on the expected outcome, that is only bounded between the two extremes a and b. This distribution is e.g. used to describe the quantization error of Analog-to-Digital converters (ADCs), where $a = -q/2$ and $b = +q/2$, with q being the resolution of the quantizer, corresponding to the ADC full scale FS, divided by the number of quantization levels, that is 2^n with n the number of bits. In the general case the mean value of a r.v. with uniform distribution is $(a+b)/2$ and the dispersion is $(b-a)^2/12$, that for an ADC correspond to 0 and $q^2/12$, respectively. Another example of a uniform distribution is related to the expression "the error of the instrument is $\pm e\,\%$", that refers to a definition of the error in the classical sense with no knowledge about its distribution, except its boundaries, normally assumed symmetric.

2.2.3.2 Normal, or Gaussian, distribution

The Normal, or Gaussian, distribution is well known and describes satisfactorily many natural phenomena, first of all thermal noise, in general affecting all electrical and electronic systems and measurements. Its distribution is

$$p(x) = \frac{1}{\sqrt{2\pi}} e^{-(x-\mu)^2/2\sigma^2} \qquad (2.2.4)$$

and is sometimes manipulated by a change of variable $z = (x - \mu)/\sigma$ to have a normalized zero-mean r.v. with PDF and CDF values tabulated in many books. What is sometimes overlooked is that a truly Gaussian distribution has arbitrarily large outcomes associated to a non-null probability, characteristic that is lost each time the original phenomenon is recorded in a digital system or passes through an electrical or electronic system, that has bounded outputs; what we obtain is a truncated Gaussian distribution. In general truncation of small noise signals by data acquisition system is not an issue because of the much larger scale setting with respect to span and dispersion. When setting the scale the operator shall be aware of the probability of missing a reading because of an out-of-scale value that saturates the measurement system: this is often addressed by considering the crest factor and the associated probability (see before sec. 2.2.1 and Table 2.2.1); values between 4.0 and 4.9 of crest factor give extremely low probabilities of missing a sample.

The *Central Limit Theorem* states that, given n r.v. x_i of arbitrary distribution (subject however to some constraints), the PDF of the r.v. z that is the sum of the said r.v. has a Gaussian distribution with mean value the sum of the mean values and variance the sum of variances. If the sum is normalized by the number of variables n and the x_i r.v. share the same PDF (e.g. are all outcomes of the same experiment as in the case when we average successive readings of the same variable to obtain a more precise measurement), then z is an unbiased estimator of the population mean μ_x and the mean square error of this estimator goes to zero with increasing n, being its variance $\sigma_z^2 = \sigma_x^2/n$. Thus, the variance of z is known, and so the error of the *sample mean estimator*, once the variance of the population σ_x^2 is known.

2.2.3.3 Rayleigh distribution

Given a two-dimensional Gaussian distribution with two independent variables x and y of zero mean and same standard deviation σ, the random variable that represents the modulus or length of this two-dimensional vector, $z = \sqrt{x^2 + y^2}$, has a Rayleigh distribution (see sec. 2.6.2.2 for a practical example). The distribution is

$$p(z) = \frac{z}{\sigma^2} e^{-z^2/2\sigma^2} \qquad (2.2.5)$$

The mean value is $\sigma\sqrt{\frac{\pi}{2}} = 1.25\sigma$ and the standard deviation is $\sigma\sqrt{2 - \frac{\pi}{2}} = 0.655\sigma$, where σ is the standard deviation of the two underlying Gaussian distributions.

This distribution is the first one of a series of unsymmetrical distributions, that are generally used to describe the behavior of combined Gaussian distributions and the effects on the overall intensity related to attenuation and propagation.

When the summed squared Gaussian components are three, the distribution is called Maxwell. Both the Rayleigh and the Maxwell may be considered a variant of the Chi-square distribution, when the considered variable is the square root of the sum of the squares, rather than the sum of the squares, of the individual normally-distributed input variables.

2.2.3.4 Student's t distribution

Another distribution that is mostly useful with problems with a low number of degrees of freedom (such as when a limited number of samples are available to make an estimate) is the "t" distribution, called — as known — Student's distribution after the pseudonym used by W.S. Gossett, when he published his work while working at the Guinness brewery[1].

$$p(x) = \frac{1}{\sqrt{\nu\pi}} \frac{1}{\left(1+\dfrac{x^2}{\nu}\right)^{\frac{\nu+1}{2}}} \frac{\Gamma\left(\dfrac{\nu+1}{2}\right)}{\Gamma\left(\dfrac{\nu}{2}\right)} \qquad (2.2.6)$$

where ν is the number of the degrees of freedom and $\Gamma(\cdot)$ is the Gamma function.

This family of curves depends on the ν parameter and for $\nu \to +\infty$ the distribution approaches the Normal distribution. The importance of this distribution is in describing the behavior of small data sets for phenomena with original Normal distribution: the t variable obtained by the following transformation

$$t = \frac{\bar{x} - \mu}{s/\sqrt{n}} \qquad (2.2.7)$$

with $\bar{x}$ the sample mean of a random sample of size n (n elements) from a Normal distribution with population mean μ and s the sample standard deviation, has a Student's t distribution with $\nu = n-1$ degrees of freedom. It is evident that it is not possible to operate with less than two elements in the sample and the interest is however in small samples, because for tens of elements the two distributions (Student's t and Gaussian) are already not easily distinguishable.

So the t distribution is useful to describe the sample mean estimator when the population mean is known (or a very good estimate of it may be obtained) and the population standard deviation is unknown. When considering the sample mean estimator with an unknown variance of the population, the best replacement for it is the sample variance estimated from the available samples; when the number of samples is very low it may deviate largely from that of the population. The t distribution is thus used to express the confidence interval for low n values[2].

---

[1] "Student" [William Sealy Gosset], "The probable error of a mean," *Biometrika*, Vol. 6, no. 1, pp. 1-25, March 1908.
Please have a look at the quite complete and insightful Wikipedia page at: `http://en.wikipedia.org/wiki/Student%27s_t-distribution`.

[2] It was proposed in an early draft of the EN 50121-2 standard, at the working document stage, to evaluate the compliance to limits of traction line emissions (caused mainly by pantograph arcing and thus intermittent) with a minimum number of recordings. It is still kept in the EN 55011 standard to evaluate compliance of production from testing on a sample set.

2.2.4 Statistical sampling and estimates

The estimation of moments based on a sample gives the so-called "sample mean" and "sample variance": the former, noted $\bar{x}$, is simply the arithmetic mean of all the N items of the sample and leads to a consistent estimate of the mean value μ of the entire population; the latter, noted s^2, is a biased estimate of the variance of the population, unless the normalization factor is brought to $N-1$ instead of N (the reason is that s^2, as a central moment, is evaluated with respect to the known base value, that is the sample mean $\bar{x}$, and not the mean of the population μ, thus losing one degree of freedom).

$$\bar{x} = \frac{1}{N} \sum_{i=1}^{N} x_i \qquad s^2 = \frac{1}{N-1} \sum_{i=1}^{N} (x_i - \bar{x})^2 \qquad (2.2.8)$$

Sample averages may be obtained by means of ensemble averages (that is "expectation") as above, or by means of time averages, when the values come from a process that shall be stationary and ergodic (see sec. 2.5.2). Thus time averages over the available record may replace ensemble averages.

For clarity we recall briefly what "consistent" means and some other terms used for estimation of statistical variables.

2.2.4.1 Bias of sample estimates

The *bias error* b is a constant difference between the estimate $\hat{\theta}$ and the — supposed known — true value θ^*; the word "error" may be dropped, using simply bias and leaving the term error for random errors. The *random error* is defined by the expected dispersion of the estimate around the true value.

$$b_\theta = E\left\{\hat{\theta}\right\} - \theta^* = \lim_{N \to \infty} \frac{1}{N} \sum_{i=1}^{N} \hat{\theta}_i - \theta^* \qquad \sigma_\theta = \sqrt{\lim_{N \to \infty} \frac{1}{N} \sum_{i=1}^{N} (\hat{\theta}_i - E\{\theta\})^2} \qquad (2.2.9)$$

It is evident that the sample mean has zero bias and its random error is the standard deviation, or the sample dispersion as a replacing estimate.

In some cases it is more informative to use the fractional counterparts of the bias error and random error with respect to the true value, that is $\varepsilon_b = b_\theta/\theta^*$ and $\varepsilon_r = \sigma_\theta/\theta^*$.

An *unbiased* estimator features a zero bias error. An estimator with a mean square error lower than any other estimators is said *efficient*. A *consistent* estimator features a zero bias error and a random error going to zero with increasing probability, as the number N of observations increases.

Depending on the underlying distribution, the best estimator may vary, so that the sample mean might not be the best choice always. As put in the NIST Handbook for Statistical Methods [118], "the optimal (unbiased and most precise) estimator for location of the center of a distribution is heavily dependent on the tail length of the distribution. The common choice of taking N observations and using the calculated sample mean as the best estimate for the center of the distribution is a good choice for the normal distribution (moderate tailed), a poor choice for the uniform distribution (short tailed), and a horrible choice for the Cauchy distribution (long tailed)." The best estimator for the uniform distribution is the simple mid-range value, obtained by averaging the two extreme values of the interval a and b $((a+b)/2)$: because of the many deviations from a clean vertical cut at the extremes of the distribution, this estimator is still subject to some inaccuracy and may be supported by some robust estimate of the two extreme values. For the Cauchy distribution, instead, the median is the best estimator of the center location. A quick check to identify the distribution is of course the inspection of a suitable plot of the data.

2.2.4.2 Bias of histogram methods

PDFs may be estimated from observations by the histogram method: let's consider again a set of N data values x_n $(n=0, 1 \ldots N)$ obtained from the recorded signal $x(t)$, stationary and with mean value μ. It is straightforward to estimate the PDF of x_n by choosing a set of intervals of width w (or w_i for generality, and thus different for each i), defined as $d_i = a + iw$, where a is the lower bound of the range of x values and i is an integer $i = 0, 1, \ldots K$, with K indicating the number of classes (or bins) chosen for the $[a, b]$ interval, so that $K = (b-a)/w$. If the number of data values falling within a single interval $[d_i, d_{i+1}]$ is indicated as N_i, such that $N = \sum N_i$, then a histogram can be built and the estimate of PDF p (and of its integral P, the Cumulative Distribution Function, CDF) may be computed as: $p_i = N_i/N$ and $P_i = \sum_{k=0}^{i} N_k/N$.

The common practice is to assign the obtained relative probability value to the center of the bin, located in X_i and embracing a range of values $[X_i - \Delta x/2, X_i + \Delta x/2]$. If the true underlying PDF has no change of slope within each bin (i.e. its first derivative is constant), then the average of all the ideal points falling inside the i-th bin gives effectively the true PDF value p_i^*; on the contrary, if there is a change of slope of the original PDF over a single bin, this means simply that the chosen width Δx is too large and that a smaller one would give better results. The resulting relative bias error of the PDF estimate depends on the higher order derivatives of the PDF and a first approximation is [12]

$$\varepsilon_b(x) = \frac{(\Delta x)^2}{24} \frac{p''(x)}{p(x)} \tag{2.2.10}$$

2.3 Time-frequency transforms

2.3.1 Fourier transform

Quite complete references for Fourier transform theory exist and report mathematical insights [21, 100, 121, 149], offering as well a more intuitive approach [90]. DFT/IDFT implementations are widely available in many math libraries and programming environments, both as Fast Fourier Transform (FFT) and recursive Fourier transform, called Goertzel's algorithm[3], that allows to limit efficiently the computation only to the few needed components with a proportional speedup.

The DFT transform pair is

$$X[k] = \sum_{n=0}^{N-1} x[n] e^{-j\frac{2\pi}{N}kn} \qquad x[n] = \sum_{k=0}^{N-1} X[k] e^{+j\frac{2\pi}{N}kn} \qquad (2.3.1)$$

with the same number N of samples in time (index n) and frequency (index k) domains.

2.3.2   Joint time-frequency transforms

Joint time-frequency representations transform a one-dimensional time signal into two-dimensional representation in the time-frequency plane. In this representation, time varying frequency can be easily displayed and studied. What a standard Fourier transforms misses is the time location, for which it is enable to distinguish two symmetrical situations of one sinusoidal tone at higher frequency preceding a second one at lower frequency within the same observation window, or viceversa: the Fourier spectrum will show always the two tones frozen in a unique spectrum picture taken over the said observation window; reducing its length or shifting it will cause one of the two tones to disappear, or its measured amplitude reduced.

Modulations are a typical example of signals featuring time varying frequency; transients also may feature different spectral signature during their occurrence, from the initial front to the decaying slope with various kinds of superimposed components.

In general joint time-frequency transforms are also more effective in detecting potentially non-stationary signals out of noise, being able to spot out local changes of the spectral signature, without smearing them out while integrating all components within the same observation window, that for the Fourier transform shall be long enough to accommodate for a satisfactorily small frequency resolution. A very good reference for the analysis of non-stationary signals by means of one of the fundamental joint time-frequency transforms, the Wigner-Ville transform, is the paper by Martin and Flandrin 30 years ago [102].

2.3.2.1   Short Time Fourier Transform and Gabor transform

A moving-window Fourier transform is the first example of frequency analysis with an adjustable time resolution: in Short Time Fourier Transform subsequent transformation windows of time duration T are taken, resulting in a frequency resolution $df = 1/T$: such windows may be overlapped or not and we have already considered implicitly this case in sec. 2.3.1 and 2.6.2.2.

---

[3] The original paper is G. Goertzel, "An Algorithm for the Evaluation of Finite Trigonometric Series," *American Mathematical Monthly*, Vol. 65, No. 1, Jan. 1958, pp. 34–35. Details are given in [131].

A weighting function may be added, so that the Fourier transform is taken of the product of the signal and the weighting function: it is exactly the role of smoothing windows, used for tapering and control of frequency leakage (see sec. 2.3.3). The rigorous application of the Fourier transform is based on a rectangular weighting function. In general, other kinds of weighting functions may be used: when a Gaussian function $e^{-\pi t^2}$ is used, then the Gabor transform results, for which the time-frequency area is minimal (in other words the Gabor inequality holds with the equal sign for Gabor transform, representing a lower bound for the joint entropy).

From the expression of the original Gabor transform it may be seen that it is a linear transform.

$$\mathrm{GT}(t, f) = \sqrt{\frac{1}{2\pi}} \int_{-\infty}^{+\infty} x(t)\, e^{-(t-\tau)^2/2} e^{-j2\pi f(\tau - t/2)}\, \mathrm{d}\tau \tag{2.3.2}$$

2.3.2.2 Energy transforms and Cohen family of transforms

Enhanced time and frequency resolution may be achieved by correctly selecting the kernel of the transform. Joint time-frequency transforms in general require the multiplication of two copies of the signal $x(t)$ suitable displaced, after a complex conjugate operation: for this reason they are often named "energy transforms", or "power transforms". This product is then multiplied by two exponentials along the two transform axes, that may be called time and frequency. Then transforms distinguish one from the other for an additional shaping of the kernel that is represented by the so called "parametrization function". This is the general scheme of Cohen's transforms, that is able to represent the largest part, not to say all, of the existing joint time-frequency transforms:

$$\mathrm{C}(t, f; g) = \iiint x(s + \tau/2)\, x^*(s - \tau/2)\, e^{j2\pi\xi(s-t)} e^{-j2\pi ft} g(\xi, \tau)\, \mathrm{d}\xi\, \mathrm{d}s\, \mathrm{d}\tau \tag{2.3.3}$$

The first to appear before Cohen's work in the '70s was the Wigner-Ville, that may be used also to express differently the whole family of Cohen's transforms. In the Wigner-Ville transform the function $g(\cdot)$ is unity.

All non-linear transforms (such as this family of bilinear transforms) suffer of the problem of the spectral cross-terms, whenever the input signal contains more than one tone [138]: cross terms may be recognized as such if the spectrum of the input signal is known or if at least its structure in terms of frequency occupation is known; from a general standpoint cross terms may be confused with real signal components.

The Choi-Williams was conceived as a smoothed version of the Wigner-Ville, giving a smaller amount of cross-modulation terms.

$$g_{CW} = e^{-(\pi\xi\tau)^2/2\sigma^2} \tag{2.3.4}$$

The constant σ controls the amount of smoothing: with a large σ going to infinite the Choi-Williams converges back to the Wigner-Ville.

$$\mathrm{CW}(t, f) = \sqrt{\frac{2}{\pi}} \iint\limits_{-\infty}^{+\infty} \frac{\sigma}{|\tau|}\, x(s+\tau/2)\, x^*(s-\tau/2)\, e^{-2\sigma^2(s-t)^2/\tau^2}\, e^{-j2\pi f t}\, \mathrm{d}s\, \mathrm{d}\tau \qquad (2.3.5)$$

SWV is another transform purposely conceived as a smoothed version of the Wigner-Ville distribution:

$$\mathrm{SWV}(t, f) = \int\limits_{-\infty}^{+\infty} h(\tau) \int\limits_{-\infty}^{+\infty} k(s-t)\, x(s+\tau/2)\, x^*(s-\tau/2)\mathrm{d}s\, e^{-j2\pi f t}\, \mathrm{d}\tau \qquad (2.3.6)$$

Using such transforms without inspecting the results for cross-modulation terms is risky, especially for those derived from the Wigner-Ville and Choi-Williams, that however offer very good performance in time-frequency resolution.

Conversely, the Gabor transform is quite reliable with no cross-modulation terms and uniform response for various kinds of signals, except those with a large chirp. A very interesting comparison for typical radar and air target signals is reported in [152].

The clarity of spectrum representation and time and frequency resolution of the Wigner-Ville have been combined to the Gabor linear transform, void of cross term products by definition: the Gabor-Wigner transform was proposed in [124].

2.3.3 Spectral leakage and windowing

Whenever capturing a time-domain signal and performing a time-to-frequency transformation, the problem of spectral leakage arises [60, 90, 121]. Harris describes the leakage phenomenon in an elegant way: "From the continuum of possible frequencies, only those which coincide with the basis will project onto a single basis vector; all other frequencies will exhibit non zero projections on the entire basis set. This is often referred to as spectral leakage and is the result of processing finite-duration records. Although the amount of leakage is influenced by the sampling period, leakage is not caused by the sampling. An intuitive approach to leakage is the understanding that signals with frequencies other than those of the basis set are not periodic in the observation window." The periodic extension of a signal not commensurate with its natural period exhibits discontinuities at the boundaries of the observation interval; the discontinuities are responsible for spectral contributions (or leakage) over the entire basis set. The form of this discontinuity corresponds to the abrupt step as shown in Figure 2.3.1.

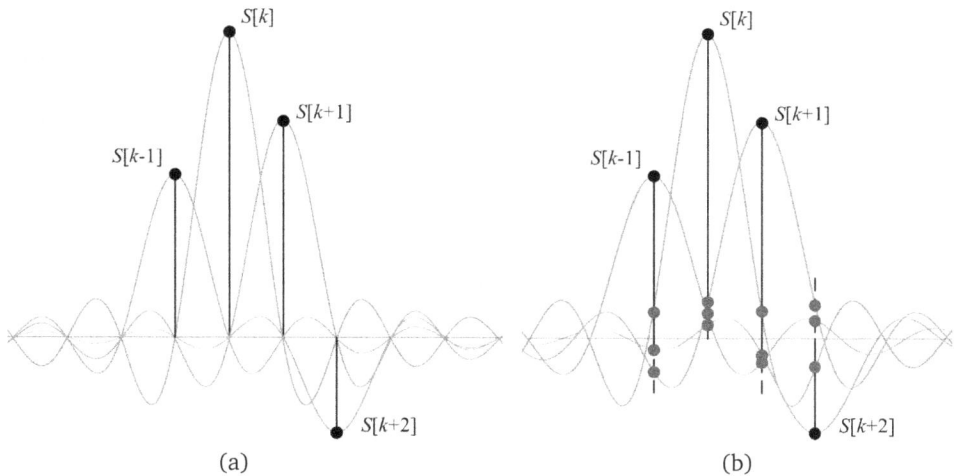

Figure 2.3.2 – Effect of a *sinc* with non-integer period compared to that of the underlying signal: (a) synchronized situation where the *sinc* of each sample do not leak onto the adjacent samples; (b) the observation period (and the *sinc* oscillation period) is increased by 20% and the leaking terms onto adjacent samples are shown as gray dots (for the smallest negative sample $S[k+2]$ this interference amounts to about 30% in absolute value, while for the other larger samples it is less than 20%).

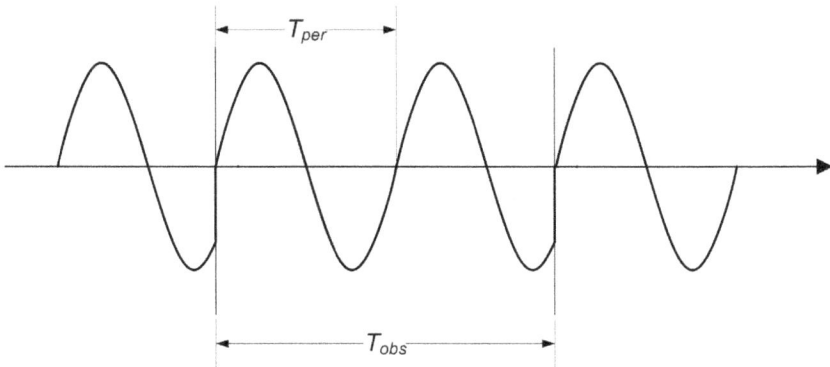

Figure 2.3.1 – Periodic signal and its extension over an observation interval (called also "epoch"), where it is not periodic.

An equivalent explanation may be given by considering the behavior of the *sinc* function that is convolved in frequency with each sample at successive frequency bins, corresponding to the implicit rectangular windowing due to the cutoff of the time-domain signal at its boundaries. The *sinc* has a maximum in the origin and zeros at multiples of $1/T$, where T is the time duration of the epoch captured with the rectangular window. If the fundamental and the harmonics of the captured signal do not occur where the *sinc* zeros are (that is to say that they are synchronous with the implicit period T set by the rectangular cut-off), at each frequency f_h there will be non-zero convolution terms that add with arbitrary sign to the main bin located at f_h, thus altering its original amplitude. This is shown in Figure 2.3.2.

Smoothing windows (or equivalently *tapering windows*) are weighting functions applied to data to reduce the spectral leakage associated with finite observation intervals.

From one viewpoint, the window is applied to data (as a multiplicative weighting function) to reduce the order of the discontinuity at the boundary of the periodic extension where the signal was cut. This is accomplished by matching as many orders of derivative as possible at the boundary of the resulting windowed data and setting them to zero, or nearly so. Thus windowed data are smoothly brought to zero at the boundaries, so that the periodic extension of the data is now continuous up to some high-order derivative.

Similarly to the interpretation of frequency leakage based on the *sinc* function in frequency domain, the action of smoothing windows may be also interpreted as a uniform reduction of the secondary lobes of the *sinc* function, thus reducing in case of lack of "synchronization" the impact of additional non-zero terms during convolution.

Using windows has impact on many attributes of the spectral analysis, including detectability, resolution and dynamic range (see Table 2.3.1).

Concerning the influence on noise level, we observe that the amplitude of the harmonic estimate at a given frequency is biased by the accumulated broad-band noise that falls inside the window bandwidth. In this sense, the window behaves like a filter, gathering contributions over its bandwidth; to minimize this accumulated noise and maximize the signal-to-noise ratio, window bandwidth shall be small, where the term "bandwidth" is the equivalent noise bandwidth (ENBW) (see sec. 2.6.1.1). Analogously, the internally generated numeric noise of the transform calculation (although very small and often negligible) is spread over a larger number of bins and the resulting noise floor is lower.

The interference, or mutual effect, between spectral components is called *frequency leakage*, or *spectral leakage*. Such phenomenon is in reality referred to in some literature as long-range frequency leakage, opposed to the other form of leakage, the short-range frequency leakage, that corresponds to the simpler fact that when the spectrum is represented as samples on a discrete frequency vector, the real underlying signal frequency may lie between two discrete points, thus worsening or biasing the frequency estimate. Radicalizing the definitions for simplicity, the two forms of leakage, despite similar names, have a different nature and behavior and are to be treated differently: smoothing (or tapering) windows are used for the long-range, while interpolation is used for the short-range leakage. It is author's opinion that the true leakage by its proper name be the long-range frequency leakage, that will be shortly addressed, while the short-range leakage is caused by a problem of limitation of the frequency resolution (see sec. 2.3.3.1 below on scalloping loss). However, the two phenomena are in reality inter-related and not so easily separable, because assuming a lack of correspondence between the true signal frequency and the DFT sample points on the frequency axis for the short-frequency leakage to occur, implies that the signal is not periodic in the observation window T, thus leading also to the other form of leakage, the long-range frequency leakage.

Even for the case of a single sinusoidal tone with a definite spectral line (not located at a DFT sample point, or in other words not periodic in the observation window), the leakage from the kernel on the negative-frequency axis biases the kernel on the positive-frequency axis; the sinusoidal tone line spreads around, ending up with a lower peak amplitude and causing an increase of the noise floor due to coherent terms, that are not suppressed by averaging. This bias is the most severe and bothersome for the detection of small signals in the presence of nearby large signals. To reduce the effects of this bias, the window should exhibit low-amplitude sidelobes far from the central main lobe, and the transition from the main lobe to the low sidelobes should be very rapid. One indicator of how well a window suppresses leakage is the peak sidelobe level (relative to the main lobe); alternatively the asymptotic rate of fall-off of these sidelobes may be used.

The use of tapering windows worsens frequency resolution according to the main-lobe width, that is usually measured with respect to the rectangular window of time duration T; as a countermeasure, the time duration of the epoch extracted from the signal may be increased proportionally beyond T to compensate for this loss of frequency resolution. Normally, for the most common windows, the main-lobe width at $-3\,\mathrm{dB}$ lies in the range 1.5 to 2.0 times that of the rectangular window, so accounting for a 30 % to 50 % loss of frequency resolution.

Since the window assigns an effective bandwidth to the spectral lines, we would be interested in the minimum separation between two equal strength lines, such that for arbitrary spectral locations their respective main lobes can be resolved and identified in the final resulting spectrum (we will see in sec. 3.1.3.5 that this is the operative definition of effective resolution). The DFT output points are the *coherent* addition of the spectral components weighted through the window at a given frequency; at the crossover points of the window frequency kernels, the gain from each kernel must be less than 0.5, or in other words the crossover points must occur at the 6 dB points of the windows or below. So, $-6\,\mathrm{dB}$ bandwidth is relevant when evaluating the effective resolution for coherent contribution.

If short time duration tone-like signals must be detected, the non overlapped DFT analysis of the signal might miss the tone if it occurs near the time window boundaries; in general overlapping is good at tracking the dynamic behavior of the spectrum for quasi-stationary slow-varying signals. In this case overlapped sub-sequences must be used and the overlap percentage is usually 50 % or 75 %, not only for practical reasons, but also to particularly reduce the correlation between the spectral components of successive sub-sequences; this is shown in sec. 2.6.2.2.

2.3.3.1 Scalloping loss

When evaluating broadband signals populated by many unrelated components, a significant issue is also the *scalloping loss*, that is the attenuation of signal components that lie mid between two bin frequencies; such phenomenon was defined above as short-range frequency leakage. The bin width and thus the bin frequency spacing is set by the chosen transformation period T; at the crossing of the window transforms located at two adjacent bins there is the maximum processing gain reduction and this amounts to 4 dB. The exact expression of the scalloping loss when the sinusoidal tone frequency f_s is not aligned to one of the Fourier spectrum bin frequencies f_k involves the $\mathrm{sinc}(\cdot)$ function (the rectangular window transform) and it is $\mathrm{sinc}^2(\pi(f_s - f_k)T)$, the maximum occurring as expected when the difference between the two frequencies is half the resolution frequency: $|f_s - f_k| = 1/2T$.

Two countermeasures are normally adopted for this phenomenon:

- increase frequency resolution fictitiously by *zero padding* (or zero stuffing), that is adding zero samples to the original sequence increasing the number of samples N involved in the calculation (bringing them always to a power of two for convenience), but keeping the captured signal power unchanged and without any improvement in the signal-to-noise ratio, except the span of the broadband incoherent noise over more frequency bins, but leaving the spectral density (i.e. per unit frequency) unchanged;

- interpolation of the exact location and height of the component by exploiting the spectral samples in the nearby bins (normally referred to simply as *frequency interpolation*); there exist closed-form expressions for some smoothing windows that give the frequency and amplitude correction terms; interpolation have been accomplished differently in the literature, depending on the used window and the attained degree of approximation (please see sec. 2.3.3.3).

2.3.3.2 Smoothing windows

The most common smoothing windows, that are also found on-board VNAs (and FFT analyzers), are the rectangular (i.e. no windowing), Hann (or von Hann), Hanning, Hamming, Gaussian, Kaiser-Bessel and Dolph-Chebyshev windows. Formulas are reviewed in [99], where windows and their transforms are also plotted in time and frequency domains.

A synthesis of the main performance indexes of the above windows is shown in Table 2.3.1. A few explanatory notes are necessary: 1) the Hanning window for $\alpha = 2$ is the von Hann window; 2) the side-lobe fall-off of the Dolph-Chebyshev is zero because all side-lobes have the same height; 3) the order of the Kaiser-Bessel [60] includes a multiplicative factor π, that is absent in the Matlab function `kaiserwin()`.

The roll-off for smoothing windows is simply dictated by the type of window as for instance obtained by some power of the $\cos(\cdot)$ function or else; as it may be noticed smoothing windows have normally a roll-off of 6 dB/octave, with increasing roll-off for the Hanning group for increasing α (as said, the exponent of the $\cos(\cdot)$ term); the Bohman has 24 dB/octave, being obtained as the convolution of two windows featuring 12 dB/octave each; finally, for the Dolph-Chebyshev window expressing roll-off is not meaningful because the side-lobes are all limited by design to the specified level.

2.3.3.3 Interpolation of frequency and amplitude

As anticipated, the problem of scalloping loss (see sec. 2.3.3.1) affects the accuracy of the amplitude estimation resulting from the Fourier spectrum of signals whose components do not fall exactly on the Fourier frequency bins for the chosen observation period T and number of sample points N (this phenomenon was called short-range

Window	Degree	Highest side-lobe [dB]	Side-lobe fall-off [dB/oct.]	Coherent gain	ENBW	-3 dB BW	-6 dB BW	Correlation 75%/50% overlap
Rectangle	–	-13	-6	1.00	1.00	0.89	1.21	0.75/0.50
Hamming	–	-43	-6	0.54	1.36 (1.38)	1.30	1.81	0.707/0.235
Hanning	$\alpha = 1.0$	-23	-12	0.64	1.23	1.20	1.65	0.755/0.318
	$\alpha = 2.0$	-32	-18	0.50	1.50	1.44	2.00	0.659/0.167
	$\alpha = 3.0$	-39	-24	0.42	1.73	1.66	2.32	0.567/0.085
	$\alpha = 4.0$	-47	-30	0.38	1.94	1.86	2.59	0.486/0.043
Bohman	–	-46	-24	0.41	1.79	1.71	2.38	0.545/0.074
Gaussian	$\alpha = 2.5$	-42	-6	0.51	1.39	1.33	1.86	0.677/0.200
	$\alpha = 3.0$	-55	-6	0.43	1.64	1.55	2.18	0.575/0.106
	$\alpha = 3.5$	-69	-6	0.37	1.90	1.79	2.52	0.472/0.049
Kaiser-Bessel	$\beta = 2.0$	-46	-6	0.49	1.50 (1.519)	1.43	1.99 (2.02)	0.657/0.169
	$\beta = 2.5$	-57	-6	0.44	1.65 (1.678)	1.57 (1.594)	2.20 (2.23)	0.595/0.112
	$\beta = 3.0$	-69	-6	0.40	1.80 (1.824)	1.71 (1.729)	2.39 (2.423)	0.539/0.074
	$\beta = 3.5$	-82	-6	0.37	1.93 (1.959)	1.83 (1.855)	2.57 (2.602)	0.488/0.048
	$\beta = 4.0$	–	–	(0.344)	(2.086)	(1.972)	(2.770)	–
	$\beta = 6.0$	–	–	(0.282)	(2.532)	(2.387)	(3.360)	–
Blackman-Harris	3-sample	-61	-6	0.45	1.61	1.56	2.19	0.610/0.126
	4-sample	-92	-6	0.36	2.00 (2.04)	1.90 (1.93)	2.72 (2.70)	0.460/0.038
Dolph-Chebyshev	$\alpha = 2.5$	-50	0	0.53	1.39 (1.41)	1.33 (1.344)	1.85 (1.875)	0.696/0.223
	$\alpha = 3.0$	-60	0	0.48	1.51 (1.537)	1.44 (1.46)	2.01 (2.04)	0.647/0.163
	$\alpha = 3.5$	-70	0	0.45	1.62 (1.65)	1.55 (1.57)	2.17 (2.197)	0.602/0.119
	$\alpha = 4.0$	-80	0	0.42	1.73 (1.76)	1.65 (1.67)	2.31 (2.34)	0.559/0.087
	$\alpha = 5.0$	-100	0	–	– (1.96)	– (1.85)	– (2.60)	–

Table 2.3.1 – Smoothing windows performance [60] verified with numeric calculations (differing values are shown between parentheses).

frequency leakage to use a common definition). The problem extends to the determination of the exact frequency of the signal component, bracketed by the two adjacent frequency bins f_k and f_{k+1}, and then in some cases to phase correction. The correction is accomplished by applying a fractional change using an opportune coefficient, determined by means of various interpolation techniques.

At a first glance interpolation might be linear between the two samples immediately at the left and the right of the unknown desired signal component. A higher order interpolation is much desirable (e.g. quadratic) to take into account the convexity of the bell-shaped peak, necessitating thus three (or more) points.

One of the first interpolation techniques [75], after simplifying the $\sin(\cdot)$ function with its own argument for small angle values, provides a linear determination of the frequency correction δ starting from the ratio of the left and right bin samples:

$$\alpha = \frac{|S[k+1]|}{|S[k]|} \qquad \delta = \frac{\alpha}{1+\alpha} \qquad f_s = (k+\delta)df \qquad (2.3.7)$$

The amplitude is then corrected by making reference to either the left or the right term, choosing the largest one to improve accuracy:

$$|A_s| = \begin{cases} \dfrac{2\pi\delta}{N}\dfrac{|S[k]|}{\sin(\pi\delta)} \\[4mm] \dfrac{2\pi(1-\delta)}{N}\dfrac{|S[k+1]|}{\sin(\pi(1-\delta))} \end{cases} \qquad (2.3.8)$$

And the phase may be obtained as

$$\varphi_s = \begin{cases} \angle S[k] - a\delta + \dfrac{\pi}{2} \\[4mm] \angle S[k+1] - a(\delta-1) + \dfrac{\pi}{2} \end{cases} \qquad \text{with} \quad a = \pi\frac{N-1}{N} \qquad (2.3.9)$$

A few years later the first extensions to tapered signals (i.e. using smoothing windows, please see previous section 2.3.3.2) began to appear, often focusing on the most widely used windows, such as the Hanning [55].

For the Hanning window (for which the resulting spectrum is analytically known), the fractional frequency correction δ is given by

$$\delta = \frac{2\alpha - 1}{\alpha + 1} \qquad (2.3.10)$$

and the corrected complex amplitude (thus including the phase information) is

$$A_s = \begin{cases} \dfrac{2\pi\delta(1-\delta)}{\sin(\pi\delta)} e^{-i\pi\delta}(1+\delta)\, S[k] \\[4mm] \dfrac{2\pi\delta(1-\delta)}{\sin(\pi\delta)} e^{-i\pi\delta}(\delta-2)\, S[k+1] \end{cases} \qquad (2.3.11)$$

2.4 Rise time and bandwidth

Information on rise time may be extremely useful both when hardware specifications and direct measurement are preferably in such terms. It occurs for example when specifying the performance of oscilloscope probes or when characterizing transients.

2.4.1 Rise time

In simple terms, considering a signal step function between two voltage levels V_- and V_+, the rise time is the time taken by the signal to change from a specified low state (or low level, V_L) to a specified high state (or high level, V_H). With the word "specified" we intend that these levels are known, as well as any related tolerance and criterion to assess that the signal has reached the high state; they are for example established by a standard, an internal procedure, or are typical of a class of digital signals (such as e.g. TTL logic levels). Think for instance to an oscillating behavior, with the signal crossing the high level threshold upward, but then crossing again in the opposite direction going farther away. A signal may be reaching the desired level in a stable way, remaining in a tolerance band around or above the level, or oscillating far from the crossed threshold line, outside the tolerance band; this is exactly how the settling time of operational amplifiers is defined.

Typically these low and high threshold pairs (V_L, V_H) in electronics and signal analysis are taken as $(10\%, 90\%)$ or $(20\%, 80\%)$, depending on the type of signal and the electrical context, or undergo more detailed specifications, such as for the input and output signals for some old logic families.

The rise time may be also defined going beyond the measurement of a time interval between two crossing points, that is yet the preferred method, that can be implemented in a very straightforward way with oscilloscopes and visually evaluating the waveshape: definitions based on bandwidth and impulse response give almost equivalent results, as it is shown in the following.

2.4.1.1 Rise time based on threshold crossing (10-90 % and 20-80 %)

Crossing a pair of thresholds may be defined as a suitable criterion, identifying the low and the high level so to mask out both superimposed noise and any waveform deformation when detaching from the lower pedestal and when reaching the upper level. Deformation, however, occurs for many reasons related to the behavior of the electronic circuits, such as variable gain, slew rate limitation, saturation, variable output impedance, ringing, etc.

The most diffused threshold pair is $10\%/90\%$ and the corresponding rise-time is indicated as t_{10-90}. When a large amount of ringing and reflections is present (also because of the wide frequency spectrum of signals), a safe identification of the beginning and the end of the rising edge is performed relying on more robust thresholds, 20% and 80%, and the rise-time is indicated as t_{20-80}.

2.4.1.2 Rise time based on bandwidth

Rise time may be also based on an agreed bandwidth, such as $-3\,\text{dB}$, $-6\,\text{dB}$ or the Equivalent Noise Bandwidth, ENBW. For the first two bandwidths the definition of rise time is quite operative: first, find in the signal spectrum the frequency at which the amplitude is $3\,\text{dB}$ or $6\,\text{dB}$ below the dc value; then, the rise time is determined with inverse proportionality using π at the denominator as the scaling factor (some authors propose instead using a multiplicative factor of 0.3 to 0.45 at the numerator, as discussed later at the end of sec. 2.4.2).

$$t_{-3\,\text{dB}} = \frac{1}{\pi B_{-3\,\text{dB}}} \qquad t_{-6\,\text{dB}} = \frac{1}{\pi B_{-6\,\text{dB}}} \qquad t_{\text{ENBW}} = \frac{1}{\pi B_{\text{ENBW}}} \tag{2.4.1}$$

It is recalled that the Equivalent Noise Bandwidth criterion is indicated by some authors as the "rms" criterion. The reason for using such definition is that it allows to pass rapidly from time to frequency domain, and back.

2.4.1.3 Rise time based on impulse response

In this case the analyzed signal is assumed as the result of an ideal step function applied to a linear system with impulse response $h(t)$, which may be determined as the first derivative of the step response. Of course such operation is correct in theory, but has the serious drawback of amplifying the noise and high frequency disturbance, so that filtering and smoothing is nearly unavoidable; in reality, the relevance of this definition is more theoretical than practical. The rise time is defined as the standard deviation of such impulse response multiplied by a scaling factor $\sqrt{2\pi}$.

$$\sigma^2 = \int_{-\infty}^{+\infty} t^2 \frac{h(t)}{H(0)}\,dt - \left[\int_{-\infty}^{+\infty} t \frac{h(t)}{H(0)}\,dt \right]^2 \tag{2.4.2}$$

$$t_\sigma = \sqrt{2\pi}\,\sigma \tag{2.4.3}$$

where $H(0)$ is the dc gain of the system (the dc value of the Fourier transform of $h(t)$).

This definition may be considered the most correct from a theoretical point of view and is often used when comparing different dynamic systems and their impulse responses: see [99] for a detailed comparison of one-pole, two-pole, Gaussian and Bessel-approximated Gaussian low-pass systems.

2.4.2 Cascaded LTI systems and rise-time propagation

If we have N linear time invariant systems in a cascaded connection with an input signal entering from the left and propagating to the right, the signal propagates following the rule of convolution of each impulse response and we will finally see that the output is the result of the overall convolution of the block impulse response. The total rise time, with a demonstration based on the Central Limit Theorem, results in the square root of the sum of the square of the single rise times.

$$t_{r,tot} = \sqrt{t_{r1}^2 + t_{r2}^2 + \cdots + t_{rN}^2} \qquad (2.4.4)$$

This relationship is extremely useful and recalled by many, but its proof is lost in the mist of the '50s: the original demonstration is attributed to Henry Wallman in a paper written for the Applied Mathematics Symposium in 1950[4], but the one-page paper simply states that "the step-function response of cascaded networks, each free of overshoot, tends to error-function integral." A more detailed examination appears in the book by Valley and Wallman published two years before [154], but no demonstration is given. On the contrary, in Johnson and Graham [77], Appendix B, the rationale of the proof is clearly stated: it is based simply on the observation that the incoming step signal convolves with the impulse response of the network; similarly convolution occurs between the probability density functions of two summed random variables and in this case their variances add; then, the square of the dispersion of the edge of the step signal and of the impulse response add in the same way. If the same definition of rise time and associated variance is used, then the proof holds and (2.4.4) keeps valid; the important thing is to adopt the same definition of rise time for all the cascaded networks: 10-to-90 %, 20-to-80 %, central slope, maximum slope, etc.

From (2.4.4) the formula for the compensation of the rise time of measuring blocks $t_{m.bl.}$ under linearity assumption may be readily derived: the DUT rise time may be estimated by rms subtracting $t_{m.bl.}$ from the measured rise time t_{meas}.

$$t_{DUT} = \sqrt{t_{meas}^2 - t_{m.bl.}^2} \qquad (2.4.5)$$

Whilst many think that the expression (2.4.4) is exact and may be applied in all cases, "it is actually a statement of trend in the limit" [154]: it is nevertheless quite accurate and already for $N = 2$ the error is only 10 %! Other considerations especially for systems with moderate overshoot are reported in [99].

2.4.3 Signal bandwidth

The most intuitive definition of bandwidth of a signal is the highest frequency component that is significant; then, to understand what "significant" really means and implies, we may attack the problem from different viewpoints.

If we take an ideal square wave and we begin to select the components of its spectrum, starting from the dc component (or 0th order harmonic), then adding the fundamental and the higher order components one by one, two facts are observed: the top of the resulting signal gets flatter resembling that of the square wave and the rising slope gets steeper, reducing correspondingly the rise time. When a threshold is fixed for either the amplitude or the rise time, then the corresponding "significant" harmonic order can be identified.

The problem with an ideal square wave is that it has infinite bandwidth and zero rise time; the components of the Fourier spectrum have an amplitude inversely propor-

---

[4] H. Wallman, "Transient response and the central limit theorem of probability", Proceedings of Symposium in Applied Mathematics, Vol. 2, pp. 91, 1950.

tional to their harmonic order, so they drop off with a $1/f$ behavior. On the contrary, for a real waveform that resembles a square wave, when again approaching the problem of identifying its bandwidth, we might ask ourselves when (i.e. for which component subtracted progressively from the reconstructed waveform) such a waveform begins to depart substantially from the reference ideal square wave of identical period. Again we need to fix a threshold and a rule might be that the power associated to the single component is equal or less to the 50 % of the corresponding square wave component (going from power to amplitude, this means that the amplitude is 70 % of the corresponding square-wave component). Or we can decide that the critical frequency is the one above which the components of the real waveform start to drop off faster than $1/f$ (and this for a nearly trapezoidal wave-shape occurs at a frequency that is the reciprocal of the rise time, as shown in the next section and commented later on in sec. 3.1.9).

Practically speaking, bandwidth is a concept related to the occupied frequency axis with respect to some characteristic or some use of the signal: we have already seen the $-3\,$dB bandwidth, as a general means to establish the frequency interval where a signal has the majority of its power concentrated, or for dynamic systems the frequency interval for which an input signal is not substantially attenuated. The $-6\,$dB bandwidth has quite a similar meaning, adding a delta interval to the previously determined $-3\,$dB frequency interval, the extension of which depends heavily on the roll-off, that is the rate of decrease of signal spectrum or system response versus frequency (usually measured in dB/decade or dB/octave). Choosing one criterion or the other depends on the application: we have seen in sec. 2.3.3.2 that the $-6\,$dB criterion is recalled when coherent composition of signals is to be accounted for and a $6\,$dB attenuation of signal amplitude corresponds to 0.5; conversely a $3\,$dB attenuation corresponds to 0.707 and we may observe also that 0.707 is the factor that comes into play when two sinusoidal tones of the same intensity are added in quadrature, that is assuming a random phase relationship, or similarly when the Power Spectral Densities of two uncorrelated noise sources are summed together. The Equivalent Noise Bandwidth ENBW is another form of characterization of the bandwidth occupation of a dynamic system, including the roll-off and the side-lobes in the stop-band: we have considered it as a performance index of smoothing windows and for rise time definition.

2.4.4   Rectangular and trapezoidal pulse train

A rectangular pulse train of amplitude A, with period T and rectangle duration τ (that corresponds to a square wave of duty cycle τ/T) is shown in Figure 2.4.1(a). The coefficients of the Fourier series are

$$c_n = \frac{A\tau}{T}\frac{\sin(\pi n\tau/T)}{\pi n\tau/T} = \frac{A\tau}{T}\operatorname{sinc}(\pi n\tau/T) \tag{2.4.6}$$

with T appearing in both fractions, in order to have a more elegant and compact expression, that involves the *sinc* function.

(a)

(b)

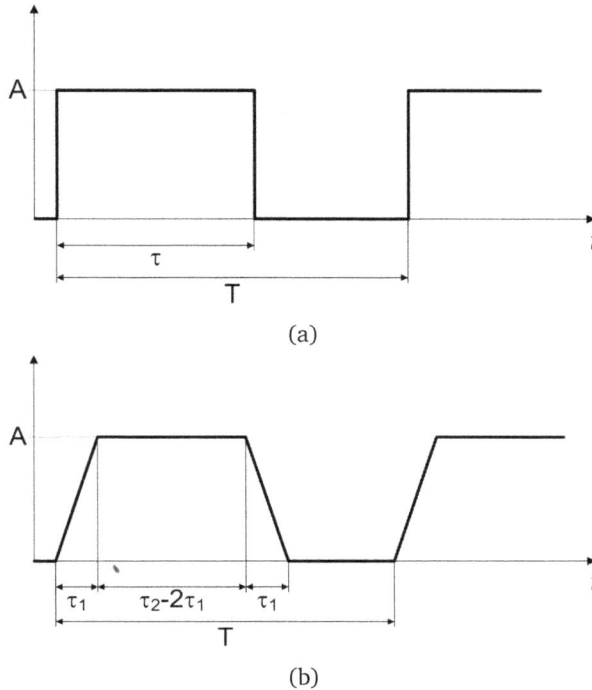

Figure 2.4.1 – (a) Rectangular pulse train and (b) Trapezoidal pulse train.

The integral of the square of the first lobe gives the power that is contained in the first components of the spectrum between 0 and $1/\tau$:

$\int_0^{1/\tau} |R(f)|^2 \, df = 0.92 A^2 \tau$ (the result being obtained by numerical integration).

The total power of the rectangular pulse is $A^2\tau$ by integration in the time domain over the entire period T (and use of the Parseval's theorem). Thus, the spectral width of the rectangular pulse encompasses the 92 % of the power, leaving outside in the high-order components only the 8 %.

Let's consider a trapezoidal pulse train waveform as a more realistic representation of a digital base-band signal or converter switching signal (see Figure 2.4.1(b)): the former time interval τ, used in the rectangular pulse train example for the active part of the wave, becomes τ_2 and the rise and fall times of the impulse flanks have time duration τ_1. The amplitude is always A and the period T. It is easy to see that the trapezoidal waveform is obtained by the convolution of a rectangular pulse $r_1(t)$ of width τ_1 with a narrower rectangular pulse $r_2(t)$ of width τ_2. Thus the Fourier transform of the trapezoidal pulse train is given by the products of the two transforms $R_1(f)$ and $R_2(f)$.

The frequency extension of the spectrum varies for different values of the rise time τ_1 (expressed as % of period T): the bandwidth is evaluated in terms of energy and of faithful reproduction of the pulse rising and falling edges (see Figure 2.4.2). The behavior may be quite different depending on the amount of rise time: any practical quick rule to estimate the bandwidth of digital signals based solely on the clock frequency must assume something on the rise time. If we agree that many clock wave-

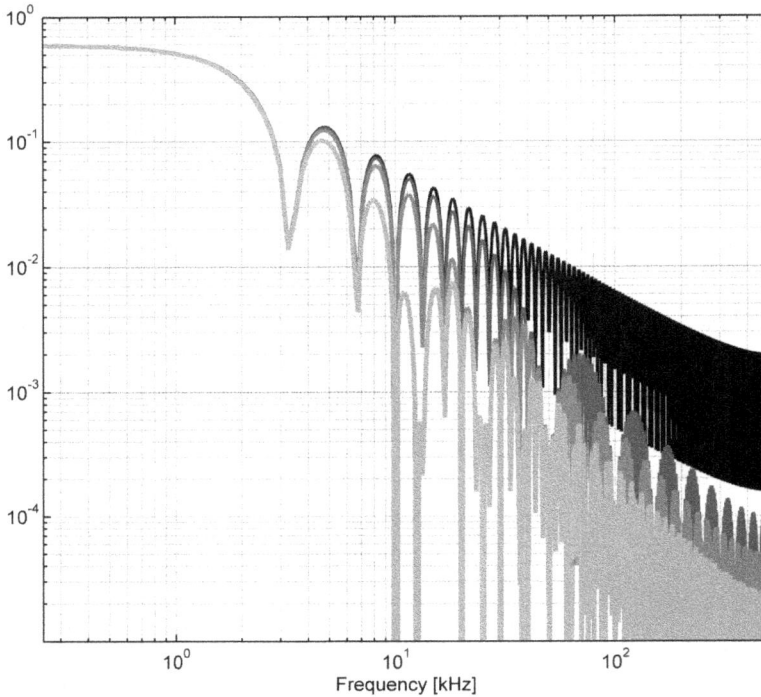

Figure 2.4.2 – Spectra of a rectangular pulse (black) and a trapezoidal pulse with varying rise time: 2%, 4% and 8% of the period in three gray tones. In this example the period is $T = 1\,\text{ms}$, the duty cycle 30%, the amplitude is $A = 1$ and 3:1 zero-padding is used to visually improve the calculated spectrum.

forms have a rise time between 4 % and 12 % of the period, then adopting the average value of 8 % leads to an estimate of the bandwidth based on (2.4.1), that ranges between $0.35/0.08\,f_{ck} = 4.37 f_{ck}$ to $0.45/0.08\,f_{ck} = 5.62 f_{ck}$, so on average 5 times f_{ck}.

The other parameter that is relevant to the spectrum shape is the duty cycle, that is the τ/T ratio. A square wave of $1\,\text{kHz}$ period is studied for varying duty cycle (0.5, 0.4, 0.25, 0.125 and 0.0625) for two cases: slow rise time, that is 2 %, and fast rise time, that is 0.2 % (not as fast as normally available waveforms from signal generators, but enough to see the difference on the chosen frequency scale). The two cases are reported in Figure 2.4.3. The spectrum is flat at the lowest frequency, so that the first value on the displayed frequency axis corresponds to the dc value: it is simply $A\tau/T$, that for $A = 1$ corresponds to the duty cycle itself. The fundamental, even if not immediately visible in the figures, is by (2.4.6) A/π, in our case 0.3183. Additionally, the smaller the duty cycle (and the shorter the time duration of the mark portion of the period), the farther the first dip in the spectrum, followed by the first lobe: the frequency location of the first dip (or zero) is the inverse of the mark duration, that is $1/\tau$. Whatever the duty cycle, all spectra stay beneath the same envelope, touching it with their lobe tips. The effect of the rise time is signified by a dip in the envelope, that for the 2 % case ($20\,\mu\text{s}$ rise time) is correctly located at its reciprocal, that is $50\,\text{kHz}$, and moves to $500\,\text{kHz}$ when the rise time is reduced by an order of magnitude to 0.2 %.

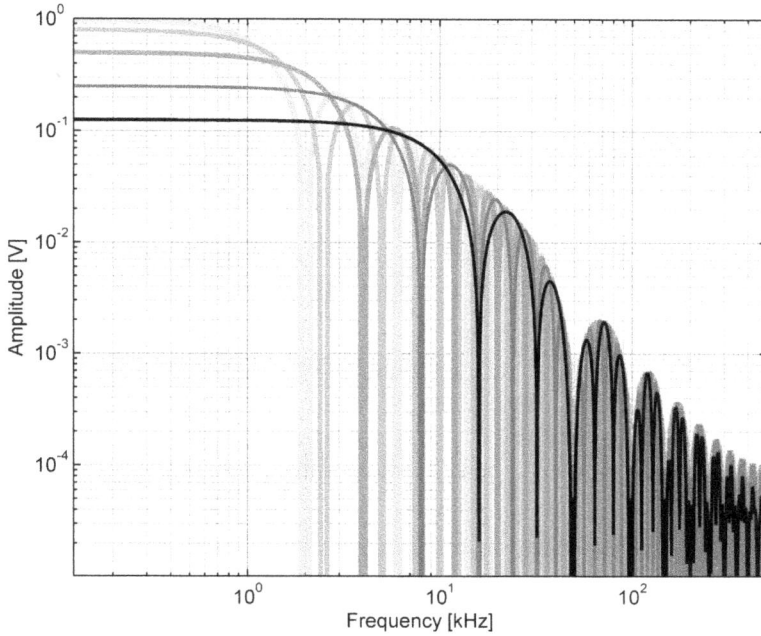

(a)

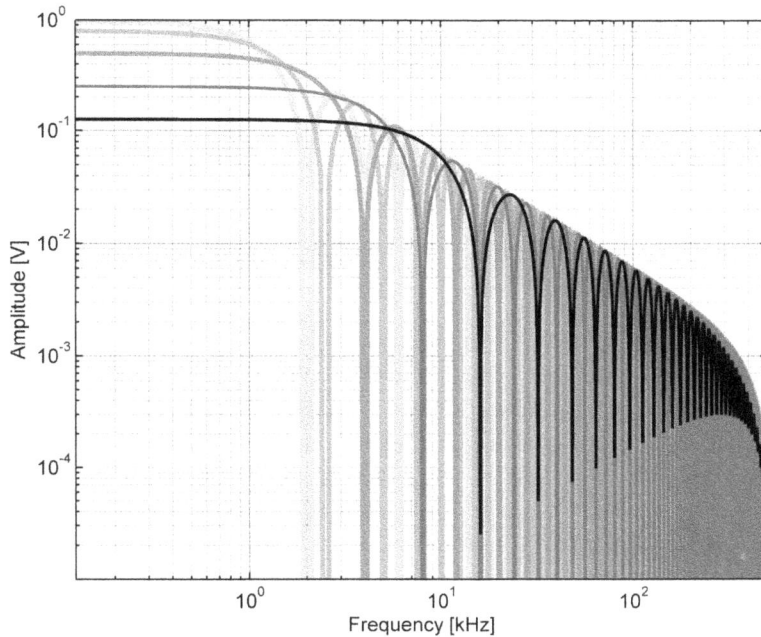

(b)

Figure 2.4.3 – Spectra of a rectangular pulse with (a) 2% and (b) 0.2% rise-time for varying duty cycle (0.5, 0.4, 0.25, 0.125 and 0.0625 from light gray to black). In this example the period is $T = 1\,\mathrm{ms}$, the signal excursion is 0-1 V and 3:1 zero-padding is used to visually improve the calculated spectrum (250 Hz resolution).

2.5 Random processes: definitions and stationarity

A random process (or stochastic process) is a collection of random values (random variables) that represent how a system evolves through time, giving such values as output. What we have when the sequence of values of some output is recorded as the time passes is a *time series*. Such a system is the copy of a deterministic system with its own output quantities to which the indeterminacy of randomness has been added: this does not mean that the new output will be highly unpredictable, or will be subject to a kind of weird behavior, but simply that its values will be characterized by a probability distribution, that may also have time dependency, that is that may vary as long as the process develops through time.

The latter is an extreme case that may be encountered often in the real world, such as for example the noise of a device that is warming up. In general, we will focus on processes with well behaved statistics, that represent already a useful tool for modeling and representing the behavior of many real systems.

Additionally, from a practical point of view the mentioned "time" in many cases might be discrete rather than continuous, such as when the observed system is e.g. a stock market index, a performance index of some quality measure (number of defects, time between two failures, etc.), a quantity that is measured, acquired or stored at given instants of time only. In many situations the underlying process is still continuous in time, but we do not have access to it or it is better to reduce the degrees of freedom and the amount of stored and analyzed information to some discrete time instants. When considering electrical and electronic systems and signals, the knowledge of the underlying physical principles tells us that the phenomena are continuous time (down to a sufficiently small scale, of course). From this, we may say that with a little flexibility continuous and discrete time are not mutually exclusive and that they may be interchanged as appropriate.

2.5.1 Basic definitions

For any experiment in which the outcomes develop through time and cannot be predicted within a reasonable error due to unavoidable experimental errors (e.g. noise and measurement uncertainty) we talk of a random process. A time history of data points recorded at some time during an experiment is only one of the possible realizations; the collection of all such realizations defines completely the random process. Being the number of collectibles and the time allowed for the experiment finite, we are bounded to work with a limited number of realizations to characterize the process or to characterize it with *a priori* knowledge, based on what we know about its statistical properties and inner mechanisms. Sometimes it may be convenient to fit a general known statistical model to the available realizations.

Examples of random processes may be a sinusoidal signal generated by an oscillator whose fundamental frequency is bound to vary because of random internal noise and to drift because of temperature variations and aging; the mains voltage spectrum whose change in time (e.g. the amplitude of fundamental and harmonics) is due to external causes, such as the varying operating conditions for the loads connected to the same supply network that are unpredictable and not under experimenter's control; additive noise alone, in principle, transforms all signals in random processes for which there are no two really identical recordings.

A random process is thus identified by a set of time recordings (time history) in the form of signals, whose time duration, amplitude resolution and frequency resolution are established by the way they were measured and recorded. Several characteristics of course may be due to the influence of test setup and the adopted measurement method: we will see that besides the well-known parameters of time-domain recording (sampling time, observation window, etc.), when recording with a more or less direct measurement in frequency domain, the dynamic nature of several phenomena shall be considered: several instrumentation settings may bias or irreversibly affect the results.

2.5.2 Stationarity and ergodicity

A random process is said stationary if its ensemble statistics do not depend on time; this gives us the certainty that statistics calculated on a data sample do not depend on time. Of course depending on which statistics may or do not vary with time, a random process may be more or less strictly stationary, with a range of degrees of stationarity. Regarding stationarity, a process is classified as [13]:

- first-order stationary, if its first-order probability density function remains equal regardless of any shift in time, so that the average value is constant;

- second-order stationary, if its second-order probability density function does not vary over any time shift applied to both values;

- strict sense stationary (SSS), if all its statistical properties do not vary over any applied time shift;

- wide-sense stationary (WSS) to relax the requirements of SSS, since it becomes evident that the strict requirements of a SSS process are more than that is often necessary in order to adequately approximate calculations on random processes; for a WSS process the average value is constant and the correlation function is defined only by the time-shift.

The fact that some statistics do not depend on time does not imply that ensemble and time statistics coincide. If they do, we say that the process is *ergodic*. Given a stationary process, we are going to define weakly ergodic and strongly ergodic processes. A process is said *weakly ergodic* if time average for mean and covariance computation give the same result as ensemble averages; if all other statistical properties of the process are calculable through time averages, then the process is *strongly ergodic*.

To interpret the two definitions with a closer look to measurements and metrology jargon, we may say that stationarity indicates that the process statistics are independent on the initial time instant and that repeated measurements at different instants of time are consistent (repeatability). Ergodicity states that ensemble and time-domain statistics may be interchanged and that the results obtained with one of the two methods may be reproduced also with the other one.

Besides the direct knowledge of the underlying physical mechanisms, from which stationarity may be postulated, suitable tests exist to evaluate the stationarity of sampled random data [12, 13]. It may be assumed for simplicity that any non-stationarity relevant for our problems is revealed by a time trend of the mean square value of data.

2.6   Auto-correlation and Power Spectral Density

Definitions and considerations open this section, establishing the basic properties for random processes evaluation.

Given the process $x(t)$, the auto-correlation function is defined in a general way as

$$R_{xx}(t_1, t_2) = E\{x(t_1)\,x(t_2)\} = \int\!\!\!\int\limits_{-\infty}^{+\infty} x(t_1)\,x(t_2)\,f(x_1, x_2)\,\mathrm{d}x_1\,\mathrm{d}x_2 \qquad (2.6.1)$$

The average power of $x(t)$ is the value $E\{x^2(t)\} = R_{xx}(t, t)$; the auto-covariance is given by the auto-correlation function with the average value removed:
$C(t_1, t_2) = R_{xx}(t_1, t_2) - \mu(t_1)\mu(t_2)$.

The cross-correlation of two random processes $x(t)$ and $y(t)$ is $R_{xy}(t_1, t_2)$:

$$R_{xy}(t_1, t_2) \;= E\{x(t_1)\,y^*(t_2)\} = E\{x^*(t_1)\,y(t_2)\} =$$

$$= \textstyle\iint_{-\infty}^{+\infty} x(t_1)\,y^*(t_2)\,f(x_1, y_2)\,\mathrm{d}x_1\,\mathrm{d}y_2 \qquad (2.6.2)$$

Here are some definitions:

- two processes $x(t)$ and $y(t)$ are mutually orthogonal if $R_{xy}(t_1, t_2) = 0$ for every t_1 and t_2; they are uncorrelated if $C_{xy}(t_1, t_2) = 0$ for every t_1 and t_2;

- a process $x(t)$ is a white noise if values $x(t_1)$ and $x(t_2)$ are uncorrelated; if these two variables are also independent, then the process is strictly white noise.

The Fourier transform of the auto-correlation function of a random process $x(t)$ corresponds to the Autospectral Density Function, ADF, also called Power Spectral Density, PSD, $S_{xx}(f)$.

$$S_{xx}(f) = \int\limits_{-\infty}^{+\infty} R_{xx}(\tau)\,e^{-j2\pi f\tau}\mathrm{d}\tau \qquad (2.6.3)$$

Analogously, the Fourier transform of the cross-correlation function between two random processes $x(t)$ and $y(t)$ is called Cross-Spectral Density Function, CDF[5], or simply Cross-Power Spectral Density, $S_{xy}(f)$.

---

[5]   Rarely used and misleading, because it is confused with the Cumulative Distribution Function.

$$S_{xy}(f) = \int\limits_{-\infty}^{+\infty} R_{xy}(\tau)\, e^{-j2\pi f\tau}\,\mathrm{d}\tau \tag{2.6.4}$$

The ADF functions are real-valued even non-negative functions, while the CDF is a complex-valued function. All these functions are defined for a frequency interval that extends from $-\infty$ to $+\infty$: remembering that they are even, it is always possible to define a so-called one-sided PSD, defined only for non-negative frequency values, so closer to reality, that is usually indicated by the letter "G":

$$G_{xx}(f) = 2S_{xx}(f) \qquad \text{for } f \geq 0 \tag{2.6.5}$$

A sufficient condition for their Fourier transformation is that the integrals of their absolute values are finite, and this holds for quite a lot of correlation functions, except some pathological cases, such as the correlation of an infinite sinusoidal signal, whose integral of the absolute value does not converge (but is anyway bounded).

These relationships are often called Wiener-Khintchin relations[6].

2.6.1 Random signals and Linear Time Invariant systems

Let's consider the input process $x(t)$ to a Linear Time Invariant (LTI) system (such as a filter or an amplifier) and the output process $y(t)$, and for simplicity let's focus on stationary processes, so to write the auto-correlation functions as $R_{xx}(\tau)$, $R_{yy}(\tau)$ and $R_{xy}(\tau)$, with $\tau = t_1 - t_2$, and correspondingly the PSDs, $S_{xx}(f)$, $S_{yy}(f)$ and $S_{xy}(f)$. Under the assumption of stationarity and ergodicity[7], the latter are related to each other through the LTI frequency response $H(f)$ as

$$S_{xy}(f) = H(f)\, S_{xx}(f) \qquad S_{yy}(f) = |H(f)|^2\, S_{xx}(f) \tag{2.6.6}$$

so that it is easy to calculate the output mean square value

$$\psi_y^2 = \overline{y^2} = \int\limits_{-\infty}^{+\infty} S_{yy}(f)\,\mathrm{d}f = \int\limits_{-\infty}^{+\infty} |H(f)|^2\, S_{xx}(f)\,\mathrm{d}f \tag{2.6.7}$$

or for the one-sided PSD functions $G_{xx}(f)$ and $G_{yy}(f)$

$$\overline{y^2} = \int\limits_{0}^{+\infty} G_{yy}(f)\,\mathrm{d}f = \int\limits_{0}^{+\infty} |H(f)|^2\, G_{xx}(f)\,\mathrm{d}f \tag{2.6.8}$$

---

[6] The relationship between the autospectrum and the autocorrelation functions was demonstrated independently by Norbert Wiener and Alexander Khintchin in the '30s and is thus usually called Wiener-Khintchin relation. Please have a look at the basic information for first reference reported in Wikipedia: http://en.wikipedia.org/wiki/Wiener%E2%80%93Khinchin_theorem.

[7] Ergodicity is necessary in the proof when exchanging the integrals (time domain averages) with the expectation operator.

As a reminder, in the following notation $S_{xx}(f)$ indicates the two-sided PSD that extends from $-\infty$ to $+\infty$ on the frequency axis, while $G_{xx}(f)$ indicates the one-sided PSD, defined only for the positive frequency axis. $G_{xx}(f)$ is thus in direct relationship with frequency domain measurements.

Last, it is useful to recall that $\overline{y^2} = R_{yy}(0)$.

The LTI action (e.g. for a filter) is to shape the PSD of the input random process. Assuming an uncorrelated white noise with a flat PSD $S_{xx}(f)$ as input, the memory that is in the LTI time constants reflects into the spectrum of the output noise $S_{yy}(f)$. The formerly white uncorrelated input noise becomes "colored" and has an auto-correlation that is no longer a Dirac delta, but shows the same time constant of the LTI system. We will shortly see, when talking about the Equivalent Noise Bandwidth in sec. 2.6.1.1, that also the total rms output noise is related to the extension of the LTI frequency response.

2.6.1.1 Noise bandwidth of a filter

The Equivalent Noise Bandwidth B_n (also indicated as ENBW) of a real system with transfer function $H(f)$ and gain $|H(f_0)|$, is the bandwidth of an ideal filter with box-like shape and with the same gain, which yields the same level of output power as the real system [63, 101]. The noise bandwidth of a filter can be defined for low-pass and band-pass filters. If white noise voltage $v(t)$ with constant flat PSD $S_{vv}(f) = \eta$ is applied to the input of the filter, after defining the mean-square output voltage for the filter in question and a boxcar equivalent one, B_n is

$$B_n = \frac{1}{H_0^2} \int_0^{+\infty} |H(f)|^2 \, \mathrm{d}f \tag{2.6.9}$$

The noise bandwidth of a band-pass filter is defined in a similar way simply changing the limits of integration.

Two classes of filters are considered as example: one with N real poles all with the same cut-off frequency and the other is a Butterworth of order N. The equivalent noise bandwidth is reported in Table 2.6.1 for N ranging from 1 to 5; f_0 indicates the pole frequency and B_{-3dB} is the frequency at which the gain is $-3\,\mathrm{dB}$ the static gain H_0 (or center band gain for band-pass filters).

2.6.1.2 Equivalent input noise

The use of the model above allows to reason on the concept of Equivalent Input Noise (EIN), a concept largely applied in radiofrequency and electronics, e.g. for instruments and active devices. The application of EIN is straightforward: measuring (or quantifying in other ways) noise at the output of an electronic system, identified as the random process $y(t)$, $n(t)$ is the EIN that going through the LTI produces exactly the same output noise power (mean square value ψ_y^2). Noise power is divided by the LTI gain H_0

Order	Real poles		Butterworth
1	$1.571\,f_c$	$1.571\,B_{-3\,\text{dB}}$	$1.571\,B_{-3\,\text{dB}}$
2	$0.785\,f_c$	$1.220\,B_{-3\,\text{dB}}$	$1.111\,B_{-3\,\text{dB}}$
3	$0.589\,f_c$	$1.155\,B_{-3\,\text{dB}}$	$1.042\,B_{-3\,\text{dB}}$
4	$0.491\,f_c$	$1.129\,B_{-3\,\text{dB}}$	$1.026\,B_{-3\,\text{dB}}$
5	$0.420\,f_c$	$1.114\,B_{-3\,\text{dB}}$	$1.017\,B_{-3\,\text{dB}}$

Table 2.6.1 – Noise bandwidth of low-pass filters: for real-pole filters the ENBW is given in terms of the corner frequency f_c or the $-3\,\text{dB}$ bandwidth $B_{-3\,\text{dB}}$; for Butterworth the ENBW is given in terms of the $-3\,\text{dB}$ bandwidth $B_{-3\,\text{dB}}$ only [101].

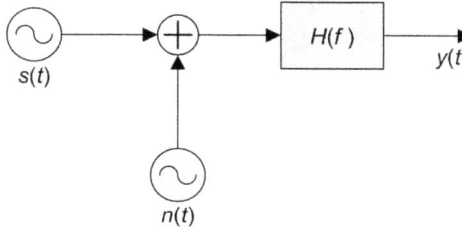

Figure 2.6.1 – Linear Time Invariant system with signal and noise source.

(the static gain $H(0)$ for low-pass systems or center-band gain $H(f_0)$ for band-pass systems); the reason behind the normalization for the LTI gain is that the interest in noise is for the operating frequency range of the electronic system, normally characterized by an almost constant frequency response $H(f)$.

$$\psi_n^2 = \psi_y^2 / H_0^2 \qquad (2.6.10)$$

2.6.1.3 Signal-to-noise ratio

We can attack the problem of propagation of noise through LTI systems evaluating the signal-to-noise ratio (SNR) and how the system response affects it. The signal-to-noise ratio is defined as the ratio of power of signal $s(t)$ and noise $n(t)$ over bandwidth B. At the input the SNR is:

$$\text{SNR}_i = \frac{\psi_s^2}{\psi_n^2} = \frac{P_s}{P_n} \qquad (2.6.11)$$

The origin of the noise may be outside or inside the system, and for the latter we have adopted the abstraction of bringing it back to the input in terms of equivalent input noise (see sec. 2.6.1.2 above). Thus, $s(t)$ and $n(t)$ are summed together and combined into $x(t)$ before entering the system; we assume that $s(t)$ and $n(t)$ are incoherent and that there is no net contribution to the cross-product that has a zero average ($E\left[s(t)\,n(t)\right] = 0$) and does not appear in the system output $y(t)$ when calculating expectation or average.

The power of the two processes going to the output is affected by the LTI frequency response; assuming that the LTI filter is conceived to have a flat gain g over the signal bandwidth and that its Equivalent Noise Bandwidth (ENBW) B_n is known, then

$$\text{SNR}_o = \frac{g\,P_s}{\eta B_n} = \frac{P_s^o}{P_n^o} \qquad (2.6.12)$$

As exemplified in Table 2.6.1, the ENBW B_n is always larger than the $-3\,\text{dB}$ bandwidth, because the out-of-band filter response (made of the roll-off portion and the stop-band) is included in the computation; so the wisest behavior to maximize SNR_o is to, first, choose the smallest bandwidth B compatible with the signal bandwidth and, second, to select the highest filter roll-off (i.e. number of poles and filter type), as technically attainable, or reasonable.

Since signal and noise cannot be really separated, but are accessible only at the LTI output, the successive measurement of the mean square value of the output, with and without the signal $s(t)$ applied, gives

$$P_y' = P_s^o + P_n^o \qquad \text{with}$$

$$\qquad\qquad\qquad\qquad\qquad\qquad\qquad (2.6.13)$$

$$P_y'' = P_n^o \qquad\qquad \text{without}$$

and the SNR_o may be calculated readily as $P_y'/P_y'' - 1$. The same is applicable when performing the rms measurement of a steady signal and aiming at correcting for the internal noise contribution: the rms of the tone is readily obtained by quadrature-subtracting the estimate noise rms (corresponding to P_y'') from the total rms (corresponding to P_y') (see sec. 3.1.8).

2.6.2 Power Spectral Density estimate

In this section the various methods for the estimation of the Power Spectral Density are considered. The subject is quite complex and the aim is simply that of showing how a random process such as background noise may be characterized and quantified without going into details: additional references are [12, 13, 21, 90, 121, 131]; further references may be found e.g. in [13], chap. 3.

2.6.2.1 Power Spectral Density from the Correlation function

A first method for the determination of the Power Spectral Density spectrum (PSD), or auto-spectrum, $S_{xx}(f)$ is based on its own definition as the Fourier Transform of the auto-correlation function $R_{xx}(\tau)$ (see at the beginning of sec. 2.6). By the adopted notation it is implicitly assumed that the random process that we consider is stationary. Analogously for the cross-correlation $R_{xy}(\tau)$ between two random processes $x(t)$ and $y(t)$, the Cross Spectral Density, or cross-spectrum, is the Fourier Transform of $R_{xy}(\tau)$.

By observing that correlation functions are always even functions of τ, it follows that the desired spectral densities are given by the real part only of the Fourier Transform. By the duality of auto-correlation and autospectrum as a Fourier Transform pair, it follows that the integral over one domain corresponds to the value in the origin in the other domain; so, the integral over the frequency axis of the autospectrum gives the value in the origin of the auto-correlation, i.e. the mean square value (that in turn is the sum of the variance and of the square of the mean value).

$$\psi_x^2 = \int_0^{+\infty} G_{xx}(f)\,\mathrm{d}f = R_{xx}(0) \tag{2.6.14}$$

$$G_{xx}(0) = \int_{-\infty}^{+\infty} R_{xx}(\tau)\,\mathrm{d}\tau \tag{2.6.15}$$

Since the auto-correlation function is biased by the constant term corresponding to the square of mean value, it results that its Fourier Transform (the autospectrum) in frequency domain shows a delta function at the origin ($f = 0$) with area equal to μ_x^2.

2.6.2.2 Power Spectral Density via Fourier Transform

Without demonstrating it [12, 13, 81, 149], we simply state that given a random process $x(t)$, for which we have J records of time length T indexed by index $j = 0 \dots J-1$, the autospectrum is given by the expectation of the modulus squared as shown below.

$$S_{xx}(f) = \lim_{T \to \infty} \frac{1}{T} E\left\{ |X_j(f,T)|^2 \right\} \tag{2.6.16}$$

A first implementation is the method commonly referred to as *Periodogram*; it owes its name to the capacity of spotting out hidden periodicity of signals as commented by Schuster[8] at the very beginning of the XX century (see [149], sec. 2.2.1).

$$S_{xx}(f) = \frac{1}{N} \left| \sum_{n=0}^{N-1} x[n] e^{-j\frac{2\pi}{N}n} \right|^2 = \frac{1}{N} |X[k]|^2 \tag{2.6.17}$$

Using the squared modulus of the DFT, the resulting kernel is not the rectangular window typical of implicit windowing over the observation epoch, but a triangle, whose transform is given by the convolution of the spectra of two rectangular windows.

The periodogram as an estimator is characterized by error bias and variance. The periodogram is an unbiased estimator, in that it goes to zero for an increasing number of samples N. Its variance, on the contrary, is the main problem and it does not reduce with N, keeping on fluctuating around the true PSD with erratic (noise-like) behavior: from this the periodogram is an inconsistent spectral estimator.

---

[8] A. Schuster, "The periodogram of magnetic declination as obtained from the records of the Greenwich Observatory during the years 1871-1895," *Transactions of Cambridge Philosophical Society*, Vol. 18, pp. 107-135.

Bartlett[9] introduced the idea of averaging the single squared transforms, while keeping the rectangular implicit window, thus maximizing the frequency resolution, without solving the problem of frequency leakage.

Welch[10] further modified the Bartlett method using overlapped data segments (or epochs) and applying windowing, offering lower variance than the Bartlett method. The Welch method, despite the non-dramatic reduction of the variance of the estimate and the existence of many other methods for spectral estimation, is quite popular in that it may be implemented easily and is applicable to signal spectra available in absolute value only, as those resulting from heterodyne measurements (e.g. using a Spectrum Analyzer) or by analog filtering with a conveniently small resolution (see sec. 2.6.2.3 later on).

The Welch method achieves a better estimation accuracy with *averaging*: Fourier spectra are computed on signal records, possibly using overlapping to increase their number and time resolution, and then averaged exploiting the benefits of sample estimates. In some cases smoothing windows are not available, as when averaging is done by the spectrum analyzer itself.

Incoherent averaging   The most common form of averaging in (2.6.16) is applied to the successive squared spectra to implement the expectation operator; this operation is called *incoherent averaging*, is performed without phase information and the time location of each spectrum is free of constraints. It is known that the resulting signal-to-noise ratio of the spectral representation improves with the number of samples J along the time axis, having used the term "noise" to identify any incoherent component superimposed to the signal or added by the Fourier mathematical operations. What is improving is the variance of the spectral noise components, resulting in a flatter spectrum noise floor, not the ratio between signal power and the average noise power [137]. In other words, averaging of successive Fourier transforms is increasing the *processing gain*.

A single transform has already an *inner processing gain* that is due to the behavior of each frequency bin as a bandpass filter, narrower as the number of points N increases and with increasing signal-to-noise ratio, that goes with $20 \log_{10}(\sqrt{N})$. This conclusion is valid for a number of transform points "larger than about 20-30" [90] and for a signal not overwhelmed by noise; to simplify, there is an increase by 3 dB of the amplitude SNR for each doubling of the number of frequency points N.

Let's consider the average of successive Fourier spectra over time, so moving between spectra rather than along the frequency axis of a single spectrum: the average noise floor is still there, but we have an improvement in terms of reduction of noise floor variance, that may be called *integration processing gain*. If the signal-to-noise ratio is measured between the peak of the sinusoidal tone and the largest noise floor value, reducing noise variance has impact on the signal-to-noise power ratio. The reduction

---

[9] M.S. Bartlett, "Smoothing periodograms for time series with continuous spectra," *Nature*, Vol. 161, pp. 686-687.

[10] P. D. Welch, "The use of fast Fourier transform for the estimation of power spectra: A method based on time averaging over short, modified periodograms," *IEEE Transactions on Audio and Electroacoustics*, Vol. 15, pp. 70-73, June 1967.

of the variance goes with J, so that a twice larger number of spectra to be averaged gives a halved variance and a 30% smaller standard deviation. For independent spectra (i.e. without correlation for example due to overlap), the integration processing gain is given in dB by $10 \log_{10}(\sqrt{J})$. Please see Figure 2.6.2 for an example.

Frequency Domain Equipment, such as Spectrum Analyzer, operates always in incoherent mode: see Figure 3.1.10 for a demonstration of noise floor smoothing by averaging.

Coherent averaging   When the time information (or phase information) is available and time records may be synchronized, then *coherent averaging* may be used. The meaning of the word "synchronization" implies that a periodic signal is to be estimated and that a reference frame for the phase can be identified, such as when triggering on a fundamental periodic waveform with oscilloscopes and data acquisition systems. In such cases averaging may be done in time domain on the synchronized (i.e. time-aligned) waveforms (or epochs) or in frequency domain, aligning the phase of the fundamental component of each epoch. If not properly aligned, the vectorial sum of sinusoidal terms with different, and random, phase relationships results in a reduction of the amplitude of the output, as in the case of a random noise components, for which averaging is thought. For independent samples, the reduction of noise power is proportional to the number of averaged samples J, while the amplitude of the coherent components keeps constant (if synchronization is provided). There is a true reduction of the average noise floor and the signal-to-noise power ratio improvement expressed in dB is $10 \log_{10}(J)$. See Figure (2.6.2) for visual comparison of the two noise floors.

2.6.2.3 Power Spectral Density by Analog Filtering

A straightforward approach to PSD estimation based on analog techniques is the use of a band-pass filter with bandwidth B_{bp}, followed by rms measurement (either performed with a high performance multimeter or oscilloscope, or with a Spectrum Analyzer); the output shall be normalized by B_{bp}. When using digital data acquisition instrumentation, the rms measurement is performed with special code for squaring, time integration and averaging, as it is for the most modern instruments: integration time and sampling time influence behavior and accuracy of rms estimate. For frequency domain equipment the role of B_{bp} is taken by the resolution bandwidth and the squaring by using a power reading.

$$\Psi_T(f, B_{bp}) = \frac{1}{T\,B_{bp}} \int x^2(t, f, B_{bp})\,\mathrm{d}t \qquad (2.6.18)$$

$$\hat{G}_{xx}(f, B_{bp}) = \lim_{T \to +\infty} \Psi_T(f, B_{bp}) \qquad (2.6.19)$$

$$G_{xx}(f) = \lim_{B_{bp} \to 0} \hat{G}_{xx}(f, B_{bp}) \qquad (2.6.20)$$

Using an infinitesimally small bandwidth B_{bp} is not practical, so for a finite B_{bp} value the estimate above is a biased estimate of the PSD and the bias is shown to be [13]

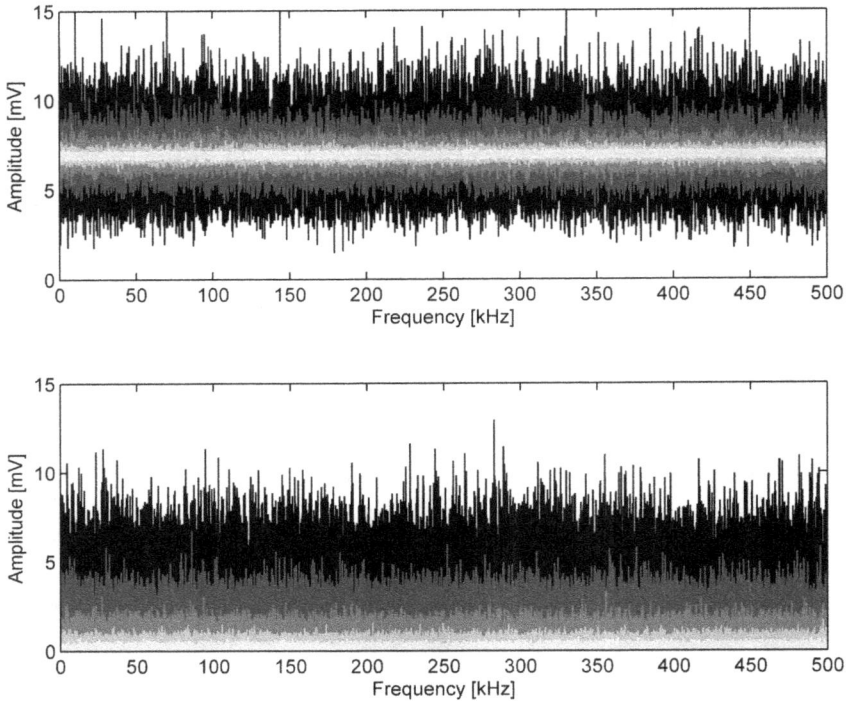

Figure 2.6.2 – Example of incoherent (above) and coherent (below) averaging of successive spectra of a normally distributed white noise; five values of $J = 4$, 16, 64, 256, 1024 were tested: incoherent averaging reduces the variance of the noise profile converging to the noise variance, while coherent averaging progressively reduces the noise floor.

$$b\left[\hat{G}_{xx}(f, B_{bp})\right] = \frac{B_{bp}^2}{24} G''_{xx}(f) \tag{2.6.21}$$

similarly to the estimate error for probability density functions. This relationship is an over-estimate of the bias error term for most real cases. It is worth underlying that the bias error acts always to decrease the dynamic range of the PSD estimate, so that "peaks are under-estimated and valleys are not so deep".

The variance of the estimate may be calculated assuming that the filtered signal over each B_{bp} frequency interval behaves like a bandwidth limited Gaussian white noise, and it is reasonable if B_{bp} is sufficiently small:

$$\sigma^2\left[\hat{G}_{xx}(f, B_{bp})\right] = \frac{G_{xx}^2(f)}{B_{bp}T} \tag{2.6.22}$$

2.7 Accuracy and uncertainty

2.7.1 Definitions

Uncertainty defines the doubt about the validity of the result of a measurement and at the same time quantifies it following one of the accepted methods: the standard deviation is commonly accepted as a quantitative measure that "characterizes the dispersion of the values that could reasonably be attributed to the measurand" [15]. Other definitions may be based on the consolidated concepts of *error* in the estimated value of the measurand or estimate of the range of values within which the *true value* of the measurand lies. Both, however, rely on an unknowable quantity, that is the error or the true value. Nonetheless, whichever definition is used an uncertainty component is always evaluated using the same data and related information.

Loosely speaking uncertainty is what accuracy was, but in a modern perspective: accuracy was strongly related to a definition of the error of the instrument, extending it to characterize an interval of variation, such as in statements referring to percentage errors $\pm x\,\%$, often including worst-case assumptions. Nothing, or almost nothing, however, was said or explicitly assumed for probability distribution of error and correlation between different error sources, internal and external. Uncertainty goes beyond and characterizes from a statistical standpoint the behavior of an instrument or a measurement, as far as the error is concerned.

The definition appearing in [15], sec. 3.1.1 and 3.1.2 couldn't be more clear: "The objective of a measurement is to determine the value of the measurand, that is, the value of the particular quantity to be measured. A measurement therefore begins with an appropriate specification of the measurand, the method of measurement, and the measurement procedure. In general, the result of a measurement is only an approximation or estimate of the value of the measurand and thus is complete only when accompanied by a statement of the uncertainty of that estimate."

Traditionally, an error is viewed as having two components: a random and a systematic component:

- random error presumably arises from unpredictable or stochastic temporal and spatial variations of influence quantities; the effect of the random error can be reduced by increasing the number of observations; its expectation or expected value is zero, the dispersion reduces for increasing observations;

- if a systematic error arises from a recognized systematic effect, the effect can be quantified and compensated by a correction factor; it is assumed that, after correction, the expectation or expected value of the error arising from a systematic effect is zero; this operation corresponds for example to the zeroing of an instrument to compensate for internal offset and we know that drift due to random components or temperature effects compromises the compensation of the systematic effect, so that the instrument shall be repeatedly zeroed.

The result of a measurement after correction for recognized systematic effects is still only an estimate of the value of the measurand, because of the uncertainty arising from random effects and from imperfect correction of the result for systematic effects.

The GUM [15] and the other standards and publications regarding uncertainty in measurements [42, 43] treat the uncertainty components arising from random effects and from corrections for systematic effects in the same way when evaluating the uncertainty of measurement results. This is justified in the GUM, sec. E.3.

2.7.2   Type A and Type B approaches

Regarding the methods used for the evaluation of uncertainty, a widespread classification divides among Type A and Type B approaches, without the intention of assigning a specific method (e.g. Type A) to a specific category, random or systematic, although there is the general tendency to interpret Type A evaluation as focusing on random errors. The GUM defines the two methods as:

- Type A: method of evaluation of uncertainty by the statistical analysis of series of observations; an observed frequency distribution is derived from the collected samples, arranged in a histogram, which a probability density function is made correspond to;

- Type B: method of evaluation of uncertainty by means other than the statistical analysis of series of observations, for example based on knowledge available on the nature of uncertainty sources and how they combine; the use of assumed probability density functions is often termed "subjective probability".

The results are substantially equivalent, both use statistical distributions and they do not imply a difference in the nature of the so estimated uncertainty components.

Regarding the combination of uncertainty terms and the estimation of the uncertainty of a measurement, although it is in principle possible to apply variations to all the quantities on which the result depends, to practically limit the size and duration of the experiment, a mathematical model of the measurement and the law of propagation of uncertainty are used.

The model consists of a relationship that relates the measurement quantities X_i to the measurement result Y:

$$Y = f(X_1, X_2, \dots X_n) \tag{2.7.1}$$

Correspondingly the estimate of the result y is based on a set of estimates of the measurement quantities x_i keeping valid the mathematical model:

$$y = f(x_1, x_2, \dots x_n) \tag{2.7.2}$$

As said, the distributions that characterize the input quantities may be based on frequency distributions (Type A) or a priori distributions (Type B). There is no preference for either approach, recognizing that a Type B estimate may be as reliable as a Type A when a substantially low number of statistically independent observations is available.

2.7.2.1 Type A method

The method consists in applying statistical analysis to obtain the best estimates on a set of samples, usually named repeated observations (see sec. 2.2.4): the best estimate of the expected value of the quantity underlying the set of observations is the average or arithmetic mean. The estimate of variance and its square root (experimental standard deviation) is a measure of the variability of the samples, due e.g. to random variations, and their dispersion around the estimated mean value. Both are described in (2.2.8), with the variance estimated with the known $(N-1)$ factor at the denominator, to take into account the small sample size. Attention shall be given to the fact that if the samples are in reality observations of a random process developing through time and they are correlated (e.g. because the sampling interval is too short), the two estimators (arithmetic mean and experimental standard deviation) may be inappropriate and the series shall be analyzed with more specific methods.

Evaluations based on repeated observations are not necessarily superior to those obtained by other means (e.g. Type B method), not only because of the influence of external factors that may adversely affect measurements used to this aim, but also by the intrinsic variability of the estimated standard deviation. Considering n samples, the experimental standard deviation of the mean s (that is the Type A estimate of uncertainty) attempts to estimate the standard deviation σ of the mean, that would be obtained by an infinite number of observations: it is known that if the samples are normally distributed, the standard deviation of s can be estimated and this represents the "uncertainty of the uncertainty estimate"; its relative value with respect to σ is approximately $1/\sqrt{2(n-1)}$ and even for a moderately large number of samples (e.g. $n=5$ and $n=10$) the resulting dispersion of the uncertainty estimate is quite large (e.g. 35.4 % and 23.6 %). Even with 50 samples, the dispersion cannot be lower than 10 %!

2.7.2.2 Type B method

Conversely, it is possible to start from a priori knowledge of distribution and scientific judgment of available information, e.g. previous Type A uncertainty estimate (this time taken directly as an input quantity), manufacturer's data or specifications, previous experience or calibration certificates. Quoted uncertainty may be thus expressed as a multiple of standard deviation or referred to a confidence interval, for which it is possible to revert back to the original estimate of the standard deviation under the assumption of the probability distribution function. The a priori knowledge comes into play to estimate the PDF for the underlying mechanisms, processes and methods:

- if the stated result stems from a previous Type A uncertainty estimate or a statistical analysis performed on a sample of observations of sufficient size, then a Gaussian distribution may be assumed (see sec. 2.2.3.2); if the size is small the Student-t distribution may be invoked (see sec. 2.2.3.4);

- when deriving an uncertainty estimate from a statement of error bound, as it is for $\pm p\%$ accuracy and class statements, then a uniform distribution is usually

Distribution	Std. dev.
Rectangular	$1/\sqrt{3}$
Triangular	$1/\sqrt{6}$
U-shape	$1/\sqrt{2}$

Table 2.7.1 – Conversion factor for standard deviations of assumed distributions for Type B uncertainty calculations.

assigned, indicating that no specific knowledge is available and that all values within bounds are equally probable;

- when values next to the boundaries are less probable, as it is physically reasonable, then the distribution is trapezoidal, driving to zero the uniform distribution above and avoiding the sharp discontinuity at its boundaries;

- when reducing to zero the inner sub-interval of lack of clue, a triangular distribution is obtained; in reality there are some situations in which a triangular distribution is physically reasonable, as for site imperfection and measured intensity for moving source.

It is clear that different distributions are characterized by a different probability of values to exceed a given threshold, such as the standard deviation; this is reasonable not only with a graphic interpretation of where the area of the distribution is located (e.g. a rectangular distribution has a consistent portion beyond a threshold, twice as much the triangular distribution), but also because the standard deviation as a second-order moment conveys information on the spread around the mean as center of gravity, not the associated probability, that depends on the distribution.

A conversion factor is needed thus between standard deviations under different PDF assumptions, for equivalent representations in terms of probability and thus level of confidence. The relationships between standard deviations of the considered distributions are reported in Table (2.7.1). The Normal distribution represents always the reference distribution, whose adoption for the final combined uncertainty is usually justified by invoking the Central Limit Theorem and the large number of contributing terms.

2.7.3 Combined standard uncertainty

To combine uncertainty from multiple sources, the variance of the terms of the model are additive; from this the justification of the need to determine standard deviations from the sample observations (Type A) or the various forms of expression of uncertainty and a priori knowledge (Type B).

2.7.3.1 Non-correlated input quantities

This is the easiest case, but often may represent only a simplification of the real measurement and system.

When considering the general equation of the model (2.7.2), the summation of variance terms of x_i is performed taking into account any multiplicative factor: terms may effectively be combined with a multiplicative coefficient inside the model or this may be a result of linearization of the in principle non-linear function f.

$$u_c^2(y) = \sum_{i=1}^{N} \left(\frac{\partial f}{\partial x_i} \right)^2 u^2(x_i) \tag{2.7.3}$$

where $u_c^2(y)$ indicates the combined variance of y. The derivative terms are often indicated as c_i and named sensitivity coefficients.

If the non-linearity of f is significant, higher-order terms shall be added to the Taylor's series, as advised in [15], sec. 5.1.2.

2.7.3.2 Correlated input quantities

The expression "correlation between input quantities" indicates combination of quantities inside the model and its equation 2.7.2: the combined uncertainty is thus calculated taking into account for generality all second order terms where interaction between quantities is indicated by mixed second order derivative terms.

$$u_c^2(y) = \sum_{k=1}^{N} \sum_{j=1}^{N} \frac{\partial f}{\partial x_i} \frac{\partial f}{\partial x_j} u(x_i, x_j) =$$

$$= \sum_{i=1}^{N} \left(\frac{\partial f}{\partial x_i} \right)^2 u^2(x_i) + 2 \sum_{i=1}^{N} \sum_{j=i+1}^{N} \frac{\partial f}{\partial x_i} \frac{\partial f}{\partial x_j} u(x_i, x_j) \tag{2.7.4}$$

where $u(x_i, x_j)$ indicates the estimated covariance for x_i and x_j

$$u(x_i, x_j) = r(x_i, x_j)\, u(x_i) u(x_j) \tag{2.7.5}$$

with $r(x_i, x_j)$ the correlation coefficient bounded between -1 and $+1$.

Covariance is usually evaluated from samples with a Type A approach:

$$u(x_i, x_j) = \sum_{k=1}^{N} (x_{i,k} - \overline{X}_i)(x_{j,k} - \overline{X}_j) \tag{2.7.6}$$

There may be significant correlation between two input quantities if the same measuring instrument, measurement standard, or reference datum is used in their determination, provided that the standard uncertainty is significant and does not disappear when combined with the others. For example, if a measurement of temperature by some thermometer is used to determine a temperature correction for several quantities and parameters, significant correlation may occur, unless a redefinition is applied, using the uncorrected quantities and the correction using e.g. a calibration curve versus temperature is added as additional input quantity.

2.7.4 Expanded uncertainty

Expanded uncertainty U is obtained from combined uncertainty u with multiplication by a coverage factor k. The coverage factor is determined to give for a given distribution a confidence interval of some width, or an associated confidence level. Quite common confidence levels are 95 % and 99 % probability: this probability is associated with a confidence interval $[L, U]$ (lower bound, upper bound), so that it may intuitively be concluded that the true value is within the confidence interval with probability $100(1 - \alpha)$ %; however, the estimate is either inside or outside the confidence interval, or, in other words, the statement is either true or false. Thus the confidence interval is to be considered using a frequency interpretation: it brackets the true value with confidence $100(1 - \alpha)$ %, thus leading to a correct guess $100(1 - \alpha)$ % of the time.

For the Gaussian distribution (that can be commonly adopted for the combined uncertainty of the output quantity if the number of input quantities is large enough, thanks to the Central Limit Theorem, even for non-Normal distributions), the confidence levels of 90 %, 95 % and 99 % correspond to confidence intervals of 1.64, 1.96 and 2.58 times the standard uncertainty (1.96 usually rounded off to 2).

2.7.5 Safe estimate and uncertainty

When the value of a measurand is reported, the best estimate of its value and the best evaluation of the uncertainty of that estimate must be given [15], sec. E.2.1. An understatement of uncertainty might lead to attribute too much trust in the values reported; embarrassing, critical or even disastrous consequences may occur when comparing results between different measurement methods for the confidence intervals not overlapping and indicating disagreement and inconsistency between reported values. A deliberate overstatement of uncertainties could also have undesirable repercussions: it could cause users of measuring equipment to purchase more expensive instruments or it could cause costly products to be discarded unnecessarily or the services of a calibration laboratory to be rejected.

When applying multiplicative coverage factors they shall be justified, they shall give the expected level of confidence and insight into combined and expanded uncertainty shall give the possibility of reversing the operation. It may be said that the concept of "safe" value does not apply, provided that all terms and contributions, including the imported uncertainties and how they are combined, are adequately described and suitable mathematical methods are adopted.

If using "safe" estimated of uncertainty, such as maximum absolute error, that gives the largest conceivable deviation from the observed results, then such estimate cannot be combined and propagated, being deemed by an ill-defined meaning. An example is the practice of expressing uncertainty interval as two values, for the lower and the upper bound, associating a *minimum* level of confidence, maybe in an attempt to stress the conservative or safe approach, but making it less useful for combined uncertainty calculation, without further information on the method used to estimate it, so that the operations can be reversed.

3

Instrumentation

Instrumentation that is used for the measurement of radiated emissions may be intuitively classified as properly said measuring, or recording, instruments and various kinds of antennas and probes to sense electromagnetic field over the required frequency range. The aim of this chapter is to give insight in the less known elements and equipment, considering both operating principles, limitations and appropriate use and settings. Spectral analysis and time-frequency domain relationships may be found in Chapter 2.

3.1   Spectrum Analyzer

The main features of Spectrum Analyzer, and by extension of frequency-domain equipment, are reviewed addressing operating principles, correct use, tradeoffs and common pitfalls. Good descriptions of SA operation and architecture may be found in manuals of major manufacturers [1, 134], together with pictures, diagrams and examples of advanced measurements, based often on options and commands of their own instruments. The focus here is on radiated emissions measurement not limited to settings and modes as per railway application standards, but covering also evaluation of radio signals for e.g. radio coverage and device characterization.

3.1.1 Architecture and block diagram

A general architecture of a super-heterodyne spectrum analyzer is considered, that, as the time goes by and numeric techniques and digital sampling are used more extensively, should be updated to resemble the so called real-time spectrum analyzers; these analyzers are in principle based on a pure time-domain capture and digital processing, but necessitate of a first mixer and down-sampling stage to extend their applicability to the highest frequency part of the spectrum. A pure numeric transform (Fourier, chirp Z or other) approach is limited by the maximum ADC sampling frequency (even with architectures with high interleaving factors) and by the huge number of samples needed; under-sampling and band-pass processing are used nowadays to push the limit to very high frequencies.

The vast majority of spectrum analyzers are still based on super-heterodyne architecture, that is more appropriate to illustrate the most common functionality and behavior of spectrum analyzers in practical situations; it's quite common that super-heterodyne architecture sets the reference to evaluate performance of other architectures and solutions. Any super-heterodyne architecture has some digital processing (including Fourier transform) applied to signals, once they have been demodulated and brought to a conveniently low intermediate frequency (IF) interval. When relevant, distinction will be made between analog and digital processing.

The typical block diagram of a Spectrum Analyzer is shown in Figure 3.1.1.

3.1.1.1 Input attenuator

Attenuators were introduced in sec. 1.5, describing them in terms of input impedance and attenuation, as well as matching capability for mismatched loads (see sec. 1.5.5).

The input attenuator is responsible for the adjustment of input signal power, in order to feed successive blocks with the correct signal amplitude; the correct operating point of the mixer, for example, is extremely important for its operation with respect to linearity, distortion, compression point, dynamic range.

Moreover, since it is a passive resistive network positioned at the beginning of the SA block chain, it ensures protection against moderate external overvoltage. Using a large input attenuation is always advisable when a measurement is performed for the first time, involving a new setup and the possibility of external disturbance; in case of more severe hazards, an external attenuator (or a filter, a transient protector, or a combination) shall be added.

Another useful feature of the input attenuator is that it ensures a better impedance matching at the input port, improving the accuracy related to reflection (VSWR) on the input line. In Figure 1.5.5 performance of attenuators featuring varying attenuation was shown, when the connected load has a VSWR variation as large as 5:1; the results have confirmed that 10 dB of attenuation ensure a VSWR lower than 1.1 for a 3:1 variation of load resistance. It is not uncommon that a calibration is performed with 10 dB of attenuation to ensure a better uncertainty (that shall be reproduced as is for certificate to be valid!); in general 5 or 10 dB of input attenuation are advised by all manufacturers somewhere in their manuals and application notes, even if for

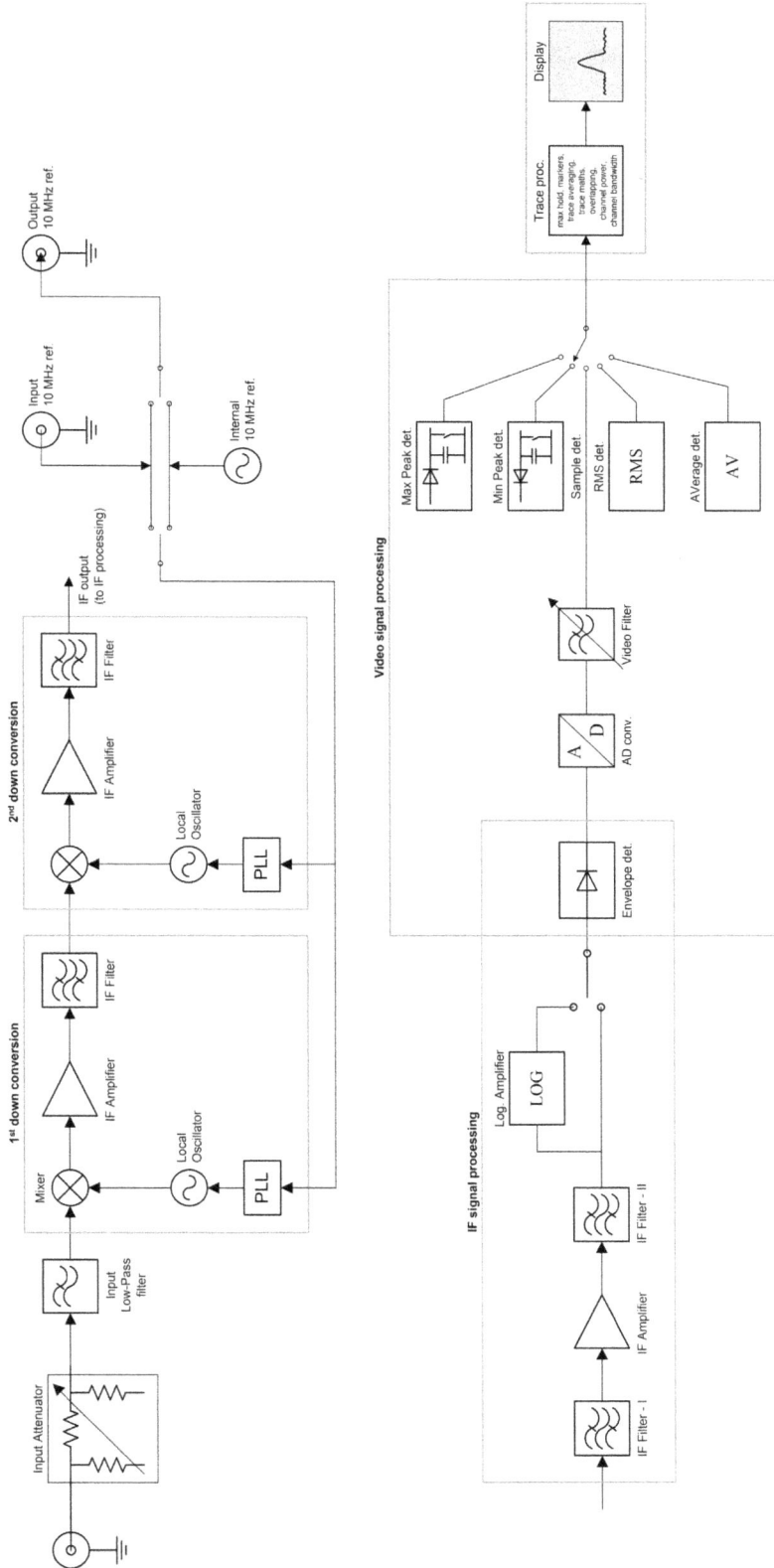

Figure 3.1.1 – Spectrum Analyzer architecture and block diagram

some particular measurements the full input dynamic range is needed, such as when measuring distortion and phase noise.

Last, when connecting the spectrum analyzer to a device with a widely changing impedance (such as an antenna and, in particular, a biconical antenna) and with far from ideal conditions (such as on site and using quite a long connection cable, exposed even to common-mode resonances with ground), the use of external attenuators at both cable ends is also advisable.

As customary, we will indicate the input attenuator setting by using the word "Att." followed by the amount in dB, as usually appears on SA display.

3.1.1.2 Pre-amplifier

The task of the pre-amplifier is to reduce the noise level by amplifying the input signal at the beginning of the SA block chain: based on the theory of noise through cascaded linear time-invariant systems (see sec. 2.6.1.2), the signal-to-noise ratio is increased if the amount of amplification before the internal noise sources is increased: some residual noise, however, will be still present due to the pre-amplifier itself and the input attenuator.

Internal noise reduction can be measured when turning on the pre-amplifier and quantifying the displayed noise floor. In Figure 3.1.2 there are the displayed noise levels for RBW = 10 kHz and with input attenuator changing from 20 dB to 0 dB, with and without the pre-amplifier. The noise level for Att = 10 dB is -116.03 dBm; stepping down to Att = 0 dB brings the noise level to -125.64 dBm. The noise contribution of the attenuator should then be estimated as the square root of the difference of the squares of the two noise levels, assuming the measurements uncorrelated. However, being the levels so different, attenuator noise can be readily estimated as the entire measured noise level, while the exact calculation gives -116.05 dBm: the displayed noise level is thus entirely caused by the input attenuator.

The use of the pre-amplifier lowers the noise level to -135.91 dBm starting from a 0 dB attenuation configuration: the effect of the pre-amplifier gain is clearly visible in the 10 dB step; assuming a perfect 10 dB power gain, the slight difference once the gain is compensated for may be used to have a rough estimate of the pre-amplifier noise term. The square root of the difference of the squares of the corrected 0 dB att. noise level and the noise level measured with the pre-amplifier on gives an estimate of -140 dBm as pre-amplifier noise contribution.

3.1.1.3 Local Oscillator

In modern spectrum analyzers the Local Oscillator is implemented with digital synthesis starting from a PLL (Phase Locked Loop) fed by a high-stability crystal oscillator (XO), generally of the temperature-compensated type, TCXO, or in high-end models with an oven-controlled crystal oscillator (OCXO), where a heating element controls the temperature around the crystal oscillator. Any kind of digital oscillator or voltage controlled oscillator (VCO) is initially fed by a crystal; fully electronic oscillators (either digital or analog) are of course possible [62], but their stability is absolutely

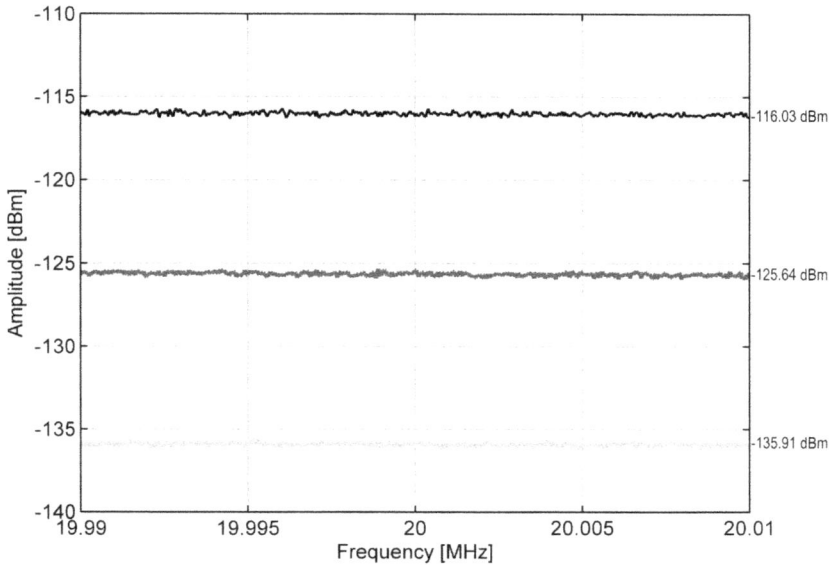

Figure 3.1.2 – Display noise level caused by input attenuator and pre-amplifier: Att = 10 dB and pre-amp off (black), Att = 0 dB and pre-amp off (dark gray), Att = 0 dB and pre-amp on (light gray); the RBW was set to 10 kHz.

worse than crystal-based oscillators. YIG (Yttrium Iron Garnet) technology may be also adopted because of the low phase-noise that characterizes it, and it was certainly widely used in the past generation of spectrum analyzers, as well as RF generators.

Information on stability is given in instrument characteristics and can be measured directly from the 10 MHz output connector, provided that an order of magnitude better stability is available in the external instrument (e.g. frequency meter). It is observed that for an OCXO LO the short-term stability is already of the same kind of the best GPS frequency references: where GPS information comes into play is for long-term stability and to limit wandering.

The Local Oscillator is adjusted in steps during the sweep, whose amplitude depends on the number of points (or pixels, see sec. 3.1.1.9) and the selected resolution bandwidth (see sec. 3.1.1.5 and 3.1.3.2). The number of points may be variable, so the user has some control on frequency resolution. The frequency step shall always be related to the adopted resolution bandwidth for a very practical exigency, that is not to miss parts of the spectrum while sweeping; when performing emissions measurements, for instance, it is always required that the frequency step is smaller than the adopted resolution bandwidth with a margin, normally staying around 50 % of the RBW value.

3.1.1.4 Demodulation and Mixer

The mixer ideally "mixes" two input signals at its ports, namely the input signal, fed by the preceding stage, and Local Oscillator signal, for down-conversion. The working principle is that of multiplying the two signals and the resulting output spectrum contains components at the difference and the sum of the individual frequencies: the principle can be explained easily by visualizing the product of two sinusoidal tones and using the prosthaphaeresis algorithm. The reader is invited to have a look at some references for details on how mixers are built and the principle used to perform the multiplication of signals (Horowitz [62]; S.A. Mass, *The RF and Microwave Circuit Design Cookbook*, Artech House, Boston, 1998, entirely dealing with mixers; Collin [45], sec. 12.7 to 12.11).

In modern spectrum analyzers, featuring an extended frequency range, there might be more than one mixer, progressively demodulating the incoming signal to lower frequency, until the IF (Intermediate Frequency) stage before log-amplifier and detectors.

Heterodyne demodulation by means of a local oscillator at frequency f_{LO}, converts an incoming mixer tone at frequency f_{in} to an intermediate frequency f_{IF}:

$$|m\, f_{LO} \pm n\, f_{in}| = f_{IF} \tag{3.1.1}$$

that for $m = n = 1$ simplifies to $|f_{LO} \pm f_{in}| = f_{IF}$. This relationship is obtained by trigonometric manipulation of mixer product (see Figure 3.1.3).

Conversely, $f_{in} = |f_{LO} \pm f_{IF}|$ and this clearly indicates that for each pair of LO frequency f_{LO} and IF frequency f_{IF}, there exist two input frequency values that satisfy the relationship, and that are thus captured and demodulated: the desired one towards dc and a second one, at higher frequency, called "image frequency". Extending this reasoning to the interval of frequencies that "nearly" satisfy the relationship, or in other words, that are close to f_{in} within a given interval, we talk of "desired frequency band" and "image frequency band". Of course at the output of the mixer also f_{LO} and f_{in} components are visible: for a sufficiently low amplitude of the input signal the mixer operates in the linear region and these four components are the only ones that populate the spectrum. This is a simplified reasoning, because for diode mixers the repeated switch on and off of the bridge diodes creates many inter-harmonics of input and LO frequencies (the explanation in sec. 10.4 of A.W. Scott, *Understanding Microwaves*, John Wiley & Sons, 1993 is very clear).

Two inconveniences may affect mixer operation:

- the signal tone that is being read is accompanied by another component that is at the same frequency distance from f_{LO} but at the opposite side of the frequency axis, i.e. this component is the image component; there is no way to separate the two resulting frequency differences if considering amplitude only, because both fall exactly at the same frequency, but have an opposite phase; if this additional information can be used, then it is possible to separate the two components with the so-called "image rejection" mixer that uses two balanced bridges at 90° displacement;

- the amplitude of the input signal increases and the mixer non-linearity creates inter-modulation products, that spread over the frequency axis: with one single input tone, the sum/difference combination of harmonics of f_{LO} and f_{in} produces frequencies that are integer multiples of the basic difference $|f_{LO} - f_{in}|$ called *single-tone intermodulation* products, or spurious terms; if more than one tone is present at the input (e.g. f_{in1} and f_{in2}), then the number of combinations increases, and sums and differences occur of integer multiples of f_{LO}, f_{in1} and f_{in2} thus producing terms that may fall next to the original components around the IF band (this is called *two-tone intermodulation*); harmonics and intermodulation may be kept under control when using the double-balanced bridge.

The mixer is the core of the spectrum analyzer and is largely responsible for its performance in terms of distortion and spectral accuracy. Key performance indexes of a mixer are the presence of harmonics and inter-modulation products, identified as *harmonic suppression* and *spurious suppression*, respectively.

Another performance index that describes the spectral properties and quantifies the amount of LO signal that leaks into the output is the *LO-to-RF isolation*: it is the ability to keep separate the components fed at each port, conveying the input uniquely towards the mixing process and the output. This is normally addressed by evaluating the leakage of the LO frequency back to the input and into the output; the overall effect is named "Local Oscillator feed-through"and is considered later in sec. 3.1.5.

Last, passives and semiconductors inside the mixer act as noise sources and add to the noise coupled through the signal and LO ports: the mixer is then characterized by its noise figure.

To avoid that strong out-of-band high-frequency components are captured and demodulated, overlapping to the correct desired components, a pre-selector filter may be added after the input attenuator: it is a tunable filter or filter bank that is optional in spectrum analyzers and mandatory in EMI receivers, and that rejects extraneous components that would otherwise be indistinguishable from the desired ones after demodulation.

For a 9 kHz - 3 GHz spectrum analyzer such IF is in the range of about 3.1-4 GHz and the LO 3 GHz above it. The IF RBW filter that accurately weights the various portions of the input signal (see sec. 3.1.3.2) can be hardly applied directly to such a high IF; the same may be said for the algorithms used for detection and signal processing (e.g. video filtering and trace processing in sec. 3.1.2). Thus, conversion to a lower IF is necessary and is done in steps, passing normally through an intermediate IF and a final IF value, the latter then fed to the Analog-to-Digital Converter (ADC). To exemplify some values, the intermediate IF might be in the range of some hundreds MHz and the final IF around some tens of MHz, compatible with the RBW values and similar to the single-IF of old spectrum analyzers [134].

When measuring wideband noise-like signals, the perceived power for the chosen RBW value may be much lower than the total power applied to the mixer by e.g. one or some orders of magnitude (it suffices to remember that RBW is a fraction of the frequency range and that the crest factor may be large for noise-like signals). It is usually advisable to set the mixer input power level around -10 dBm or lower, so that displaying a -20 dBm or -30 dBm power level with a narrow RBW ensures that the ideal limit of $+10$ dBm is never reached. Too a high level before damaging the mixer will drive it into its 1 dB compression point with signal compression and distortion by-products (harmonics of the input signal) as a consequence. Modern spectrum analyzers, not

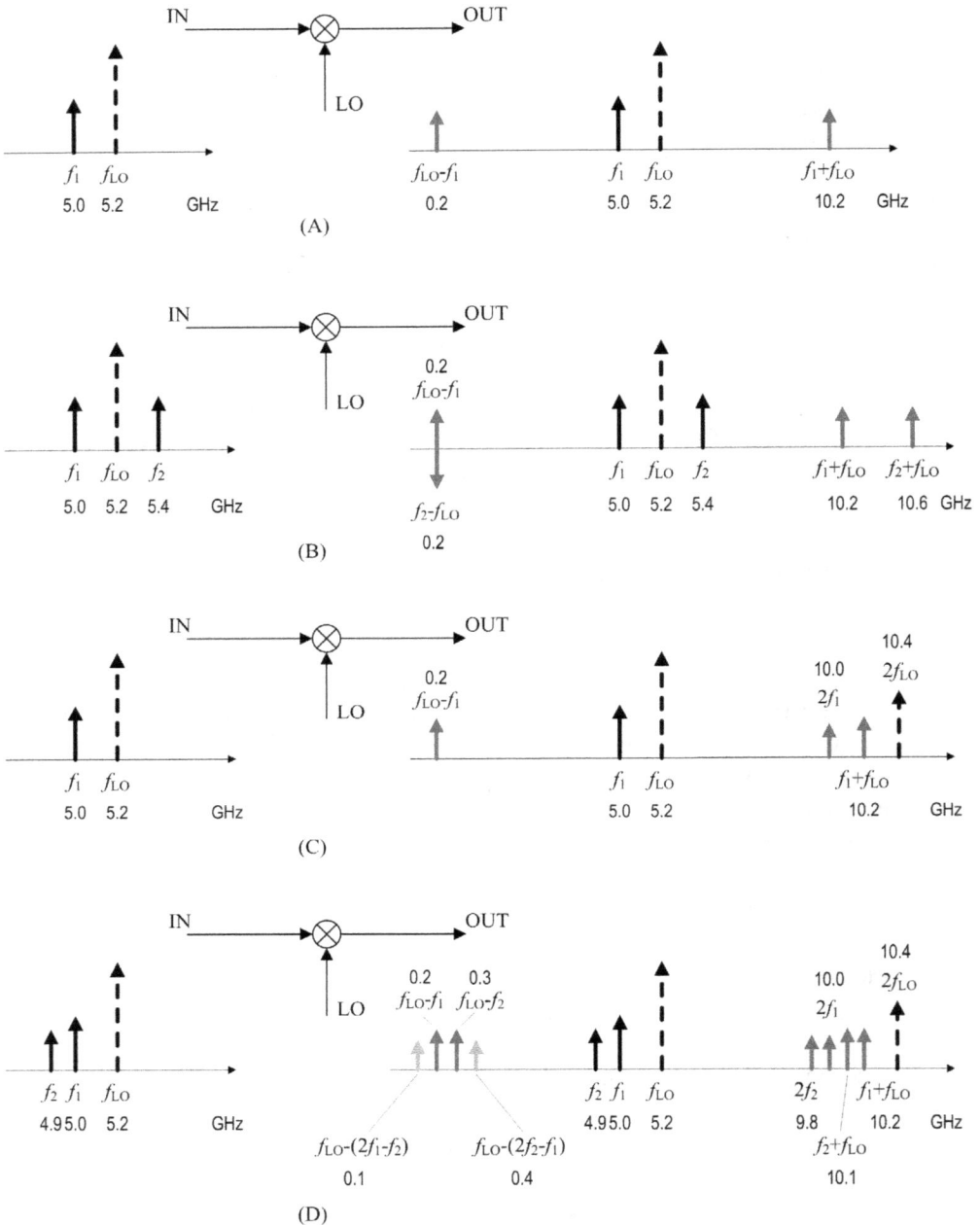

Figure 3.1.3 – Mixer operation with ideal line spectrum of input signal (solid black, made of either one or two tones at f_1 and f_2), local oscillator tone (dashed black at f_{LO}) and mixer output (including original components, in black, and mixing products, in gray): (A) simple mixing under linearity assumption and one input tone f_1; (B) two input tones mirrored around f_{LO} (f_2 is the image frequency) create concomitant IF components with same frequency and different phase; (C) first-order intermodulation products with one input tone at f_1; (D) second-order intermodulation products with two input tones at f_1 and f_2 (only the main second-order intermodulation terms in the IF band are shown in light gray).

only warn the user for exceeding the advised maximum operating point, but often have hardware or software protections.

3.1.1.5 IF filter

IF signal processing is performed at the last intermediate frequency (at some tens of MHz, it was said); here the signal is amplified again and resolution bandwidth and envelope detector are applied. The gain G_{IF} at this last IF can be adjusted in small steps (often 0.1 dB steps), so the maximum signal level can be kept constant when passed to the subsequent signal processing, regardless of attenuator setting and the mixer level. Of course, wrong settings compromise the signal-to-noise ratio, the dynamic range and may introduce some distortion; this operation is generally performed automatically within the spectrum analyzer and the user has no control on it. In case of more than one mixer, and correspondingly more than one IF processing block, there will be several IF gain terms G_{IF1}, G_{IF2}, etc., as well as different mixer levels.

The IF filter is used to define that portion of the IF-converted input signal that is further analyzed and then displayed; the IF filter is thus responsible for the accurate definition of the Resolution Bandwidth (RBW), that weights many spectral characteristics and features a very wide range, normally from few Hz to a tens of MHz.

Implementation choices are obviously those of analog and digital filtering, but also the use of the Fourier transform is an option. There is a major tradeoff in IF filter construction, combining the opposite requirements of selectivity and fast transient response (so to sweep the frequency axis by changing the LO frequency at a convenient pace). The tradeoff has converged onto the Gaussian architecture for the analog/digital implementation. For details regarding filter performance and for Fourier transform characteristics please see sec. 3.1.3.2 and 3.1.6.

Analog filters are often used for the largest RBW, let's say above some hundreds kHz for the difficulty to implement digital filters with the necessary sampling rate; analog implementation is conceived to have similar performance in terms of selectivity and shape factor at least down to 20 dB with respect to center band. Since analog filters are in any case built assembling different stages to obtain the desired roll-off, it is customary to split the filter into two parts, one soon after the IF mixer output, followed by the *IF amplifier* and then by the second part of the filter.

Digital filters have a much better selectivity than analog ones, that comes at hand when selecting the narrowest RBW values. Digital filters are also more stable with respect to temperature, drift and aging, and thus more reliable and preferable as for uncertainty and the need for periodic calibration. Moreover, since they exhibit faster transient response than their analog equivalent, they allow slightly faster sweep times. However, they shall be preceded by the ADC stage, in turn preceded by an analog anti-aliasing pre-filter, and this slowed down their adoption at the beginning.

Last, when reaching low RBW values and/or requiring fast sweep times, it is much more convenient to rely on Fourier transform (FFT) and direct frequency filtering. The use of windows for frequency leakage suppression affects the effective frequency resolution and requires a slightly longer *transformation window* (named also *observation time*). However, selectivity is much better than analog and digital implementations:

e.g. with a flat-top window a shape factor of 2.6 is attainable. The *shape factor* or *skirt selectivity* is defined as the ratio between two bandwidths, one at high attenuation, e.g. down by 20, 40 or 60 dB, and the other taken as reference, usually the $-3\,\mathrm{dB}$ bandwidth (normally used to define the RBW value itself); further details and values for the most common implementations can be found in sec. 3.1.3.2.

3.1.1.6 Log-amplifier

The log-amplifier implements an ideal logarithmic function (linear on a log-scale) of the incoming power between the two extreme points of a range that may span several decades. When years ago the log-amplifier could be built starting from a good diode (or a transistor in diode connection), attention was focused on the excellent logarithmic law needed for the base-emitter junction and datasheets of the most suitable transistors were expressively reporting the deviation from the pure log law. For a spectrum analyzer this performance hides behind the figures of "scale fidelity" or "linearity error" appearing in the datasheet.

The log amplification acts as a compressor for large noise (or signal) peaks; a peak of ten times the average level is only 10 dB higher on a power log-scale. Instantaneous near-zero envelopes, on the other hand, contain no power, but are expanded toward negative infinity decibels, and this might bias averaging (see sec. 2.1.1), while for display purpose they can be removed quite easily. The combination of these two aspects of the logarithmic curve affects the noise power statistical distribution and leads to read values lower than the true noise power for some settings (see sec. 3.1.3.4) [2]. By the way, when spotting sinusoidal tones out from noise, log compression ensures a reduced noise floor and thus an increased display dynamic range, in particular if used in conjunction with Min Peak detector.

3.1.1.7 IF envelope detector

The IF signal after band-pass filtering is passed to the envelope detector block. This block removes the IF frequency and takes the low frequency modulating signal, that is the base-band image of the incoming signal. This is like when recovering the modulating signal from the envelope of the incoming signal in AM modulation; the basic scheme for AM demodulation is that of a diode followed by a RC circuit that establishes the time constant of demodulation. With envelope detection only the magnitude of the envelope is used, while the phase information is lost; on the contrary, when processing is entirely performed by Fourier transform, as for real-time spectrum analyzers, the full information is retained.

The power content of the incoming signal is centered around the IF frequency like if it were the carrier, split towards dc and higher frequency (at twice the IF frequency), so that the desired band may be isolated and further processed. The Video Filter (see sec. 3.1.1.8) low-pass filters the detected IF output signal and eliminates the high frequency sub-band. When processing is performed digitally for the last IF block, the envelope is determined from the samples of the IF signal and a numeric implementation of the video filter is applied. For the first "digital" spectrum analyzers appearing in the '80s and for some models still in use today, the digital sampling was not operated on the IF signal directly, but on the signal at the output of the envelope detector fed to the video filter. This of course largely reduces the requirement on ADC sampling frequency, but limits the availability and flexibility of post-processing functions, especially needed today with sophisticated modulations featuring a wide dynamic range and demanding time-frequency performance.

The dynamic range of the envelope detector is very important because it largely affects the overall dynamic range of spectrum analyzer; modern spectrum analyzers feature normally 100-120 dB of dynamic range of the IF signal (see sec. 3.1.7). The user modifies the dynamic range by changing the reference level: when using this option, it is possible to widen the display dynamic range, and correspondingly to change the IF gain (but it is SA-model dependent); mixer operating point is not influenced, because down-conversion occurs before the IF envelope detector block. So, at this stage we can amplify the signal stretching it onto the displayed range, but we have no influence on signal distortion and compression, for which the input attenuation and the mixer operating point are the key settings (see sec. 3.1.7).

The IF signal before being passed to the envelope detector may undergo amplification and logarithmic compression by means of the log-amplifier described in sec. 3.1.1.6.

3.1.1.8 Video filter

The video filter in old analog spectrum analyzers was a simple RC filter following directly the output of the envelope detector (as for the basic AM demodulation circuit mentioned before) and drove directly the signal for the cathode ray tube. The name has been preserved, but all modern spectrum analyzers use analog-to-digital conversion (ADC) to implement digital processing functions, even if the video filter is still a low-pass filter with the aim of smoothing the trace, removing the jagged noise profile and increasing somewhat the displayed signal-to-noise ratio. The combined operation of detectors and video filter is quite articulated for the many options of modern spectrum analyzers, so that a clear all-comprehensive detailed connection scheme is not easily attainable.

The ADC also has its own recommended input power level specified by the manufacturer, that corresponds to its full scale range probably with some margin to avoid signal clipping at the two extremes; the margin necessary to avoid clipping must again take into account the crest factor of superimposed noise: considering Gaussian noise a crest factor of 2.6 is met 99 % of the time (see sec. 2.2.1), resulting in 3 to 5 dB of margin for the equivalent power.

If the Video Bandwidth VBW is smaller than the Resolution Bandwidth RBW, then the sweep time is further increased with respect to the bare requisite regarding the transient response of the RBW filter (see sec. 3.1.3.6).

3.1.1.9 Detectors

Detectors available in a modern spectrum analyzer are Peak (as Max Peak and Min Peak), Average, RMS, Sample. Their description and the explanation of the basic operation follow.

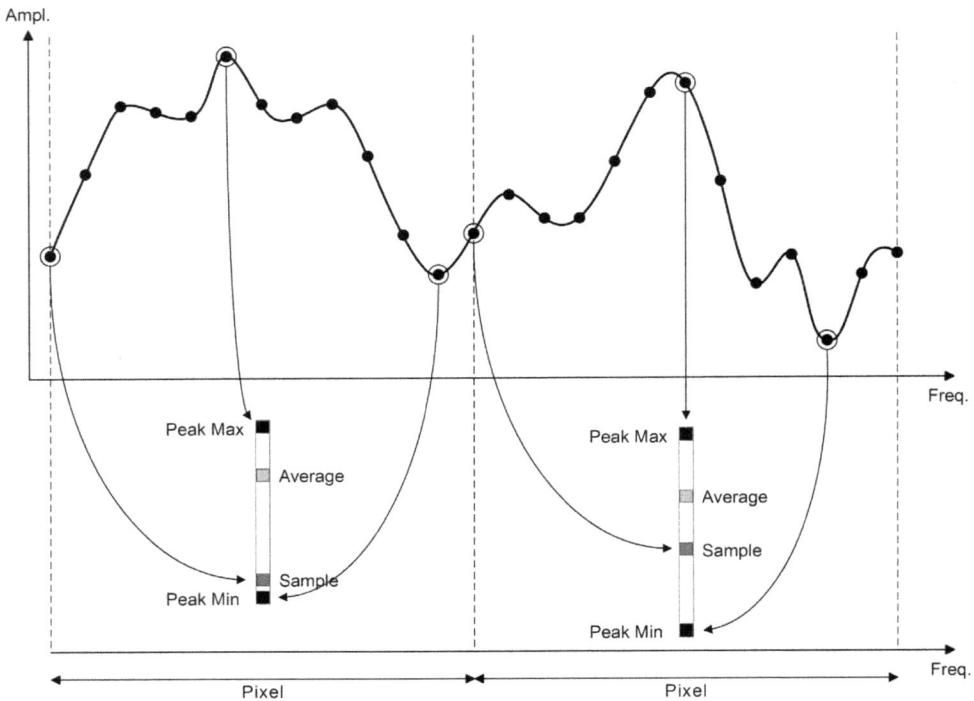

Figure 3.1.4 – Operation of detectors exemplified.

The explanation of their operation is based on the correspondence between a portion of the spectrum and its representation on the display, that is made of a certain number of pixel columns (that we call for brevity "pixel"): a single pixel may contain the spectral information of a relatively large frequency interval; in this case the information contained in the interval, after being processed with the detector, is synthesized in a single value that is then displayed.

The operation of detectors is graphically described in Figure 3.1.4.

When considering random signals, such as noise, detector performance may be quite different and may depend also on the number of samples fed to the detector at each step: each detector operates on a pixel basis and collects the samples pertaining to that pixel in the chosen frequency span and resolution; when the number of samples per pixel is limited, this might bias the final spectral estimate and influence detector performance.

Max and Min Peak detectors   The Max Peak detector displays the maximum value. From the samples allocated to a pixel the one with the highest level is selected and displayed. Even if wide spans are displayed with very poor resolution bandwidth (this occurs when span/RBW is greater than the number of pixels on the frequency axis), no input signals are lost. Therefore this type of detector is particularly useful for EMC measurements and to capture at best transient signals with a wide frequency occupation.

The Min Peak detector selects from the samples allocated to a pixel the one with the minimum value for display. Its usefulness alone is limited, but together with the Max Peak detector it is able to implement a maximum span detector.

Sample detector   The Sample detector samples the IF envelope for each pixel of the trace to be displayed only once. That is, it selects only one value from the samples allocated to a pixel, e.g. the first one, to be displayed. If the span to be displayed is much greater than the resolution bandwidth (again for a span/RBW greater than the number of pixels on the frequency axis), input signals are no longer reliably detected.

RMS and Average detectors   The RMS (root mean square) detector calculates for each pixel the power of the displayed trace from the samples allocated to that pixel. The result corresponds to signal power within the span represented by the pixel. For RMS calculation samples of the envelope are required on a linear level scale. The reference resistance value ($50\,\Omega$) may be used to calculate power from voltage readings.

The AV detector calculates the linear average for each pixel of the displayed trace from samples allocated to a pixel. For this calculation samples of the envelope are required on a linear level scale. Averaging on a log scale wouldn't be useful because the largest values that are relevant to the result would be log compressed; it is however possible to enable this function by selecting a narrow Video Filter bandwidth (VBW) after processing the IF envelope on a log scale. In all cases averaging is also obtained by using the Video Filter after the Sample detector. Needless to say, averaging of AV detector is performed on subsequent samples of the IF envelope signal, while *trace averaging* option (see sec. 3.1.2) performs averaging of trace pixels, far apart in time (each separated by a sweep time ST interval).

Quasi-Peak detector   The Quasi-Peak (QP) detector "yields a measure roughly correlated to the subjective annoyance effect on AM broadcast services", and "has also found restricted use for measuring interference in television" [72]. So, its intended use is for interference measurement applications, adopting defined charge and discharge times that weight severity in terms of amplitude and rate of occurrence for periodic and intermittent disturbance. These times are established in Table 2 of the ANSI/IEEE C63.2 standard [72] and in Table 2 of CISPR 16-2-1 [40] (see Table 3.1.2 later at the end of sec. 3.1.3.6). QP response is compared to that of other detectors in CISPR 16-1-1 [38] at sections 5 through 7: with respect to the Peak detector the measurement is done by applying a pulse train and assuming a pulse repetition rate of $25\,Hz$ for frequency interval A and $100\,Hz$ for the other frequency intervals; the resulting difference is $6.1\,dB$ for interval A, $6.6\,dB$ for interval B, and $12\,dB$ for intervals C and D. The European Directive 2004/104 for EMC of automotive applications [51] and CISPR 12 [37], conversely, give a straight $20\,dB$ correction factor for measurements at $30\,MHz$ and above, that is overestimating in most cases.

3.1.2 Trace processing

Modern spectrum analyzers allow some post-processing of acquired traces, such as averaging, almost always implemented as a running average of new samples since the last reset: for each pixel new traces are accumulated and averaged arithmetically or using an aging factor (see sec. 2.6.2.2). The averaging operation is performed taking into account display settings, so linearly or logarithmically, depending on the chosen display units and scale. Averaging in post-processing faces thus the same problems already considered for video filtering (a form of averaging by low-pass filtering) and for AV detector itself: the use of Sample detector and linear-scale setting yields correct results, while the log scale setting underestimates by 1.45 dB; with Min Peak, Max Peak or Sample detectors using Video Filter (with low enough VBW/RBW ratio) helps the running average to converge.

The *max hold* option is again applied in trace post-processing and consists in keeping the envelope of the maximum value for each pixel; the result represents thus a worst-case estimate of the measured amplitude and may be useful e.g. to obtain an upper bound of the background noise, as it is done during measurements of electromagnetic emissions. Additionally, when sweeping over the variable-frequency carrier signals of digital modulations, such as OFDM, the overall spectrum and the graphical representation of the band occupation, it is useful to capture signal portions on the fly and accumulate them on the display trace.

Channel power measurement is another example of trace processing that requires the definition of the channel bandwidth (see sec. 3.1.11.1). Many spectrum analyzers have built in parameters for many standard formats, such as cellular and wireless network transmission protocols. The possibility of covering channel bandwidth with several adjacent samples, rather than reading it altogether with a single measurement at the largest RBW, allows to measure even modern wideband spread-spectrum modulations without spending for expensive equipment: depending on modulation characteristics and statistics, it may be necessary to accumulate several readings.

Trace saving and recall allows to superpose past and new traces for visual comparison or to apply markers.

Trace markers are quite useful in spotting out numeric values accurately, more than the display and resolution limitation. They allow also to calculate the difference in frequency and/or amplitude between two points in the same trace or with respect to a saved trace. Of course, with modern spectrum analyzers that allow to save and download to an external computer the raw numeric values, all kinds of post-processing are possible, using general purpose math packages, such as Matlab, Octave, Mathcad, NumPy.

Enhanced marker functions implement more complex operations and calculations (as in the case of phase noise), saving time and preventing mistakes.

3.1.3 Performance and operations

This section considers the operation of SA elements presented previously, verifying and quantifying performance reporting also manufacturers data. Statistical analysis is heavily used squeezing information from repeated tests and the analysis of distributions and expectations.

3.1.3.1 Input VSWR and reference impedance value

The input Voltage Standing Wave Ratio is often specified in terms of "less than" (so an upper bound over a frequency interval is often given) and the upper bound of the inequality ranges usually from about 1.2 to 1.8. It is first to note that 1.2 is in any case a pretty large VSWR when accuracy and low uncertainty are required, resulting in a 20 % difference between the maximum and the minimum of the voltage wave at the RF input connector, or in other terms $\pm 10\,\%$ (or ± 1 dB) around an average point. Most important, the VSWR specification is always accompanied by attenuator setting, that is almost always set to 10 dB. Higher attenuation values may further improve the VSWR (see sec. 1.5.3), but less desirable for the noise floor; smaller attenuation in turn has a bad impact on input VSWR.

The reference impedance is $50\,\Omega$ for almost all spectrum analyzers, some with the option for $75\,\Omega$ input setting. In any case measurements at any other reference impedance are possible with a $50\,\Omega$ spectrum analyzer: the possible solutions are i) adding a resistor in series (matching approximately the measured line impedance), ii) using a coupling transformer or iii) inserting an attenuator.

Attenuators are commonly used to correct impedance mismatch: a typical example is an antenna with highly variable input impedance connected to the spectrum analyzer, where the attenuator normalizes the VSWR for changing frequency. Of course the power dissipation in the attenuator is not an issue when measuring low-level signals; it is on the contrary for when large RF power levels are applied for immunity testing, so that 3 dB and 6 dB attenuators are commonly used.

3.1.3.2 Resolution bandwidth

The setting of the Resolution Bandwidth (RBW) may span over about six or seven decades, from one or few Hz to tens of MHz; in old spectrum analyzers the possible choices where just two or three per decades (distributed in 1-3-10 or 1-2-5-10 scales), while in modern ones digital processing in principle allows almost any value, limited by practical reasons (e.g. pre-calculated values of digital filter coefficients or a limited number of installed analog filters).

Resolution bandwidth, impulse bandwidth and skirt selectivity Let's focus on the definition of RBW as IF filter bandwidth: for example, the -3 dB bandwidth is the frequency spacing between two points of the transfer function at which the insertion loss of the filter has increased by 3 dB relative to the center frequency; analogously a very similar -6 dB bandwidth definition may be adopted. Why so much interest in the definition of 3 dB and 6 dB bandwidths? Since the definition is based on the frequency response we are talking of "amplitude dB", so with a factor of 20 in front of the log function: the 3 dB and 6 dB insertion loss correspond to 0.707 and 0.5 the center band gain. It is clear also that the -6 dB definition corresponds to half the amplitude gain and the $-3\,\mathrm{dB}$ definition to half the power gain. From a different viewpoint the $-6\,\mathrm{dB}$ bandwidth for Gaussian filters can be made correspond roughly to the *impulse bandwidth*.

The impulse bandwidth is defined as

$$B_i = \frac{1}{H_0} \int\limits_{0}^{+\infty} H(f)\,\mathrm{d}f \qquad (3.1.2)$$

where $H(f)$ is the amplitude (or voltage) frequency response and H_0 is the dc or center band value. The Equivalent Noise Bandwidth of a filter was defined based on the square of the frequency response, because this definition is based on power (see sec. 2.6.1.1).

For EMC and EMI measurements, where the spectrum of pulses, clicks and transients is often measured, 6 dB bandwidths are exclusively specified and the CISPR 16 standards make reference to 6 dB bandwidth for the standardization of IF filters of EMI receivers.

The definition of the $-20\,\mathrm{dB}$ bandwidth is sometimes added (that corresponds to a decade in amplitude, or an insertion loss of 10) as a first characterization of the roll-off (or skirt) of the filter. The *shape factor* (or *skirt selectivity*) is specified down to the $-60\,\mathrm{dB}$ bandwidth as $\mathrm{SF}_{60/3} = B_{-60\,\mathrm{dB}}/B_{-3\,\mathrm{dB}}$.

Depending on the analog or digital implementation, the effective bandwidth measured at $-3\,\mathrm{dB}$ and $-6\,\mathrm{dB}$ points may vary and, as a consequence, also the ENBW is subject to change. Furthermore, even when specifying the IF filter as a four-pole, five-pole or Gaussian, different manufacturers and implementations have slightly different values.

In [134] Rohde & Schwarz gives a complete set of values for the relationship between the $-3\,\mathrm{dB}$, $-6\,\mathrm{dB}$, ENBW and pulse bandwidth for the IF filters used in their spectrum analyzers; they are reported in Table 3.1.1.

Agilent [2] gives very similar figures for the 4-pole and 5-pole synchronous filter implementation (less clearly called "4 filter" and "5 filter" by Rohde & Schwarz in [134]): ENBW $= 1.128\,B_{-3\,\mathrm{dB}}$ and $1.111\,B_{-3\,\mathrm{dB}}$, respectively. For FFT-based processing Agilent gives ENBW $= 1.056\,B_{-3\,\mathrm{dB}}$, not so different from the Gaussian filter specified in Table 3.1.1 (1.065). In the same Agilent application note the impulse bandwidth of a 4-pole analog and a Gaussian filter are declared $1.62\,B_{-3\,\mathrm{dB}}$ and $1.499\,B_{-3\,\mathrm{dB}}$ respectively: while the latter matches the 1.506 value shown in Table 3.1.1, the 1.62 value is much different from the 1.806 one. It is thus observed that while 4-pole and 5-pole implementations may differ remarkably between manufacturers (as we have seen), there is a general agreement on the Gaussian response filter performance.

3.1.3.3 Input attenuation correction

The input attenuator is one of the most relevant sources of error, and, if corrected, of uncertainty, in absolute amplitude measurements: the error of combined attenuator and IF filter gain has systematic components, so it is not always true that the same attenuator setting gives the same amplitude error whatever the chosen IF RBW.

Multiples of $B_{-3\,\mathrm{dB}}$	Analog 4-pole	Analog 5-pole	Gaussian
−6 dB bandwidth	1.480	1.464	1.415
ENBW	1.129	1.114	1.065
Pulse bandwidth	1.806	1.727	1.506

Table 3.1.1 – Relationships between -3 dB, -6 dB, ENBW and pulse bandwidths (Rohde & Schwarz [134]).

For example, in the tested unit there is a certain amount of over compensation of input attenuator, so that the 0 dB attenuation (exclusion of input attenuator) gave an amplitude reading of $189.27 - 189.33$ dBµV that resulted always, or almost always, in the lowest reading. At RBW = 1 MHz the error for the 10 dB and 20 dB attenuation settings is +1.8 dB with respect to 0 dB attenuation; at RBW = 100 kHz the spread that contains all the tested attenuator settings is only 0.25 dB and it is 0.18-0.22 dB at the remaining RBW values from 10 kHz down to 10 Hz. There is no unique correction for the attenuator setting alone, but the error is related to the input attenuation-RBW setting pair. Such an error, if not compensated, affects the absolute amplitude accuracy by about ±0.1 dB; uniform distribution might be assumed, having found no specific correlation with the RBW setting except at 1 MHz. Moreover, it was observed that the FFT operating mode gives slightly larger amplitude estimates of about 0.15 dB.

Except extreme cases, such values are within the declared absolute accuracy. The best use of Spectrum Analyzer is for relative amplitude measurements (e.g. attenuation and frequency response), while a power meter is always a much better choice as reference for absolute readings.

It is thus advisable to verify always SA response by using a reference source that for the case of the internal IF filter response may be the 10 MHz reference itself; as a matter of fact we are looking for inconsistencies in the correction of attenuator and RBW combined gain settings, so trying to compensate for a systematic error. On the contrary, when stability and phase noise are questioned, an external reference shall be used.

3.1.3.4 Detector output for different signals

Depending on the type of input signal, different detectors may provide different measurement results: when dictated by standards the choice of the detector is compulsory; sometimes there is the possibility of choosing between detectors, provided that compensation factors are applied; in other cases, e.g. for scientific investigations and evaluations other than strict compliance to limits, the operator is free of choosing (choosing the wrong or inappropriate detector leads thus to a biased or erroneous measurement or to worsened accuracy and performance) .

Sinusoidal signal   Assuming that the spectrum analyzer is tuned to the frequency of the input signal (span = 0 Hz), the envelope of the IF signal and thus the video voltage of a sinusoidal input tone with sufficiently high signal-to-noise ratio are constant.

Therefore, the level of the displayed signal is independent of the selected detector, since all samples exhibit the same level and since the derived average value (AV detector) and RMS value (RMS detector) correspond to the level of the individual samples.

Being different the amount of resulting noise for different detectors, if the original signal-to-noise ratio is not sufficiently large, this contribution will be relevant and there will be differences in the displayed amplitude. It is easy to calculate the expected deviation by adding the difference in detector readout D (see the next paragraph) to the signal-to-noise ratio SNR, both expressed in dB, and reversing the log-relationship to extract the deviation around the tone amplitude (see sec. 3.1.8).

Noise The behavior is much different, however, taking random signals, such as noise or noise-like signals, in which the instantaneous power varies with time. If we consider the crest factor of the noise signal, then it is clear that the Max Peak and Min Peak detectors will report values as much as different from the RMS values as 10-12 dB if a Gaussian noise is observed for a sufficiently long time (please see sec. 2.2.1 for crest factor and related probability for a Gaussian distribution, and sec. 3.1.11 for a quantitative estimate), with the Max Peak and Min Peak detectors respectively over-weighting and under-weighting the noise level.

When measuring noise with low span/RBW ratios, the number of samples per pixel will be low (down to only one for extreme settings) and the Sample detector will give the same result of the Max Peak detector. In reality, using Sample detector, the spectrum analyzer may have lower effective video bandwidth than in Max Peak detection mode, because of limitations of the sample-and-hold circuit that precedes the A/D converter; examples include the Agilent 8560E-Series spectrum analyzer family with 450 kHz effective sample-mode video bandwidth, and a substantially wider bandwidth (over 2 MHz) in the Agilent ESA-E Series spectrum analyzer family.

The RMS detector calculates for each pixel the total rms of data samples pertaining to that pixel. The use of additional video filtering is not necessary and not advisable, since it would lower the displayed values down to a theoretical maximum error of 2.51 dB for Gaussian noise input.

The Average detector performs the average of samples pertaining to a pixel. If a linear envelope detector is used (no log compression), then applying video filtering by narrowing the VBW value brings a similar result and the two settings are compatible. If the log-amplifier is used, on the contrary, video averaging leads to a lower value than the true average value. In the case of Gaussian noise this difference is 1.45 dB.

We may say that performing an RMS detection of signal envelope is substantially equivalent to evaluate the average of signal power; however, there is a difference for the power calculated respectively as RMS average and the square of AV detector output.

The demonstration of the difference between RMS and AV detector output can be achieved by considering the Rayleigh distribution f_{Rayleigh} of the amplitude of noise at the output of the IF filter (considering the distribution of the magnitude, with in-phase and in-quadrature components, both Gaussian): the AV envelope detector gives the average envelope voltage v_{IF}, while the RMS gives the average power of the envelope voltage with respect to a reference resistor value R.

$$\bar{V} = \int_0^\infty v_{IF}\, f_{\text{Rayleigh}} dv = \sigma \sqrt{\frac{\pi}{2}}$$

(3.1.3)

$$\bar{P} = \int_0^\infty \frac{v_{IF}^2}{R}\, f_{\text{Rayleigh}} dv = \frac{2\sigma^2}{R}$$

Thus the ratio of the power estimated as square of the average voltage $\bar{V}$ over a reference resistor R and the power indicated by the RMS value is

$$\Delta P = 10\, \log_{10} \left(\frac{\bar{V}^2/R}{\bar{P}} \right) = 10\, \log_{10} \left(\frac{\pi}{4} \right) = -1.05\,\text{dB}$$

(3.1.4)

Considering the Peak detector, its response may be quite variable, depending on the sweep time capturing the peaky values expected for the large crest factor of Gaussian-like noise: it is thus understood that the compensation factor of 20 dB indicated by CISPR 12 commented at the end of sec. 3.1.1.9 is somewhat arbitrary and may lead in many situations to overcompensate measurement results obtained with Peak detector. The variability of Peak detector output for different span, sweep time and VBW values was extensively reported in [99]. In EMC measurements VBW is normally set equal to or larger than RBW to exclude it, but this accounts for nearly 5 dB of variability. The influence of sweep time is marginal provided that it is above a minimum value that in the performed tests was about 1 to 3 seconds (see Figure 3.1.5).

3.1.3.5 Frequency resolution and effective resolution

The displayed trace is stored digitally as a vector of points that can be downloaded to a USB memory, a local hard disk or through a cable connection. The points are separated on the frequency axis by the so-called *step* (or *frequency step*), that results simply from the selected span divided by the number of points the SA uses to store the trace; the latter may be varied going to system settings and normally ranges between 400 and 2000 (several thousands in EMI receivers, that cover wide frequency ranges in one single sweep). The step represents the frequency resolution of the internal sample representation, but it is not the effective resolution, so does not indicate the ability to discriminate between spectral elements really present in the input signal.

The IEEE Std. 748 [71] defines the *effective resolution* as "the ability to display adjacent responses discretely; the measure of resolution is the frequency separation of two responses which merge with a 3 dB notch."

The effective resolution is to be evaluated on the final result of the measured spectrum, so including RBW and the IF filter selectivity — as it might be expected —, but also the sweep time and the influence of the video filter and its transient response. For every combination of frequency span and sweep time there exists a minimum achievable resolution value, called optimum (or minimum) resolution, defined as

$$R_{opt} = K \sqrt{\frac{\text{SPAN}}{\text{ST}}}$$

(3.1.5)

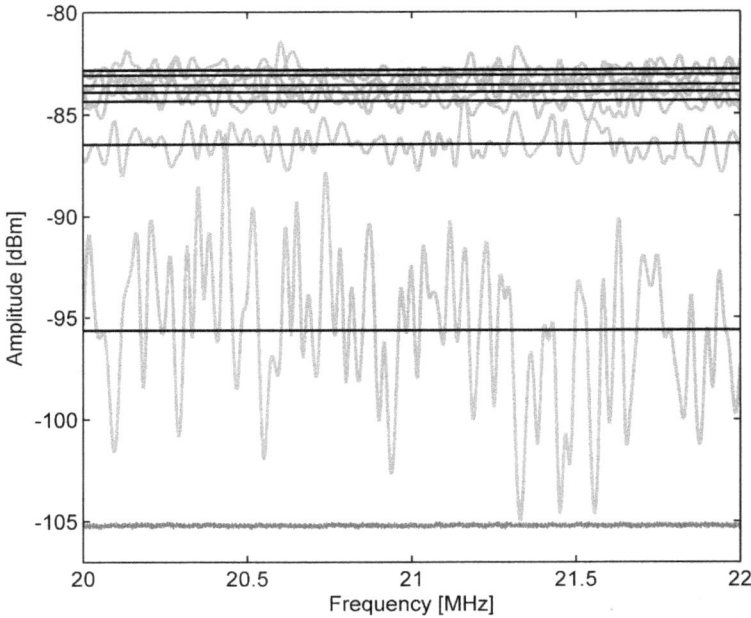

Figure 3.1.5 – Peak detector behavior for increasing sweep time: RMS detector reference curve (thick gray), Peak detector curves from bottom to top for increasing sweep time (ST = 0.1, 0.3, 1, 3, 10, 30, 100, 300, 1000 s); the horizontal black lines are the mathematical average of each curve.

where the frequency span, or simply the span, is the whole frequency interval swept by the SA, and the sweep time ST (that is considered below in sec. 3.1.3.6) is the time needed to perform the measurement from the start to the stop frequency. This expression is derived from (3.1.8), used to determine the minimum sweep time; the parameter K here shall be distinguished from k there: $k \propto K^2$ and they refer to resolution bandwidth and frequency resolution, respectively. The resolution in this context has an operative definition, as soon as we go to the next paragraph, where a measurement method is proposed; in (3.1.8) the resolution is the Resolution Bandwidth and it is considered a fixed characteristic, not a measurable performance.

The impact of effective resolution on spectrum interpretation may be verified in various situations:

- if amplitude modulation (AM) is considered, with carrier at f_1 and an applied modulation tone of frequency f_m, then sidebands and carrier are separated on the display as soon as the RBW gets down to the modulation frequency value and below it; an example is shown in Figure 3.1.6, where the same carrier with 50% AM modulation at 5 kHz is measured with both RMS and Peak detector, varying the resolution bandwidth around the modulation frequency (300 Hz, 1 kHz, 3 kHz, 10 kHz): the spectra are perfectly equivalent for the evaluation of the amplitude of the main components, that appear clearly separated when the RBW is 1 kHz or less; the noise floor is as expected larger for the Peak detector and af-

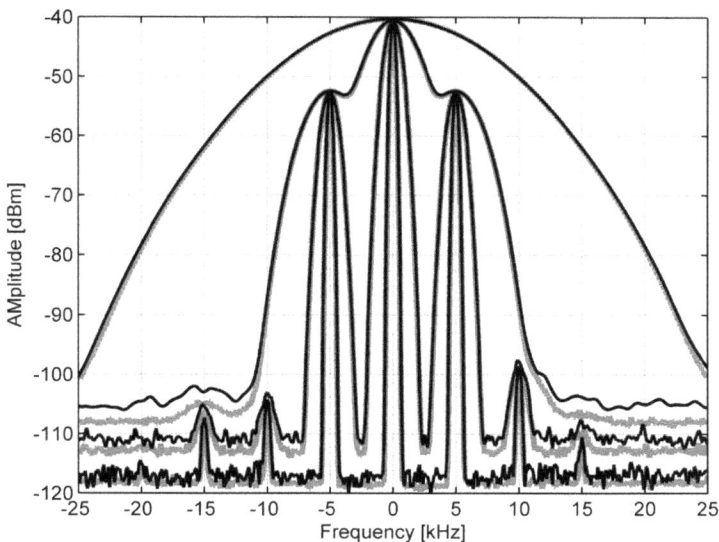

Figure 3.1.6 – Amplitude modulation (20 MHz carrier at $-40\,$dBm, 50% modulation at 5 kHz) as appearing for different RBW values (10 kHz, 3 kHz, 1 kHz, 300 Hz, RBW/VBW$=10$) and using RMS (gray curve) and Peak (black curve) detectors. The sweep time was always adjusted at about five times the minimum value (particularly relevant for the RMS detector, when measuring modulation peaks, not the noise floor).

fects the accuracy with which the smaller harmonic byproducts of the generator are measured;

- a pulse train, that is a discrete spectrum signal (or line spectrum signal), may result in a continuous or discrete spectrum depending on the RBW value compared to its fundamental f_1 (corresponding to repetition interval T), that is also the frequency separation of harmonic components (see sec. 3.1.9 for a more complete analysis of spectrum analyzer behavior with discrete-spectrum signals).

3.1.3.6 Sweep Time

The Sweep Time (ST) defines the time it takes to sweep the frequency range for the N points set from the start to the stop frequency (the span); each point (or pixel) has thus a fraction of the total ST (that may be called *dwell time*), inversely proportional to N. Being the chosen number of points in close relationship with the span/RBW ratio, it is said that the time T' apportioned to measuring at one frequency point with a bandwidth RBW is directly proportional to RBW[1]:

$$T' = \frac{\text{ST}}{\text{SPAN}}\,\text{RBW} \tag{3.1.6}$$

---

[1] The use of RBW rather the number of points N aims at deriving a compact expression for the minimum sweep time that involves IF filter quantities and not the display size, that is N. It is understood that N depends on the frequency step and not on the IF resolution filter RBW.

The main constraint for the ST setting is the transient response of the IF filter, that depends not only on the chosen RBW value, but also on the type of filter (analog, digital, FFT) and its order (see sec. 3.1.3.2). For a correctly damped filter, with no remarkable overshoot, taking a readout of its output before the transient response has vanished results in under-estimating the real output value and it is necessary to wait for the rise time to pass (see sec. 2.4.1). The rise time of the filter is thus inversely proportional to the RBW value, provided that all other details and corrections are conveyed in a suitable coefficient k:

$$t_r = \frac{k}{\text{RBW}} \tag{3.1.7}$$

Combining the above quantities and making $T' = t_r$, a compact expression for the minimum sweep time $\text{ST}_{\min}$ for the used RBW is obtained:

$$\text{ST}_{\min} = k \frac{\text{SPAN}}{\text{RBW}^2} \tag{3.1.8}$$

where k hides all details on filter structure and trade-off for transient response error. There is some disagreement on the value of k, as reported by different manufacturers and stemming out from different assumptions, with k ranging from $k = 1$ to $k = 2.5$.

The transient response may be corrected, if the decision is taken to shorten the sweep time beneath the limit established by (3.1.8): of course this needs to be implemented automatically in the spectrum analyzer by the manufacturer, who knows well the details of the implemented filters and their complete time response (this results in the so-called desensitization, unavoidable when measuring fast signals).

If the FFT option is used instead, ST is dramatically reduced: in theory for smaller RBW values the advantage is orders of magnitude, but in practice, due to the computational overload, the ST reduction may be no more than about two orders of magnitude. The transient response time is replaced by the *observation time*, that is the duration of the transformation window (see sec. 2.3.1 and 2.3.3), not directly related to span, but to frequency resolution.

3.1.3.7 Comparison with experimental results

In Figure 3.1.7 a comparison for sweep time is reported of theoretical expressions and real evaluations for different span/RBW values. It is evident that the FFT mode is about two orders of magnitude faster than the sweep mode, as expected. Moreover, it may be observed that the sweep time in Sweep Mode set by the manufacturer as the minimum value is always much longer than the minimum value established by (3.1.8), thus ensuring very accurate readings.

3.1.3.8 CISPR 16-2-1 specifications

Last, requirements in the CISPR 16-2-1 standard [40] are considered, i.e. the minimum required sweep time (called "scan time") to perform EMC measurements with the various detectors already examined in sec. 3.1.1.9: Table 1 and Annex D of the

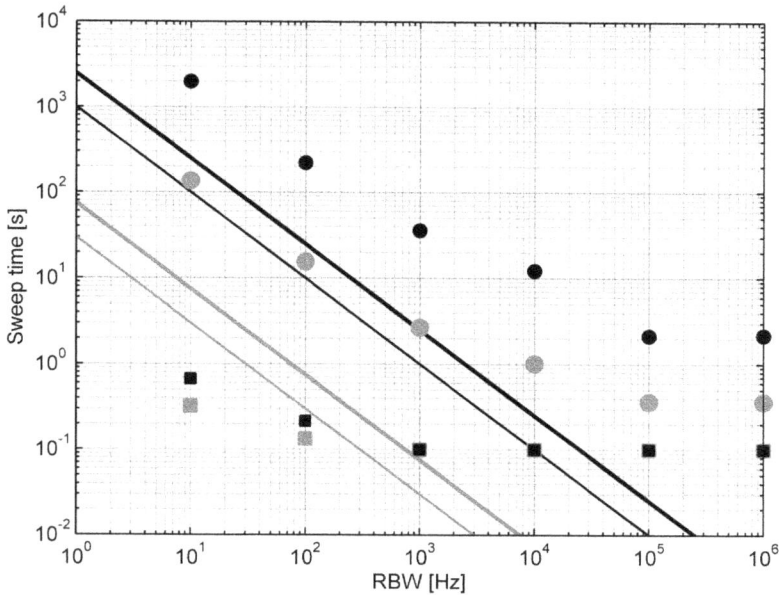

Figure 3.1.7 – Required Sweep Time as a function of the Resolution Bandwidth for a given span/RBW excluding Video Bandwidth (VBW/RBW set to 10). The sloped lines correspond to estimations: eq. (3.1.8) for a span/RWB=1000 with $k = 2.5$ (black thick) and $k = 1$ (black thin) and for a span/RBW=30 with $k = 2.5$ (gray thick) and $k = 1$ (gray thin). Both sweep mode (circle) and FFT mode (square) were tested in the two span/RBW ratios (black for span/RWB=1000, gray for span/RWB=30).

	Frequency band	Peak det.	Quasi-Peak det.	Average det.
A	9 - 150 kHz	14.1 s	$2820\,\text{s} = 47\,\text{min}$	$1255\,\text{s} = 20.9\,\text{min}$
B	0.15 - 30 MHz	2.985 s	$5970\,\text{s} = 99.5\,\text{min}$	$5134\,\text{s} = 85.6\,\text{min}$
C-D	30 - 1000 MHz	0.97 s	$19400\,\text{s} = 323.3\,\text{min}$	$8051\,\text{s} = 134.2\,\text{min}$

Table 3.1.2 – Minimum scan times as specified by CISPR 16-2-1 [40].

standard report the scan times required for the four frequency bands A, B, C and D and for the Peak, Quasi Peak and Average detectors, that are the detectors accepted for EMC measurements and for which limits are specified by various EMC standards.

3.1.4 Zero span

The zero-span mode consists of a direct reading of IF filter output after the envelope detector, having the LO frequency held to a fixed value defined by the center frequency setting. In all modern digital spectrum analyzer the signal is digital and taken at the ADC output, so that it is no longer possible to tap the IF output and send it to external equipment[2]. In old spectrum analyzers reading the IF output is a safe and easy way to

---

[2] Reconstruction by a dedicated DAC reconversion of the digital ADC output would be needed.

have a tunable band-pass filter: the signal may be sent to a time-domain equipment, such as an oscilloscope, a multimeter, a data acquisition system, and band limited measurements become possible, to record transients in noisy environments, to perform crossing rate distribution and amplitude probability distribution (APD) measurements etc.. Nowadays, signal processing functions are replacing this approach, but it comes at a price, so that a modern digital sampling oscilloscope with a moderate analog bandwidth of let's say 60 to 100 MHz and an old cheap spectrum analyzer may be a good alternative.

Zero span is quite useful to track time behavior of modulations (e.g. setting the center frequency to one of the lateral bands carrying the modulation information, or measuring the instantaneous power over a channel bandwidth), to measure weak signals of known spectral distribution, but hidden by noise in time domain, and to measure the time response of the IF filter itself. Regarding the former, channel power measurements may be performed in swept or zero-span modes (see details in sec. 3.1.11). Also in zero-span mode the limiting factor to follow fast RF pulses is the IF filter rise time.

When operating in zero span on very long time intervals, observing phenomena that are narrowband (such as a lateral band), the local oscillator stability and drift may have impact: small fluctuations of the measured amplitude may appear caused by the change of the portion of signal spectrum intercepted by the IF filter centered at the supposed fixed center frequency, in reality probing around with erratic behavior.

3.1.5 Local Oscillator feed-through

The phenomenon of feed-through from the local oscillator has consequences that influence somewhat measurement results and might be interpreted as a limitation of the spectrum analyzer performance, or even worse as a true genuine measured signal indicating e.g. an excess of disturbance or band occupation at the lowest frequencies. It is also a not-so-often documented phenomenon and shall be tested by the user.

When performing swept measurements over an extended frequency band while starting from low frequency (as in the case of conducted or radiated emission measurements starting at 9 kHz), there is an evident increase of the noise floor at the lowest frequency values. At the input mixer, operating at the largest local oscillator frequency, there might be a certain degree of coupling and as a consequence some leakage of LO signal in the IF path. So, if a low enough start frequency is selected, such a low frequency input signal is converted to an IF frequency that is practically the same of the LO, that feeds through and is only slightly or at all attenuated by the IF filter. The resulting noise floor has a peculiar shape, that is larger towards dc, with a profile like flicker noise or what in the FFT calculation is called "snow-drift profile". The consequences are reduced dynamic range and sensitivity, a wrongly suggested presence of excess noise, and low-order harmonics leaking in the incoming signal.

The phenomenon is well described in Figure 3.1.8, where several sweeps between 5 and 145 kHz are shown in both sweep and FFT mode: the curves start from below for RBW = 100 Hz and move up as RBW is set to 300 Hz, 1 kHz, 3 kHz, 10 kHz and 30 kHz; the video bandwidth VBW was set to one tenth of RBW to smooth the curves just for a graphical exigency. It is apparent that for the smaller RBW values the curves are spaced by about 5 dB each, meeting the assumption of broadband incoherent noise, but for the largest RBW there is no further increase because of the coherent LO leakage, that adds to the underlying noise. All the curves are not horizontal except for the lowest RBW values and only for the last data points right-hand side: this is again a symptom of the LO leakage and of a typical phase noise profile. The behavior is roughly the same for sweep mode and FFT mode to proof that the phenomenon is not related to the processing method, but occurs before the IF filter.

LO leakage in the lowest part of frequency range is particularly important for EMC measurements: band A indicated by CISPR [40] extends right from 9 to 150 kHz; MIL-STD-461 prescribes a minimum frequency of 10 kHz for both radiated and conducted measurements; EN 50121-2 [27] and EN 50212-3-1 [28] for railway applications in their 2006 version started from 9 kHz. The adopted RBW values are those prescribed or tolerated by the EMC standards and range between 100 and 1000 Hz: 1 kHz is sometimes allowed to reach faster sweep times when measuring non-steady phenomena, as it is for rolling stock in the EN 50121-3-1; the MIL-STD 461 also allows the use of a larger bandwidth, but application of corrective factors is forbidden. In addition, as commonly specified in EMC standards, no Video filter will be used, setting the RBW/VBW ratio to 10, but it is of no relevance to evaluate the impact of LO feed-through.

It is quite evident, then, that LO feed-through phenomenon is able to impair a test campaign if not properly considered: at RBW = 1 kHz the slope is still evident, with an increase of 4-5 dB; the use of a larger RBW, at 3 kHz, is unacceptable.

3.1.6 Swept versus Fourier analysis

Modern spectrum analyzers that apply analog-to-digital conversion at the output of the IF block allow the user to choose between a traditional "swept analysis" and a frequency-domain image, obtained by Fourier transform of digitized IF data. It is underlined that FFT (or Fourier) analysis as implemented in spectrum analyzers or receivers is performed on the IF signal not on a baseband digitized version of the original signal, as it would be sampling directly with an oscilloscope.

With the publication of Amendment 1 to the third edition of CISPR 16-1-1 [38], the use of FFT-based measuring instruments has been allowed for EMI compliance measurements [143]. A number of product standards have then included this method in their latest modifications (for example, CISPR 13, CISPR 15 and CISPR 32), other standards such as CISPR 11, CISPR 12 and CISPR 25 are following. As said, time domain analysis and FFT may be applied to the RF signal at different stages: working on the IF signal has the advantage of keeping the advantages of heterodyne demodulation in terms of limiting the input bandwidth and enhancing the dynamic range; presently commercial RF equipment with time domain and FFT option operate in this mode with various degrees of optimization with respect to the number of points, the sampling hardware and the complexity of FFT and other mathematical functions, in order to ensure real time operation [22, 95]; a direct time domain approach may be far more cumbersome for the involved amount of data and for the unfavorable signal-to-noise ratio, but has the advantage of a ready implementation in all modern digital

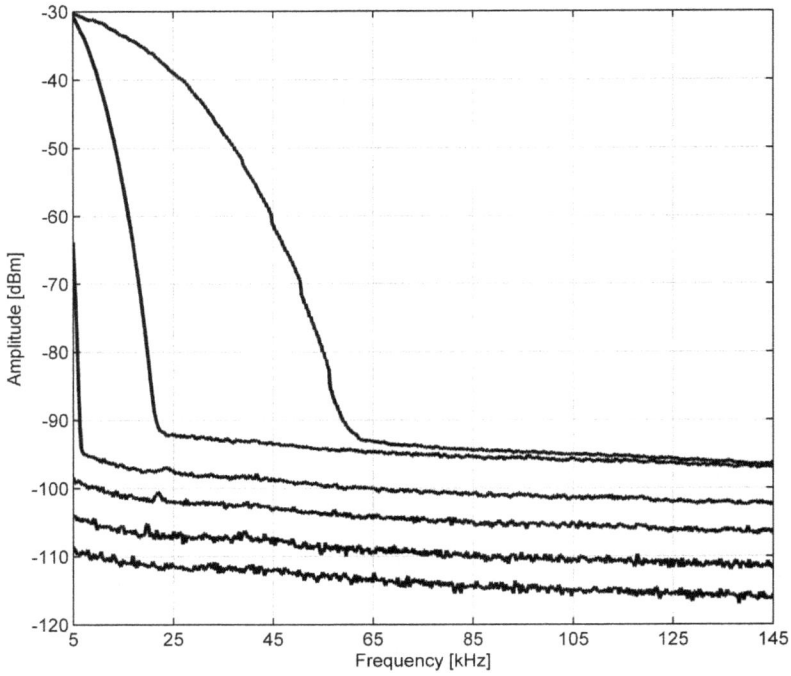

(a)

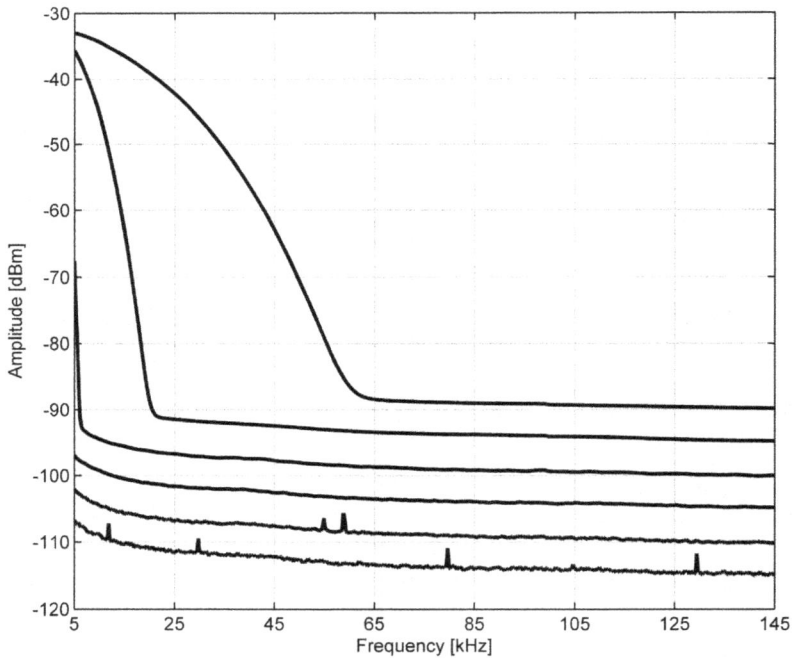

(b)

Figure 3.1.8 – Example of Local Oscillator feed-through for different RBW choices (100 Hz, 300 Hz, 1 kHz, 3 kHz, 10 kHz and 30 kHz, from bottom to top): (a) sweep mode, (b) FFT mode.

storage oscilloscopes, using possibly the on-board computer for further processing. For transients such as those originating from the electric arc at the sliding contact [20] this approach is attractive in that the full waveform of the transient is captured and may be further processed, including FFT and spectrum evaluation, even simulating various kinds of detectors.

The Fast Fourier Transform (FFT) allows to reduce the minimum sweep time and to keep it almost unaltered while changing settings; sweep times of only hundreds of ms may be achieved for many combinations of settings. FFT, however, is not universally applicable to all measurement conditions and some extreme settings that in Sweep Mode analysis lead only to some inaccuracy and biasing of the trace, in FFT mode may turn out into incomprehensible traces and weird behavior: steps and discontinuities, asymmetric traces, fluctuations.

Since FFT analysis is applied to the digitized IF envelope signal (as RBW filter would do), the FFT resolution is adjusted for the same effective resolution of RBW filter setting in Sweep Mode; the number of spectra and any internal averaging operation are adjusted to match VBW setting and the desired sweep time value (a sweep time input value is still indicated as reference). The combined effect of smoothing windows for frequency leakage suppression and possible internal averaging over successive transforms displays smoother spectra with almost no apparent noise-jagged profile.

When discussing variance reduction by averaging N points of the same trace, the independence of the data points was assumed (see sec. 3.1.2). This independence is ensured in sweep mode, due to the time required to sweep from one measurement point to the next one; in FFT mode the degrees of freedom may be less, due to the correlation resulting from the shorter time record and the influence of windowing, thus impacting slightly on the expected variance reduction. To author's knowledge there is no literature reference that gives quantitative results, so this problem is addressed experimentally in sec. 3.1.10.

When measuring transients, it is necessary to select the right method for the fastest sweep attainable: normally, at lower RBW values and moderate spans, FFT mode is the fastest; however, for large RBW and spans the answer is not straightforward and needs some reasoning. Transients, such as steps and pulses, have a wide spectrum with a narrow time location and they cannot thus be considered narrow-band; from this observation the choice of a large frequency span and a correspondingly large RBW to keep the sweep time as fast as possible (please see sec. 3.1.3.6).

3.1.7 Dynamic range and mixer operating point

The dynamic range should not be confused with the display range, in relationship to the *reference level* and the selected scale. The dynamic range provides information about SA capability to simultaneously process signals with very different levels, including e.g. a signal and its harmonics; of course, for the evaluation of harmonics and in general harmonic distortion, the dynamic range is not the only limiting factor: also mixer linearity plays a role, followed by other less relevant non-linearity issues.

The limits of dynamic range depend on the measurement to be performed. The lower limit is determined by thermal noise and phase noise, both contributing at a different extent to the noise floor: while the former is more or less flat, the latter decreases when the carrier offset increases (see sec. 3.1.5). The upper limit is set either by the 1 dB compression point or by distortion products occurring in case of over-driving. When the input signal at the first mixer approaches its 1 dB compression point, the mixer will operate non-linearly and there will be very high levels of distortion terms. Operating at small resolution bandwidth and reducing the noise floor, distortion products may become visible and the user should undertake a number of checks: their amplitude is plausible? does it reduce drastically reducing the input signal amplitude? or, similarly, do they reduce by increasing input attenuation? When the spectrum analyzer is operated automatically, the reference level and the combination of attenuator and IF amplifier are such that the mixer operating point is monitored and the 1 dB compression point is never reached.

Trying a raw estimate of the dynamic range in modern spectrum analyzers, simply comparing the minimum noise and the maximum input levels, we may conclude that amplitudes that differ by as much as 160 or 180 dB may be displayed. However these range limits cannot be reached at the same time for the same SA settings: internal blocks such as mixer, log-amplifier and ADC have a much smaller dynamic range and their operating points shall be carefully decided, in order to maximize the available dynamic range, that normally result in a smaller range up to a hundred of dB[3].

Considering first the maximum input level, values around $25-35$ dBm are normally encountered, that shall be reduced to 10 dBm or less, not to damage or over-drive the first mixer. A compromise shall be found in the selection of mixer level. If RF attenuation is high and mixer level is low, the levels of distortion and intermodulation products are also low, but at the same time the signal-to-noise ratio is small. In this case the dynamic range is limited at the lowest end by the inherent noise. If, on the other hand, the mixer level is high, then distortion and intermodulation products are generated, whose levels exceed the inherent noise level and therefore become visible.

3.1.8 Accuracy of amplitude measurements

3.1.8.1 Scalloping loss

A first source of error in the measurement of amplitude of a peaky spectrum, such as a sinusoidal tone, is due to the scalloping loss for a spectrum analyzer, that is the fact that for too a large span/RBW ratio, there is no overlap between the IF filter frequency response ideally located on each trace pixel and during the sweep some parts of the real spectrum underneath will be unavoidably lost. Reducing the span/RBW ratio and thus the step size to less than RBW greatly enhances the amplitude estimate, leaving the possibility that a tone located midway between two trace pixels will undergo the maximum error (exactly for the same reason that was presented for the FFT scalloping loss, see sec. 2.3.3.1). Thus the scalloping error is

---

[3] This is why for phase noise measurements instruments such as phase noise analyzer can reach better performance than a general purpose spectrum analyzer.

$$\varepsilon_{sclp,\max}[\text{dB}] = 3.01 \left(\frac{\text{span}}{\text{RBW}} \frac{1}{N-1} \right)^2 \tag{3.1.9}$$

Setting the step to half the RBW as required by some EMC standards, the maximum scalloping error is $0.75\,\text{dB}$. Using a step of 1/3 brings the error down to $0.33\,\text{dB}$. Negligible scalloping errors (below $0.03\,\text{dB}$) are ensured for a 10:1 ratio between step and RBW.

3.1.8.2 Internal noise

When taking into account the effect of internal noise (be it phase noise or thermal noise), amplitude readings of externally fed input signals may always be corrected, provided that a reliable estimate of internal noise is available. It is intuitive that the "true" value of the input signal level P_s corresponds to the measured one P_m minus the effect of noise. Since the internal noise (its power being P_n) is assumed uncorrelated from the input signal (identified by P_s) with a random phase relationship, subtraction occurs in the sense of root mean square, or incoherent subtraction:

$$P_s = P_m - P_n \tag{3.1.10}$$

Subtraction shall be performed going from dB to linear quantities, and once the subtraction of power terms has been done, going back to the original dB quantities. The procedure starts from the original P_m and P_n values and to estimate the correction that is multiplicative in linear scale (dB to subtract), it normalizes the difference $P_m - P_n$ by P_m itself, thus leading to

$$\Delta P_s[\text{dB}] = 10 \log_{10} \left(1 - \frac{P_n}{P_m} \right) \tag{3.1.11}$$

that depends only on the ratio of P_n and P_m, that is their distance in dB, not on their absolute values; the resulting correction values ΔP_s to subtract to the measured value for the corrected estimate $\hat{P}_s$ of input signal power are reported in Table 3.1.3. It is extremely important that SA settings are such to smooth the noise floor and to reduce its variance (see sec. 3.1.1.9 and 3.1.10), for a consistent correction and to reduce uncertainty. At a first approximation the confidence interval that brackets the noise floor distribution reflects directly in the dB correction and is transferred as uncertainty onto the estimated signal power $\hat{P}_s$.

3.1.8.3 Absolute vs. relative reading and related errors

Amplitude measurements may be absolute or relative. It is intuitive that most error terms related to amplitude quantification are ruled out when performing a relative amplitude readout, because many internal blocks hold the operating point with comparable errors, when two measurements are taken one after the other, at a short distance in time (this is what in many instruments is called "short-term uncertainty" or "short-term variability").

$P_{meas}[\text{dB}] - P_n[\text{dB}]$	$\Delta P_s[\text{dB}]$ (additive)	$\Delta P_s \%$ (multiplicative)
1	-6.87	20.6
2	-4.33	36.9
3	-3.02	49.9
4	-2.20	60.2
5	-1.65	68.4
6	-1.26	74.9
7	-0.97	80.1
8	-0.75	84.2
9	-0.58	87.4
10	-0.46	90.0
15	-0.14	96.8
20	-0.044	99.0
25	-0.014	99.7
30	-0.0043	99.9

Table **3.1.3** – Correction values in dB for the difference-of-power estimation of measured signal power.

Regarding absolute reading, i.e. the measurement of an amplitude and its expression as a voltage or power, it is instructive to review the most relevant error terms and the way they are expressed and quantified in manufacturer's datasheets. First of all it is recalled that any quantification of an error (called *calibration*) is subject to change and drift with temperature and time; from this the need of warm-up and a specified operating temperature interval, the increase in the uncertainty for the time past the last calibration and the need for periodic calibrations. In any case calibration can reduce the error term by quantifying corrections that may be stored in the spectrum analyzer for later use, but is deemed by an intrinsic residual uncertainty, due to the used references, test setup and adopted procedures. What is normally stated in manufacturer's datasheet is the residual uncertainty due to all these factors and declared by the laboratory for the specific calibration, after all systematic errors have been removed; so the declared uncertainty is the least attainable uncertainty in ideal conditions. Declared uncertainty is deemed to increase with the time past the last calibration, until a new calibration is necessary.

On the contrary, relative reading is quite immune to many of the uncertainty issues considered for absolute reading: the difference between two points of the measured spectrum (or between two successive spectra) may be accomplished by using delta markers or other external trace processing. While long-term variability and drifts, as well as any offset or deviation, vanish in the difference, residual uncertainty and short-term variability are the only factors that matter: what is really "short term" depends not only on operating characteristics, but also on our knowledge of such characteristics and the choice of the best settings. In an attempt to simplify, we may say that short-term variability is made mainly of noise terms, internal temperature changes, linearity error (that may affect two points of the spectrum at different power levels), and short-term drift, usually expressed as %, ppm or dB over square root of unit of time (e.g. hours, days, months ...). Once the time interval over which measurements are taken is established, some of these terms may be estimated with the information appearing in the instrument manual. Putting altogether, uncertainty budget may be estimated using a Type B approach [15].

3.1.8.4 Uncertainty budget

Uncertainty estimation is considered with a closer look to the terms influencing the accuracy of amplitude measurement. The aim is identifying the sources of information for the factors that we have reviewed in the previous section and how to characterize their uncertainty. Depending on manufacturer's approach and quality system, uncertainty may be declared with a more or less explicit statement of distribution (e.g. uniform, Gaussian, U-shaped) and coverage factor (possibly only standard uncertainty is given, i.e. $k = 1$, but most often $k = 1.96$, or 2, will be given, not to say $k = 4$, as Agilent did for many instruments). Even if not clearly stated, nor commented in the literature, the impression is that the coarsest uncertainty statements are quite overestimating: while this sentence wouldn't make sense if the uncertainty terms were coming from a Type A estimate, if they are in turn the result of a Type B method applied to single components or blocks, it is possible that margins are introduced. This is the impression whenever there is the possibility of confirming Type B statements with repeated measurements and a Type A approach.

Frequency response is relevant if samples to compare are distributed over a large interval: whereas the frequency response is nominally flat (zero error on average), depending on frequency and the adopted RBW, the manufacturer will declare a varying uncertainty; frequency response tends to worsen in the very low frequency range and gives its best between a few MHz and some GHz; generally speaking, uncertainty may vary between a fraction of dB to up to 1 dB.

Linearity is always quite good and for the advised mixer operating interval (e.g. below $-10\,\text{dBm}$ down to about $-70/-90\,\text{dBm}$) linearity error is usually below 0.1 dB.

The input attenuator might be a source of uncertainty when switching between steps: some manufacturers declare $0.1 - 0.2\,\text{dB}$ and we have seen, when comparing averaged noise floors, that the correspondence is in fact quite good and that practically uncertainty is lower than 0.1 dB.

Amplitude accuracy rather than being estimated based on single uncertainty terms (as discussed before) is often presented with a catch-all statement based on some assumptions, e.g. a RBW value, some frequency points, a range for the input signal and a minimum input attenuation (e.g. 10 dB). It is evident that in this case the declared uncertainty shall be quite large and it is, being normally above $1 - 2\,\text{dB}$, and in the end not so useful.

Uncertainty related to input VSWR is always dramatic: VSWR values are always declared "less than" with quite a large upper limit; while a minimum of input attenuation is always advisable (except when maximizing the dynamic range, e.g. for distortion and phase noise measurements), when connecting troublesome sources such as long cables and antennas, additional external attenuation may be needed, and this adds to the overall uncertainty budget.

When performing power measurements (as will be considered in sec. 3.1.11) RBW accuracy is also relevant: ENBW directly weights noise-like signal power; apart from any RBW accuracy declaration, we saw that IF filter imaging allows an accurate quantification of the overall IF filter response, thus leading to accurate RBW determination, including ENBW.

3.1.9 Repetitive pulse signals, discrete and envelope spectrum

Pulse base-band signals and pulse modulated RF signals may be exemplified as used in data networks and digital links (such as Ethernet, USB, SATA to cite a few known technologies) and in digital modulations, such as modern cellular communication and wireless network formats (GSM, UMTS, WiFi) and radar signals, respectively. The reason for considering these two families of signals is that they are a good model for many real world signals.

The energy of periodic pulse signals is concentrated at discrete frequencies $n \cdot f_1$, where f_1 is the fundamental given by the inverse of the repetition period T. The envelope $\mathrm{sinc}(\cdot)$ function has nulls at integer multiples depending on the mark-to-space ratio (duty cycle) τ/T. If the pulse signal is used for modulation of a carrier, the spectrum will be symmetrically distributed above and below the carrier frequency. Depending on the adopted resolution bandwidth, three cases are possible:

- if resolution bandwidth RBW_1 is small relative to spacing of spectrum lines $\Delta f = 1/T$, then individual spectral lines can be resolved and a *line spectrum* is obtained; if resolution bandwidth is further reduced to RBW_2, amplitude and number of spectral lines do not change (i.e. they are narrowband), but the noise level reduces and thus the signal-to-noise ratio improves as $10 \log_{10}(\mathrm{RBW}_1/\mathrm{RBW}_2)$;

- if the resolution bandwidth RBW_1 is larger than the spacing of spectrum lines Δf, but smaller than the first null of the $\mathrm{sinc}(\cdot)$ envelope function at $1/\tau$ from the carrier frequency, spectral lines cannot be resolved and the amplitude height of the envelope depends on bandwidth; the amplitude within the selected RBW_1 depends on the number of underlying spectral lines collected within RBW_1. In this case we speak of an envelope display; the envelope amplitude decreases with decreasing bandwidth as $20 \log_{10}(\mathrm{RBW}_1/\mathrm{RBW}_2)$.

- if the resolution bandwidth RBW is larger than the envelope null spacing $1/\tau$, selectivity is no longer effective and the amplitude distribution in the spectrum cannot be recognized any more; increasing RBW the IF filter impulse response approaches the time function of the pulse-modulated carrier.

The change of measured spectrum shape is shown in Figure 3.1.9 to give a first graphical description of the phenomenon: the RMS detector is used and we will see that for correct quantification of the pulsed waveform a Peak detector should be used instead.

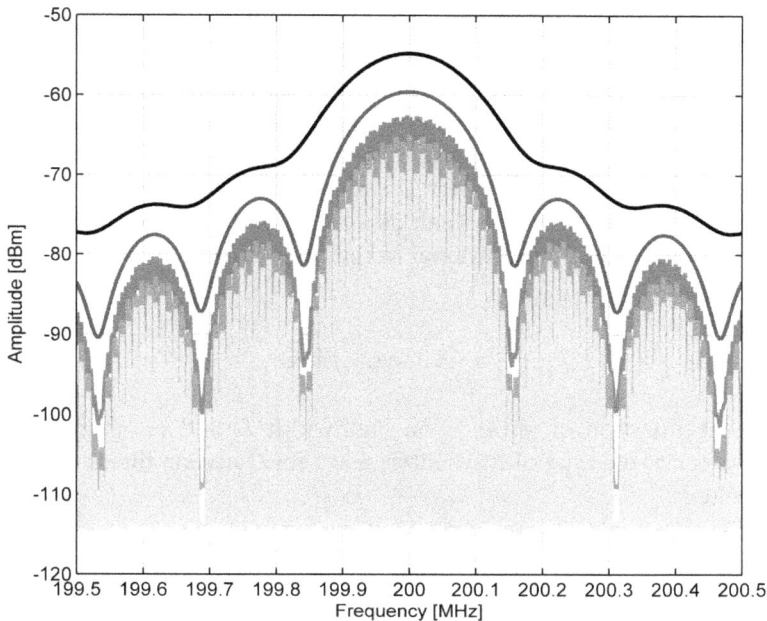

Figure 3.1.9 – Example of desensitization for a 200 MHz carrier at −40 dBm, modulated by a square pulse of 6.5 µs time duration and 10 kHz repetition frequency; measurements are done with RMS detector and RBW=1, 3, 10, 30 and 100 kHz from light gray to black.

A $\tau = 6.5\,\mu s$ rectangular pulse with $f_0 = 10\,\text{kHz}$ repetition frequency, modulating a 200 MHz carrier of $E = -40\,\text{dBm}$ (2 mV peak) amplitude[4] is considered, using RBW values ranging from 1 kHz to 100 kHz, and setting VBW=RBW. The results are plotted in Figure 3.1.9. The effect of larger RBW values is an increase of the spectrum, a smoother curve and a loss of frequency resolution. The IEEE Std. 376 [69] confirms that the amplitude in linear units of spectrum components around the sine wave frequency is $f_0 \tau E$; in the same standard we read that the first null of the spectrum (provided that the frequency resolution is sufficiently narrow to identify it) occurs at $2\pi/\tau$ from the center frequency, and this is wrong, unless we revert to use pulsation instead of frequency: the correct expression for the first null is $1/\tau$ and indeed it gives 156 kHz in the present case. The increase of the spectrum measured at the center frequency of the 200 MHz carrier gives a first indication of the desensitization phenomenon: with RBW = 100 kHz the marker at the center is −54.77 dBm, at 30 kHz it is −59.56 dBm, then −62.47 at 10 kHz, −64.93 at 3 kHz and −67.02 at 1 kHz. These measurements and the readings at the center peak were made using the RMS detector with 1 minute sweep time: the "true" value by analysis of the signal spectrum is $20\log_{10}(\tau/T)$, that is −11.87 dB below the −40.0 dBm value of the amplitude of the non-modulated 200 MHz sinusoidal tone (so −51.87 dBm) for the 6.5 % duty cycle used in the tests, neglecting cable losses and amplitude errors.

The reduction of the envelope amplitude with decreasing bandwidth is called *pulse desensitization*. We have said that the amplitude of spectrum lines does not change with

---

4 The used symbols for the rectangular pulse and the sinusoidal carrier are those of the IEEE Std. 376 [69].

bandwidth, provided that it is small enough with respect to line spacing Δf. Pulse desensitization, often indicated by parameter α, refers to the envelope spectrum and to the coarse line spectrum that is obtained when resolution bandwidth is of the same order of spectrum line spacing. The usual application of pulse desensitization is when an accurate estimate of carrier amplitude is necessary, but the carrier is pulse modulated and "visible" thus for a short time. In order to correctly return the amplitude of the underlying sinusoidal carrier, a Peak detector shall be applied; RMS detector will always give a lower output, proportional to the duty cycle.

The pulse desensitization factor

$$\alpha = 20 \, \log_{10}(\tau k B) \qquad (3.1.12)$$

depends on the duration of pulse τ, on bandwidth B and on shape factor k, that weights the effect of the type of RBW filter: $k = 1$ for Gaussian filters and $k = 1.5$ for a rectangular filter.

As easily seen in the expression above, pulse desensitization is a matter of IF resolution bandwidth and IF filter response to pulsed RF waveforms.

To this aim a Peak detector is best suited, provided that the sweep time is long enough for a stable maximum reading: in the example that follows the pulse repetition rate was fast enough (10 kHz) to hit the IF filter more than once at each sweep step (pixel); a sweep time $ST = 1$ min ensures that the scan of one pixel point lasts for about 50-150 ms for the most common choices regarding display resolution and that a minimum of 500 hits is ensured for each display point. The use of Peak detector gives a higher trace than the one obtained with RMS detector, well above the noise floor even for the largest RBW values. Reading the peak value at each pixel with enough hits per pixel, the trace is nearly flat, losing resemblance with the expected spectrum (the one in Figure 3.1.9, obtained with RMS detector, was better for illustration); its plotting does not bring any additional information or further insight, and for this reason is omitted. The result of the experiment with a $0.4 \, \mu s$ pulse modulating a $-40 \, dBm$ $200 \, MHz$ RF carrier are reported in Table 3.1.4.

The application of the desensitization factor is quite common when measuring fast pulses as for radar signals: modern SA with RBW values as large as 10 or 30 MHz are barely sufficient to follow a $100 \, ns$ rise-time RF pulse, still with a significant amplitude error, that necessitates compensation for desensitization; when measuring ultra-wideband (UWB) signals, rise time values are even shorter and desensitization compensation is absolutely necessary.

A compromise shall be found since small RBW values reduce the amplitude too much, whereas large RBW values degrade resolution and ability to identify spectral lines. The optimal relationship between τ and B was found empirically as

$$\tau B = 0.1 \qquad (3.1.13)$$

Spectrum analyzers operating in FFT mode are not suitable for pulse measurements.

Analogous behavior with respect to the quantities characterizing the modulating pulse (repetition frequency and pulse width) is expected when considering a base-band sig-

RBW [MHz]	Desens. factor α [dB]	Measured [dBm]	Estimated [dBm]
10	0	-40.00	-40.00
3	0	-43.15	-43.15
1	-9.12	-50.40	-41.28
0.3	-19.58	-61.72	-42.14
0.1	-29.12	-69.95	-40.83

Table 3.1.4 – Estimated values of the desensitization factor α and corrected amplitude when measured with Peak detector for a $0.4\,\mu s$ pulse modulating a $-40\,dBm$ $200\,MHz$ RF carrier: calculated α values for RBW=$100\,kHz$, $300\,kHz$ and $1\,MHz$ are bracketed by an uncertainty interval of about $\pm 0.25\,dB$ due to the uncertainty of the pulse generator thresholds for logic 0 and 1 when driven by the pulse digital control signal (about $15\,ns$ on the time axis); the effect of desensitization is visible already for 3 MHz RBW, that is however not included in the correction of (3.1.12) and results in a substantial error. The applied k value is 1, as for Gaussian filters.

nal: two square waves of $100\,mV$ amplitude and $10\,kHz$ repetition frequency with small 6.4 % duty cycle (thus resembling a rectangular pulse of $6.4\,\mu s$ width) and a balanced 50 % duty cycle (thus giving a line spectrum of odd harmonics), respectively, are considered in the example shown in Figure 3.1.10. Both cases demonstrate what we know already about the correctness of amplitude measurement with respect to resolution bandwidth: a larger RBW (upper half of figure for $3\,kHz$, lower half of figure for $1\,kHz$) gives a higher reading of peak amplitude.

Moreover, a narrower resolution bandwidth lowers the noise floor, as it is apparent by comparing the depth of notches: not only the noise floor is lower, but spectrum lines are well separated and there is a negligible contribution from the adjacent ones (stopped by skirt selectivity), so that the base floor appears with a noisy look.

For the short $6.4\,\mu s$ rectangular pulse, the spectrum is characterized by a first notch of the envelope correctly located at the inverse of its time duration, that is $156\,kHz$. For the 50 % duty cycle the envelope goes as the expected $1/n$, with n the harmonic order, if amplitude is considered; the spectrum was measured in dBm, so a $1/n^2$ envelope reduction is correct for power (passing from the first to the second peak, that is from the fundamental to the third harmonic, the amplitude reduction is nearly $10\,dB$, that is an order of magnitude).

3.1.10 Averaging and noise power

A spectrum analyzer has many options that implement some sort of averaging or smoothing, as we have already seen in the previous sections:

- IF filter RBW implements merging and integration of signal components over RBW itself, with larger RBW values giving smoother spectra; using zero-span combined with a large enough RBW allows measurement of power spectral density, by applying the concept of analog filtering followed by rms detection (see sec. 2.6.2.3);

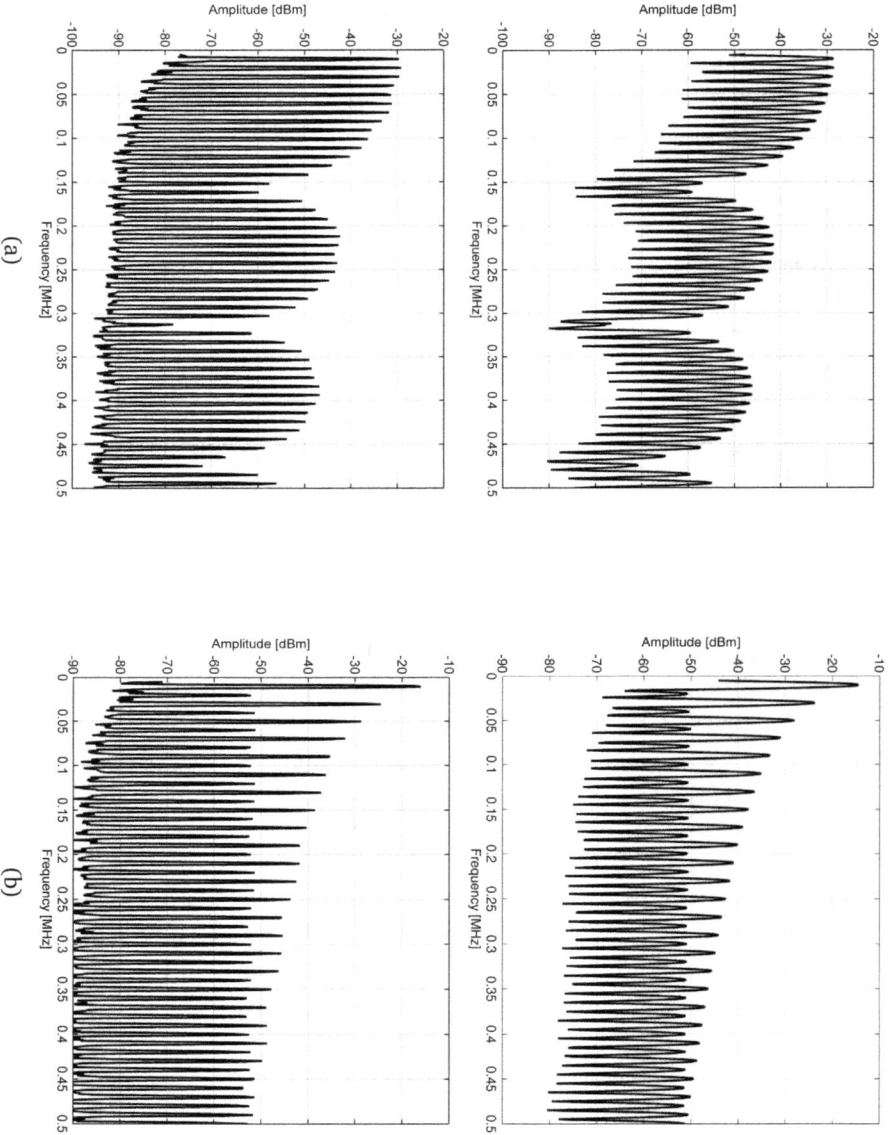

(a)

(b)

Figure 3.1.10 – Line spectra of square waves for (a) 6.4 % and (b) 49.6 % duty cycle; the effect of reduced resolution bandwidth (RBW = 3 kHz above, = 1 kHz below) is reflected in the lower noise floor between spectrum lines and a slightly lower (and more correct) height of spectrum lines.

- a slower ST (Sweep Time) is able to "accumulate" on the same frequency for a longer time, thus allowing potentially a stronger integrating action; the words "accumulate" and "potentially" aim at indicating that the final result (estimate error) depends heavily on the chosen detector and signal statistics;

- AVerage and RMS detectors are the best choices for noise power measurements: the expectation may be implemented explicitly as average or implicitly in the calculation of the rms itself; the problem is the choice of the correct representation, i.e. voltage or power, linear or log scale;

- setting Video Bandwidth VBW to a fraction of RBW reduces noise variance, but keeps the noise floor unchanged.

Considerations behind true power measurements and reduction of variance of the estimate for noise and noise-like signals are reviewed once more.

Regarding the use of average and rms, it is useful to consider more in detail on which quantities the average is operated. Averaging log data (i.e. data expressed in log scale, namely in dB) not only causes the aforementioned 2.51 dB under-response, it has also a higher than desired variance. Negative spikes of the envelope resulting from taking the log of small power levels add significantly to the variance of the log average, even though they bring very little power. Conversely, large power spikes are compressed by the log operation, and two orders of magnitude for example translate into 40 dB. We already observed that the result of a power measurement made by averaging power is lower than that made by averaging the log of power by a factor of 1.64 (2.51 dB).

Average may be performed on a number of channel power measurements x_i corresponding to different traces acquired successively: it is known that if the traces are independent and their number is large enough, then the Central Limit Theorem may be invoked to demonstrate the Gaussian distribution of the average random variable, for which we know that the variance is inversely proportional to the number of collected samples N, that is σ_x^2/N. Assuming noise-like statistics ensures that the distribution of samples is already Gaussian. This is strictly true if they are represented on a linear scale, otherwise log-scaled power readings have no Gaussian distribution any longer. However, being each sample (display pixel) the result of many underlying real measurements summed together and averaged, the power level has no longer a distribution like the "logged Rayleigh", but rather looks Gaussian, and there is no need to convert dB-format values back to absolute power. Also, their distribution is sufficiently narrow that the log (dB) scale is linear enough to be a good approximation of the power scale.

The just mentioned averaging operation that ensures that the Central Limit Theorem holds for trace samples refers to two underlying averaging operations: the average of all measured values belonging to the same pixel performed by the selected Average/RMS detector and the average performed by the resolution bandwidth filter, when collecting the signal power within its measurement bandwidth RBW. Moreover, for the Central Limit Theorem to strictly hold, samples shall be independent.

If the number of samples is low (a few samples, or just one, are contributing to each pixel) and the independence assumption is weak (as when the measurement time is

# avgs J	span/RBW=200 Sweep mode	span/RBW=20 Sweep mode
1	-104.4872 (0.7781)	-104.4977 (0.7179)
4	-104.4577 (0.4138)	-104.3919 (0.4110)
9	-104.3931 (0.2922)	-104.3865 (0.2405)
25	-104.4060 (0.1607)	-104.3838 (0.1420)
64	-104.4008 (0.1043)	-104.3762 (0.0870)
100	-104.3944 (0.0817)	-104.3984 (0.0715)

Table 3.1.5 – Noise averaging results for large and small span/RBW ratios: sample mean (and sample standard deviation between round brackets) of the noise floor (each trace is more than 500 points).

short using FFT mode instead of sweep mode), the variance reduction is less pronounced, reducing slightly the theoretical speed advantage of Fourier transform mode to attain the same signal-to-noise ratio.

The noise mean value is substantially unmodified as the number of averaged traces is increased (a small reduction is observed only for the lowest number of averaged traces, however equal to about 20% of the standard deviation). The reduction is more evident, as expected, when the span/RBW ratio is small, so that more samples fall into the same pixel increasing the significance of the average operation; this means also that at least one of the two cases (large or small span/RBW ratio) is not following the expected rule of reduction of variance with the number of averages. By calculating the ratio of each pair of standard deviation values, it is possible to verify that

$$\frac{\sigma_{J_k}}{\sigma_{J_{k+1}}} = \sqrt{\frac{J_k}{J_{k+1}}} \qquad (3.1.14)$$

and this expression meets accurately the measurements for $J_k \geq 9$.

Reduction of variance is plotted in Figure 3.1.11 for the two cases, graphically compared to the expected reduction given by the reciprocal of the number of averages.

3.1.11   Channel power measurement

Resolution bandwidths offered by spectrum analyzers are limited to some MHz or a ten of MHz, not suitable to cover the entire channel bandwidth of modern modulations; moreover, the IF filter is not selective enough to reject power terms in adjacent channels and to give an accurate measurement. For these reasons the measurement of power across the entire channel bandwidth is mostly carried on by a suitable narrower RBW and successive integration. Selecting a narrow RBW around a few % of the channel bandwidth ensures that the noise floor of the spectrum analyzer (the DANL) is kept conveniently low and that, when measuring at the edge of the channel bandwidth, we are not letting in too much of the adjacent channels. The total channel power P_{ch} is obtained by integration of readouts corresponding to trace pixels P_i [dBm] falling in-

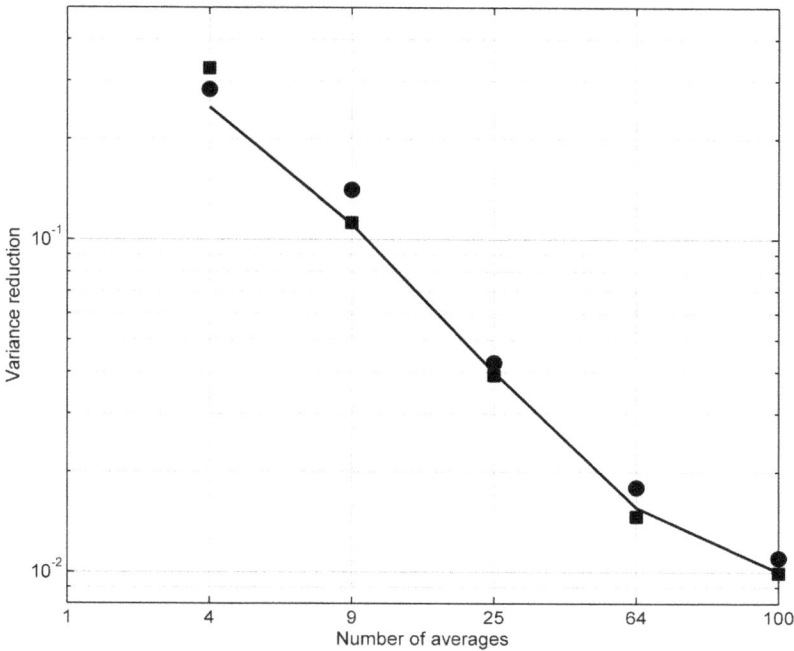

Figure 3.1.11 – Reduction of variance for trace averaging in the cases of large span/RBW (circle) and small span/RBW (square).

side the channel bandwidth (that we assume go from index N_{lo} to N_{up}); power values are obtained with a much smaller bandwidth (RBW) with respect to channel bandwidth B_{ch} and thus shall be normalized to the latter; finally, the average of the true power levels of each trace pixel on a linear scale gives the estimate of channel power.

$$P_{ch}\,[\text{dBm}] = 10\log_{10}\left[\frac{B_{ch}}{B_n}\,\frac{1}{N_{up}-N_{lo}}\sum_{i=N_1}^{N_2}10^{P_i/10}\right] \qquad (3.1.15)$$

where B_n is the ENBW of the used RBW, B_{ch} is the nominal channel bandwidth, $N_{up}-N_{lo}$ is the number of points that make the trace that falls in the channel bandwidth and effectively used for the average. The transformation from dB to the linear power scale avoids the problem related to the log influence on the average result.

A similar operation may be accomplished using the Video Filter to perform the average. If operating on dB power values distributed logarithmically, there will be the known underestimation: in the case of a noise signal with Gaussian distribution this causes the already said 2.51 dB underestimation. If the average is performed with the average option of the acquired traces represented again in dB on a log scale, then the result is the same. On the contrary, using the AV detector, it is possible to operate directly on the sample points, represented in power linear scale rather than dB.

Since the 2.51 dB correction holds only for Gaussian distribution of noise and it is thus not universally applicable, it is advisable not to perform any averaging before the traces are fed to the channel power algorithm and that VBW is set to minimum 3 times RBW, so to exclude it (with a maximum error of 0.05 dB on channel power).

Now the question is "how the power sample points P_i shall be taken in terms of the correct detector to use?" As said, the AVerage detector applied to the true power samples is a correct choice to obtain the average power reading. If using eq. (3.1.15) to estimate the average power, the use of the Peak Max detector might slightly bias the final result; the Sample detector gives unbiased results.

Regarding the biasing of Peak Max detector, we observe that the peak of noise will exceed its power average by an amount, that increases (on average) with the length of time over which the peak is observed (see sec. 3.1.3.4). A quite accurate formula that has been checked against some experimental observations in the literature [2] puts in relationship the observation time T (that is the sweep time divided by the number of points) and the impulse bandwidth of the chosen RBW setting (the exact relationship between the two depends on the IF filter type, see sec. 3.1.3.2[5]) with the observed ratio between the average power obtained with peak detector $P_{av,pk}$ and with sample detector $P_{av,smp}$.

$$\frac{P_{av,pk}}{P_{av,smp}} = 10 \log_{10} \left[\ln\left(2\pi T B_n + e\right)\right] \qquad (3.1.16)$$

Agilent in [2] indicates a better fitting of measured values to the theoretical curve of (3.1.16) if the multiplying 10 coefficient is replaced by 10.7, even if they underline that there is no theoretical support for this change.

Regarding the use of FFT mode or Sweep Mode, the time behavior of the measured signal and its modulation shall be considered.

Let's consider a constant-envelope modulation, such as FM (analog) or Frequency Shift Keying (digital), used for GSM cellular phones, satellite communications and noise-protected transmissions. The constant-envelope makes the measured power constant when that power is measured over a bandwidth wide enough to include all the relevant spectral terms, that is the channel bandwidth. Using the FFT mode over a wide enough frequency interval to cover the channel width will lead to channel power measurements with very low variance. On the contrary, sweep mode, that is typically performed with a RBW value that is a fraction of the span and narrower than the symbol rate, will show large fluctuations and the channel power measurement will thus have a larger variance.

3.1.11.1 Carrier power

The measurement of amplitude (or power) of a sinusoidal tone with steady characteristics was already considered, when talking about the compensation of the internal noise contribution by subtraction (see sec. 2.1.1 and 2.6.1.3) and when observing that with sinusoidal tones several detectors give the same result (see sec. 3.1.3.4).

---

[5] The impulse bandwidth is 1.6-1.8 the -3 dB bandwidth for older four-pole synchronously tuned filters, but more commonly in modern digital spectrum analyzers it is 1.5 the -3 dB bandwidth for a digitally implemented Gaussian filter; see sec. 3.1.3.2 and Table 3.1.1.

When the sinusoidal tone, or the entire modulated carrier, is not steady, but occurs in bursts, then a zero-span mode measurement is advisable: the largest RBW compatible with the modulation band is applied to the center frequency, which the LO is locked onto; then the time domain dynamics of the signal are tracked on the display. The measurement of the power is done by averaging samples, provided that each point corresponds to a true power measurement by selecting the RMS detector. Since there are intervals of time when the carrier is on and others when the carrier is off, the time intervals for the computation are selected manually, by applying a criterion of minimal amplitude of the measured rms, e.g. at least -20 dB (10%) of the full carrier rms (that if unknown may correspond to the largest recorded value). With modern spectrum analyzers automatic triggering is possible by setting such an amplitude value.

Regarding the use of Video Filter, it is underlined that the characteristics of the transient carrier signal are not those of a pure sinusoidal tone and the effects of the modulation may make it look like a noise signal, but probably with much different statistics: it is thus advisable to exclude the Video Filter by setting the VBW above the chosen RBW (e.g. with the already suggested minimum ratio of 3:1). This might turn into a problem, especially when spectrum analyzer software locks the minimum VBW/RBW ratio to 1, for high data rate protocols (e.g. hundreds kbps or some Mbps), where bursts occur in µs or hundreds of ns.

3.1.11.2 Channel power statistics

Besides the measurement of carrier power alone, it might be necessary to characterize the behavior of the modulated signal, e.g. in terms of peak power, peak-to-average power ratio, etc.

The peak power may be interpreted as the maximum of either the total channel power or the power over a portion of the frequency interval; in any case it is a measurement that can be made using the same settings for carrier power measurement, provided that the spectrum analyzer is able to track signal dynamics.

Analogously, when estimating peak-to-average power ratio, the value may be variable depending on bandwidth and sweep time used to catch the peak envelope power. Normally, a zero-span measurement setup with a large resolution bandwidth is the only possibility; it is advisable to verify result sensitivity to the used RBW value.

Given the probabilistic nature of the measured signals, a curve of probability for the different peak-to-average levels is quite common to adequately describe the statistical behavior of a modulation: the required curve is the cumulative distribution function that gives the probability that some peak-to-average level is overcome during measurements; of course the curve shall be accompanied by the number of acquired samples to give an indication of the statistical significance. Such characterization is not only relevant to possible interference to adjacent channels and to third-parties, but also to correctly sizing electronic circuits (most of all amplifiers) in transmitter and receiver paths, quantifying the headroom with respect to thermal power sizing and the number of times the signal amplitude is near the maximum.

3.2 Antennas and probes

Just for convenience and to follow usual conventions, in this section it is distinguished between "antennas" and "probes", to mean large devices measuring the field in one (or two, in some cases) polarizations and more compact devices, used to measure the spatial distribution of the field along three orthogonal axes.

Attention is focused on the working principles, usual characteristics, any limitation or advisable provision. In many cases available information for products consist of the characteristics we have reviewed in sec. 1.3.1; when thinking of antennas and probes as pieces of equipment that shall be reliable other characteristics may be relevant, such as influence of temperature, robustness, duration of batteries, etc., and they are very product dependent.

Practically speaking, antenna sensitivity (and its suitability for measurements) may be determined easily once the antenna factor and the noise floor and the used RBW values of the connected spectrum analyzer, giving an equivalent noise term n_{SA}:

$$E\,[\mathrm{dB\mu V/m}] = AF + n_{SA}\,[\mathrm{dB\mu V}] \qquad (3.2.1)$$

3.2.1 Magnetic field antennas

Antennas for magnetic field are mainly of the loop type, passive and active: the former is probably less expensive and does not incur in the problem of saturation, whereas the latter is far more sensitive. Sensitivity is particularly relevant when measuring at large distance and where the limits (and the expected emissions) are very low, such as the very low $-5\,\mathrm{dB\mu A/m}$ appearing at $30\,\mathrm{MHz}$ for the stationary test in Figure 4.2.1.

Active loop antennas feature negative antenna factor values indicating a significant amplification; antenna factor curves are also significantly flat, avoiding dramatic changes of sensitivity versus frequency.

3.2.2 Electric field antennas

Electric field may be measured by different antennas depending on frequency, required directivity, size.

3.2.2.1 Rod antenna

In the low frequency range a common antenna is the rod antenna, mostly useful for testing as per MIL STD 461 [110]: such an antenna measures only the vertically polarized E-field component that is in general the preponderant component near the soil. The antenna uses a ground plane to normalize its reference plane and remove variability due to different floors and soils. This antenna was conceived for use in the low frequency range, namely between $9\,\mathrm{kHz}$ and $30\,\mathrm{MHz}$: given the relatively low E-field intensity, it is an active antenna, with a sensitive FET transistor connected to the rod element; the role of the FET is to increase the input impedance to values that are compatible with the already large impedance of the rod element at low frequency, far from resonance. Care shall always be taken not to damage the FET gate by electrostatic discharge (ESD) brought in by operator's fingers.

3.2.2.2 Biconical antenna

The biconical antenna is a troublesome antenna for several factors: the large input VSWR in the lowest portion of its frequency range, the moderately large Antenna Factor (except around antenna resonance) and the susceptibility of both parameters to antenna positioning with respect to ground and orientation.

Antenna factor   The significant capacitive coupling of the antenna to the ground is due to the large surface of the conical expansions at the two antenna ends; it is also known that in vertical polarization the coupling of the two ends is different, one being very close to the ground and the other end almost uncoupled, thus creating the problem of a significant current unbalance and possibly common mode resonance, with common-mode-to-differential-mode transformation, kept under control only if the antenna balun has very good performance. Conversely it looks like the presence of the ground affects more the horizontal polarization if the radiation patterns are considered [6], and the antenna factors shown in Figure 3.2.1.

Input impedance   Regarding the input VSWR variations due to the change of input impedance, in [6] the mutual terms of the input impedance are calculated taking into account the effect of the ground; by definition the self terms are unaffected.

$$Z_{in} = Z_{11} + \frac{I_2}{I_1} Z_{12} \tag{3.2.2}$$

where Z_{11} is the self impedance in free space, Z_{12} is the mutual impedance between the antenna and its image, and the terms I_1 and I_2, that are the feeding point current into the antenna and its image, are such that $I_1 = -I_2$ in horizontal polarization and $I_1 = I_2$ in vertical polarization.

In Figure 3.2.2, for a biconical antenna at 1 m above ground with both orientations, it is possible to see that the absolute value of the mutual impedance is approximately constant with frequency for the vertical orientation (real and imaginary parts keep oscillating with frequency as expected), whereas for horizontal polarization it changes dramatically from very low values at 20 MHz to a maximum of about 100 Ω in the center of the band and then going back to small values around 300 MHz.

3.2.2.3 Logperiodic antenna

The logperiodic antenna is also named logperiodic array, made of dipoles of reducing length, starting from the lowest frequency of the operating interval and proceeding to the shortest one at the antenna tip that is active at the largest frequencies. While the frequency increases the active elements that have a tuned electrical length move along

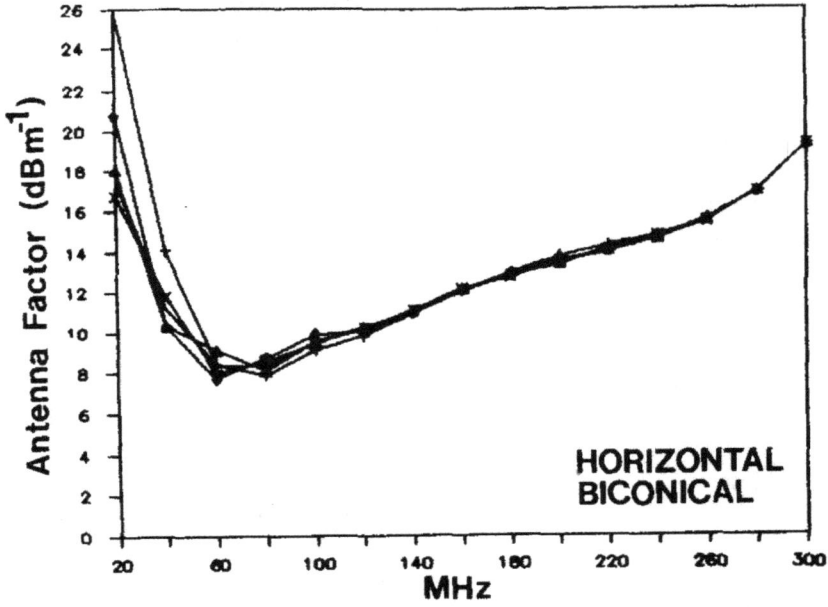

(a)

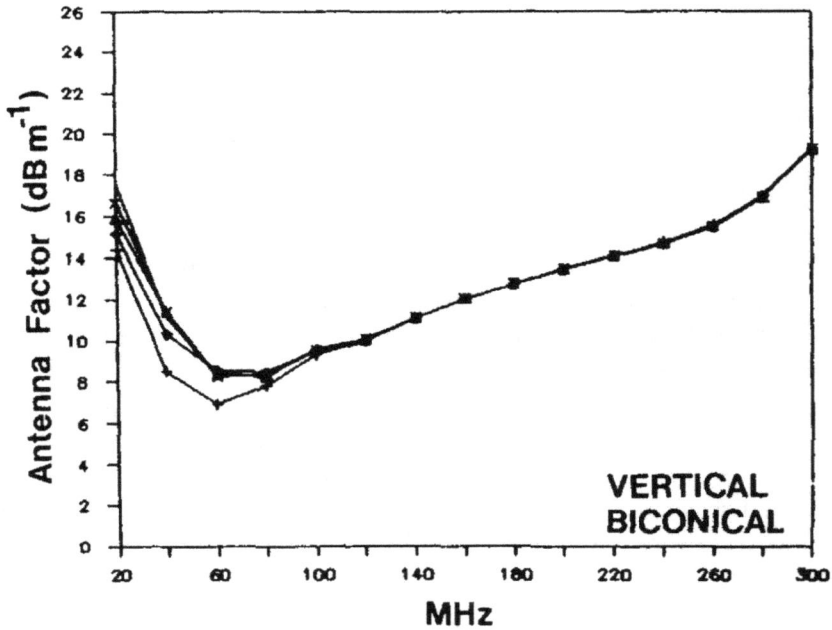

(b)

Figure 3.2.1 – Computed antenna factors for (a) horizontal and (b) vertical orientation at various heights above the ground plane (plus +: 1 m; diamond ◇: 2 m; triangle △: 3 m; cross ×: 4 m) [6].

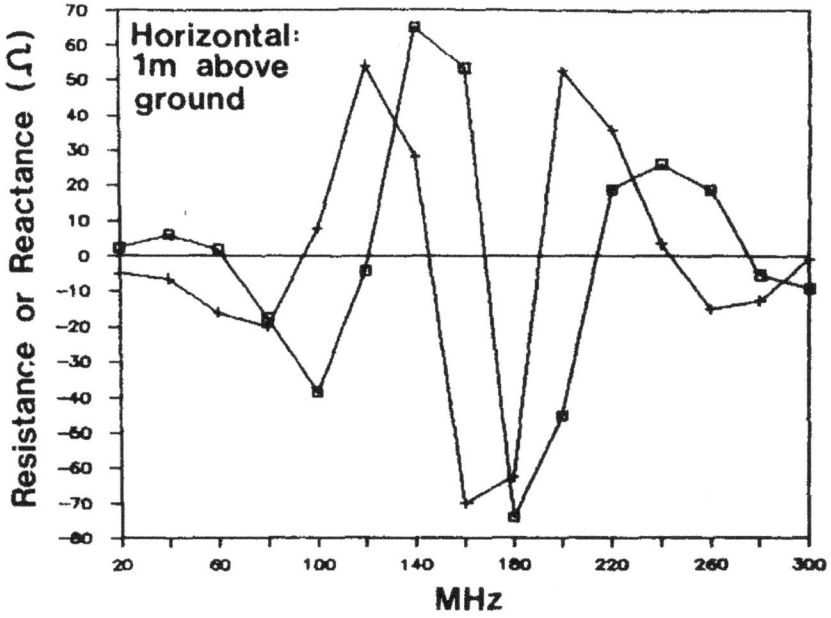

(a)

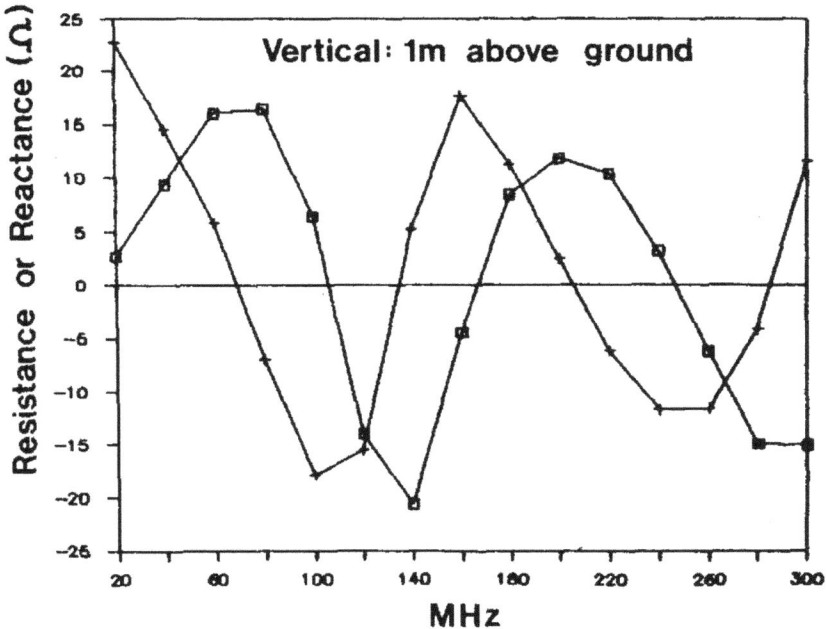

(b)

Figure 3.2.2 – Mutual impedance Z_{12} between the skeletal biconical antenna and its image for both horizontal and vertical polarizations; antenna feedpoint is 1 m above the ground plane (square $\square$ $R_{12} = $ real $\{Z_{12}\}$; plus + $X_{12} = $ imag $\{Z_{12}\}$) [6].

the antenna: for not extreme frequency values more than one element is active and the directivity is increased by additional rear and front elements; it is evident that at the lowest frequency the behavior is nearly that of a tuned dipole. Not only directivity, but also input impedance benefits of this behavior and the VSWR of this antenna is quite moderate over almost the entire frequency range. The side effect is that the electrical center of the antenna is moving when the frequency is changing, so that statements regarding the distance between the "antenna" and the EUT are not accurate any longer: the EN 50121-2 standard to this aim indicates the use of the geometrical center of the antenna; by the way the error introduced by taking the geometric center as electric center is about ± 0.3 m maximum, that for a 10 m measurement distance corresponds to $\pm 3\%$; in any case it would be possible to correct for such distance round off that is deterministic with frequency, provided that the $1/r$ assumption holds and that the influence of reflected rays keeps unaltered.

Logperiodic antenna is small enough not to have an appreciable coupling with ground already at the minimum height of 2.5 m required by the standard; similarly the coupling is not changing significantly in the two polarizations.

3.2.3 Magnetic field probes

Magnetic field may be sensed by different physical principles: coils using induction and thus limited to ac fields with a minimum frequency; Hall effect devices, sensing dc and ac and suitable for intense magnetic field; magneto resistive devices, covering dc and ac with more sensitivity and a more extended frequency range than Hall effect devices, but exposed to saturation for moderate field intensity; fluxgate are the most sensitive but with limited frequency range, due to the internal control system.

Sensors sensitive to dc component of course are biased by the earth magnetic field whose value and direction are not exactly known nor constant: buried metallic object and installations, as well as metal deposits, may affect the distribution and shape of magnetic field lines. Similarly, in the presence of strong dc magnetic field on-board in dc traction systems, accuracy of ac readings may be largely affected by non-linearity and saturation.

3.2.3.1 Coil

The physical principle is that of induction due to varying magnetic field, so that these sensors are suitable only for ac fields. The application of the Faraday-Neumann-Lenz law gives, after manipulation of the pulsation argument for sinusoidal components

$$V = -\omega BSN \cos(\omega t) \qquad (3.2.3)$$

where B is the magnetic field intensity as peak value, S is the coil area, N is the number of turns (for solenoids or multi-turn coils) and $\omega = 2\pi f$. The voltage V is the no load voltage at coil output that shall be read ideally using an infinite impedance voltmeter.

For increasing frequency parasitic capacitance may reduce slightly the output voltage, with an increasing effect up to the first resonance: to improve coil accuracy, an impedance matching network may be used that operates a compromise between the no-load reading exigency and the damping of the approaching resonance [97]. Additionally, being relationship (3.2.3) derivative in nature, integration by suitable low-pass filtering is necessary to normalize the coil output: in this case integration and matching may be performed by the same RC network [97].

The design of such coil shall trade off between sensitivity (e.g. area and number of turns), dynamic range, need of amplification, reduction of parasitic capacitance, optimization of losses to shape the frequency response when approaching the resonance. The frequency response of such coils may be quite extended both down to low frequency (e.g. less than 1 Hz with a properly designed amplifier [97]) and up in the hundreds of kHz or MHz range.

Coils are much less sensitive to external and operator's influence: a shielded coil may be built to avoid that strong electric field coupling on the curly path along the coil wire may compromise the accuracy of the magnetic field reading.

3.2.3.2 Hall effect devices

The original Hall effect occurring in a bulk material comes directly from the Lorentz's force of electrons flowing (applied biasing current) undergoing the external magnetic field: the orthogonal force displaces the electrons and causes a difference of potential that is measurable at two terminals located on an axis orthogonal to the current terminals and the applied field. This clearly establishes that the sensed magnetic field is the one orthogonal to the plane where the terminal pairs are located. Three-axis probes are made of three independent monoaxial Hall effect devices.

All semiconductors show Hall effect behavior: the induced voltage is proportional to material mobility that for the most commonly used ranges between $80000 \, cm^2/Vs$ for InSb, to 33000 for InAs, and 8500 and 1500 for GaAs and Si, respectively. Of course, since mobility is largely influenced by temperature, also Hall effect has a remarkable temperature coefficient.

Another issue is the large noise that characterizes devices: not only thermal noise (as it is for GMR sensors), but also shot noise and flicker, in common to semiconductor devices. The problem of noise and its impact on sensitivity was considered in [94]: the basic sensitivity, also named "gain", is in the range of a few to a hundred of $\mu V/\mu T$; the noise Power Spectral density has the typical flicker profile, increasing remarkably at very low frequency, so that values of $100 - 1000 \, nT/\sqrt{Hz}$ are very common from 1 Hz down to 0.1 Hz; the Ohio Semiconductors sensor model HR36 analyzed in [94] had the lowest internal noise, approximately halving the minimum detectable field.

Hall effect devices generally have a large saturation point (e.g. several hundreds mT for some commercial devices, see [94]), much larger than for instance GMR devices, and no memory or magnetization effect.

3.2.3.3 Anisotropic (AMR) and giant magnetoresistive (GMR) devices

The GMR effect, discovered independently by Albert Fert and Peter Grünberg in 1988, distinguishes from the simple magneto-resistive effect (called also "anisotropic magneto-resistive effect"), studied by Lord Kelvin more than a hundred years before. AMR consists in a change of resistance due to the alignment of orbitals under the effect of the external field, so that a reduction of resistance is expected with respect to the zero-field situation: the composition of the external field component with respect to the material magnetization H_m (expressed by angle φ, indicating the amount of $H_\parallel$ component along magnetization axis and $H_\perp$ orthogonal to it), taking into account the direction of the magnetic field created by the excitation current (indicated by angle ϑ with respect to the material magnetization axis); simplifying orientation of the current strip to lie along the magnetization (or anisotropy) axis, $\vartheta = \varphi$, and the change of resistance is given by

$$\frac{\Delta R}{R} = -k \sin^2 \vartheta = -k \frac{H_\perp}{H_m + H_\parallel} \qquad (3.2.4)$$

where k is the magneto-resistivity coefficient.

GMR effect was observed in multi-layer metal structures for which, when no external magnetic field is applied, the alignment of domains is anti-parallel and the resulting resistance is large, whereas with magnetic field applied the mutual interaction of these very thin layers (at nanoscale) may bring the thinnest ones into saturation, with a remarkable drop of the resistance value.

In some good conductors the d levels are empty, because the Fermi level occurs close to s and p energy levels. In ferromagnetic metals the d levels are partially occupied with an uneven distribution of the spins with different number of electrons and their statistics for the so called up and down spins. The less populated, or minority spin species, is more scattered with a Fermi level lying within the d band that sees a high density of states. The free mean path is correspondingly smaller and so is the electrical conductivity. With very thin layers a mix of the up-spin and down-spin electron channels is observed, that otherwise would flow separately [34].

AMR sensors are much simpler and cheap, if considering a large-scale production viewpoint. The most performing ones have sensitivity as large as $10 - 20\,\mu V/\mu T$, but in any case less sensitive than GMR, since they feature a smaller resistance change when field is applied: roughly 1-2 % against several tens of % for GMR. As a known issue they exhibit also poor "directivity", in that they are quite sensitive to the orthogonal component of the external field.

As many other resistive sensors, GMR sensors may be built around a single element, a half-bridge configuration or a full-bridge configuration, the elements being measured from outside as in a Wheatstone bridge. Their frequency response is quite extended, up to several MHz, and the sensitivity is much better than e.g. Hall effect devices. Whereas the latter feature internal noise sources typical of semiconductors (with flicker and shot noise mechanisms), in GMR devices thermal noise mechanism prevails, directly related to device resistance (usually low, in the range of hundreds Ω up to a few kΩ).

Saturation may occur with fairly low fields of a few mT, so that degaussing is absolutely necessary to remove remnant field component: auxiliary windings are placed around the sensors and driven by a local source that feeds degaussing ac waveform cycling between positive and negative values of reducing amplitude.

3.2.3.4 Fluxgate

Fluxgate principle has been studied since after the Second World War (Felch *et al.* in 1947, as reported by Primdahl [130]) and many realizations were created, even anticipating theoretical works, since the '30s. The pioneering work led to preliminary results that still showed mutual discrepancies and with experimental results, as synthetically pointed out in [130]. Primedahl collected several results in the '70s and [130] may be considered a very complete overview of physical principles, main properties and architectures of fluxgates and experimental results, of course up dating back to more than forty years ago.

Parallel and orthogonal fluxgates are available: the former have the excitation field parallel to the external field to measure, the latter, as it goes, orthogonal. The core where the magnetic flux created by the windings is summed may have a bar or a ring shape, single or twin construction and various dispositions of the excitation winding (see Figures 5 and 6 in [130]).

Although fluxgates are very accurate and sensitive for short-term measurements, offset and its drift are indicated by some as the weak side of this technology, as pointed out by Ripka, Pribil and Butta [136] [6]. On the contrary, having a look at the Bartington datasheet offset error is quite limited (less than 100 ppm for various models) as well as its drift with temperature (from ± 0.1 nT/°C to ± 0.6 nT/°C, that for a range of 20 °C limits the sensitivity for low-intensity fields to a few nT, despite the well promising power spectral density value used for the sensitivity).

Effectively, performances declared by manufacturers indicate a very sensitive sensor (with internal noise limiting it to up to 20 $pT_{rms}/\sqrt{Hz}$), with bandwidth of some kHz, a dynamic range before saturation that is satisfactory for railway applications (between several tens of μT up to 1 mT), a very good linearity error (in the range of tens of ppm), orientation error (limited to a fraction to a few degrees).

Temperature drift reduction is usually achieved exploiting differential schemes, where one of two sensors is subject only to temperature effect, that is causes a signal drifting from zero due to temperature changes and is subtracted from the complete signal of the other sensor, subject to temperature and external magnetic field.

3.2.4 Electric field probes

Electric field may be sensed by different physical principles (capacitor plates, Pockels effect, short monopoles and dipoles), but in general its measurement is more trouble-

---

[6] In their work the first two sentences of the abstract leave no doubt: "Offset and its long-term stability is a weak point of fluxgate sensors. Even the ultrastable sensors kept at no vibrations and stable temperature at magnetic observatories show offset drift."

some than for magnetic field due to the low sensitivity of such devices and the large impedance of the electric field.

Often electric field probes are sold installed inside handheld equipment or in a form suitable to be handheld. Electric field at low frequency (e.g. power supply frequency) features a large wave impedance and is heavily disturbed by conductive parts, including the operator her/himself and grass, if minimum clearance requirements are not met: the accurate measurement of electric field at power supply frequency should be performed by a suspended probe using a long stick of dielectric material (for example a fiber glass pipe of about $2\,\text{m}$ length). The IEEE Std. 644 [70] recommends 2.5 m of distance between the probe and an 1.8 m tall operator or observer to ensure that the proximity effect error is smaller than 3 %. Of course it is extremely helpful that the operator or observer stand in the region of lowest electric field strength if the source is known (e.g. catenary voltage).

On the influence of the operator or surrounding environment datasheets and manuals are almost always silent: keeping electric field sensors close to the operator's body influences dramatically the results; some tests may be made using an as most constant as possible source of electric field and getting closer and closer to the probe and moving around it.

3.2.4.1   Capacitor plates

The simplest E-field detector suitable for low frequency is that made of two conducting halves, electrically separated, that form the two plates of a capacitor that charges due to induction of the E-field in which it is immersed. The charge is proportional to the electric field intensity E, to the air dielectric constant ε_0 and a form factor k:

$$Q = k\varepsilon_0 E \qquad\qquad (3.2.5)$$

For ac fields a current flows through the plates, proportional to the derivative of the charge. Thus, whereas for dc fields a charge amplifier is necessary to read the accumulated charge, for ac fields a weak current amplifier (e.g. femtoampere) may be used (not so different from the techniques used for charge amplifiers). Rectification of the ac signal may be used, but in this case the information on the original field components is compromised.

Regarding field uniformity between plates, different geometries may be used: parallel plates shall have a large surface to increase probe gain and compared to their separation to ensure uniformity. Conductive objects external to but near the probe may exert their influence perturbing the field distribution and increasing edge effects, thus compromising accuracy: normalizing guard rings may be included to shape the field at the periphery.

3.2.4.2   Monopole/Dipole elements

Based on the same principle of the rod antenna (see sec. 3.2.2.1) monoples if a reference ground plane is provided, one for each axis. Considering also the need of amplifying a very weak signal from a high-impedance source, it is apparent that dipoles that are balanced and do not need a reference plane are a better choice. The amplifying stage can be designed completely floating.

Given the fairly large radiation pattern of dipoles, mutual coupling between the three orthogonal dipoles aligned along the chosen system of coordinates is one of the design problems of this antenna: characterization and calibration may get rid of such issue, provided that the three elements are accurately orthogonally positioned. Otherwise, not only asymmetric coupling may give rise to transformation of components, but also isotropicity is compromised.

For the small dimensions of dipole elements (few cm), equations for the Hertzian dipole antenna (see sec. 1.3.2) can be satisfactorily used up to some GHz.

3.2.4.3 Pockels effect

The electro-optic effect consists of a change of refractive indices of a material due to an applied electric field. The dependency may be linear or quadratic: the former is known as Pockels effect, while the latter is known as Kerr effect, less useful for accurate measurements; higher order dependency is possible, but usually negligible. KH_2PO_4 (KDP) and $LiNbO_3$ are materials with a significant Pockels effect.

For materials transparent at some wavelength the Pockels effect consists of a change of the refraction index: with a field E_z applied along one of the axis of the crystal, the indexes along the other two orthogonal axes change (*birifrangence*) as follows (a cubic crystal with uniform refraction index n_0 at rest is considered):

$$n_x = n_0 - \frac{1}{2}n_0^3 r_{41} E_z \qquad n_y = n_0 + \frac{1}{2}n_0^3 r_{41} E_z \qquad n_z = n_0 \qquad (3.2.6)$$

where r_{41} is one of the coefficients of the electro-optic tensor matrix of the crystal and n_0 is the ordinary index of refraction of the crystal (without electric field applied).

The change of refraction index creates a delay, or, in other words, a difference of phase between the components of the beam of light traveling through the crystal; this phase difference is called *retardance* and is usually indicated by the symbol Γ:

$$\Gamma = \frac{2\pi n_0^3 r_{41} E_z L_z}{\lambda_0} \qquad (3.2.7)$$

where L_z is the length of the crystal in the z direction and λ_0 is the wavelength of the applied beam of light.

Crystals for the measurement of Pockels effect instead of measuring the phase, after the first polarizer that selects the desired light polarization for the Pockels effect to occur with a specific orientation of the applied electric field, add another polarizer and measure the change of amplitude of two beating beams: the amplitude is determined by a transmittance function T expressed as a function of *retardance* Γ:

$$T = \sin^2\left(\Gamma/2\right) \qquad (3.2.8)$$

Light is applied to the crystal through a fiber optic and a collimating lens. Several construction details and techniques are detailed in [74].

The operating principle ensures a very large bandwidth, up to tens of MHz; depending on other factors, such as electrode and crystal size, limiting power dissipation or increasing electrical resistance, the maximum useful frequency may be lower. The internal noise is typically of the shot type due to the granularity of the photons arriving at the receiver.

Increasing the length of the crystal increases its sensitivity; however, the maximum length of the crystal is limited by its optical quality: for crystals of ammonium dihydrogen phosphate or deuterated potassium dihydrogen phosphate the maximum length may be in the order of one or few hundreds of mm, whereas for cadmium telluride, lithium niobate, and lithium tantalate the maximum length may be ten times smaller.

The main field of application is that of optical modulators or optical switches, e.g. for lasers. The sensitivity is low, so that they are used as sensors mainly to sense very high voltages and very large electric field, where the fiber optic ensures galvanic isolation and very large insulation levels.

3.2.5 Elements of uncertainty

Sources of error and uncertainty are many when the setup is complex, when elements of different nature are involved and human factors are also determinant. The latter is addressed later on in sec. 5.4.4 focusing on synchronization issues and driving style. The following sections consider some relevant uncertainty contributions related to probes and antennas and their use.

3.2.5.1 Supports, holding elements and connecting cables

In general masts, shafts and holding elements are required to be non conductive: the latter may contain sometimes small conductive elements, such as bolts and screws; in general when using non-conductive plastic materials for large elements, the effect of the alteration of the dielectric permittivity cannot be disregarded. Errors up to 1 dB, or 2 dB exceptionally at some frequency, may be due to the holding system, both in horizontal and vertical polarization (probably slightly larger in vertical polarization due to the vertical orientation of the supporting mast).

Conversely, connecting cables are and shall be conductive: antennas, in particular dipoles and biconical antennas, have a metallic support that hosts a transmission line connecting the balun an keeping the input connector farther away, about $0.5 - 0.6$ m; this metallic pipe is included in the calibration of the antenna and it is good practice to prolong ideally it when connecting the cable, getting farther away from the antenna, before turning to the vertical direction and going to the ground. The error due to the cable is much bigger when the antenna has the same polarization of the cable, that is the vertical one: the cable behaves like a frequency-dependent scatterer, coupling for same polarization and affecting significantly the antenna radiation pattern. For vertical polarization, the effect of the cable is much more significant with some resonances occurring in the lowest frequency interval up to about $200 - 300$ MHz, depending on

	$d = 3$ m		$d = 10$ m		$d = 30$ m	
	Hor.	Ver.	Hor.	Ver.	Hor.	Ver.
Biconical [30 MHz, 200 MHz]	0.0	±0.5	0.0	±0.25	0.0	±0.1
Logperiodic [200 MHz, 1 GHz]	±1.0	±3.2	±0.2	±0.5	±0.1	±0.15

Table 3.2.1 – Directivity difference error as per CISPR 16-4-2 [43], Table D.1 through D.4.

the distance between the descending cable and the antenna axis; increasing the distance of the vertical cable section to about 2 m (that for many biconical and bilog antennas is 1.2-1.6 the overall antenna length), reduces significantly the number of resonances and the attained maximum.

3.2.5.2 Height scanning and angle of arrival of reflected rays

When using height scanning for measurement of emissions, such as in EN 55011 or 55022 measurements, the direct ray and the ray reflected from the ground both reach the antenna from two different angles, that depend on the height difference between the source and the antenna and the height with respect to the ground level. Antenna directivity is given for the direction of maximum emissions, that is the maximum of the radiation pattern; with omni-directional antennas (or antennas with a broad radiation pattern) directivity values for some offset angle with respect to the direction of maximum are not so relevant. In general, the directivity function shall be adopted instead and may be profitably used for a correction if the angle of arrival of the reflected ray is known. Since the pattern or the angle of incidence is unknown, a correction of the field strength is not feasible or at least impractical. So an uncertainty contribution is introduced for all antennas, but in particular if a directive antenna is used. The standard CISPR 16-4-2 [43] reports a thorough analysis of the uncertainty contribution for the antennas used in EMC measurements, of course under the assumption of a reflecting ground, that for our railway emissions measurements is replaced by the soil, that may be assumed only partially reflective and with a non-negligible loss. The directivity difference error that may be applied as correction is expressed in terms of the two values of an interval over which uniform distribution is assumed: these values may be similarly adopted for railway emissions measurements when estimating the uncertainty budget, underlying the lack of detailed information for soil behavior and knowing that the height of antennas is limited as per EN 50121-2 and -3-1 with respect to the full height scanning between 1 and 4 m. The spread of directivity difference error for the two frequency intervals typical of biconical and logperiodic antennas are shown in Table 3.2.1 for the three measurement distance values of 3, 10 and 30 m. A very clear and detailed analysis of the problem may be found in [85].

In [16] the problem of measuring large objects at different antenna distances assuming a wide range of angles of arrival was considered and an estimate of antenna factor span and uncertainty for a bilog antenna was derived. The considered frequency range is 30 to 1000 MHz with measuring distances of 3, 10 and 30 m: for an EUT that is 1 m high and applying the height scanning technique between 1 and 4 m, then the angle of incidence may vary from 0 to 17°; reducing the measuring distance to 3 m, the maximum angle may increase up to $45°$. The bilog antenna has thus an antenna factor AF_b given by the manufacturer versus frequency but in free-space conditions and an antenna factor that depends on the direction of arrival, $AF_b(\vartheta, \varphi)$, as it is for the directivity function with respect to the directivity. Since in the paper the bilog is compared to the dipole taken as reference by almost all standards for emissions measurement, also for the dipole we may express an antenna factor AF_d in free-space conditions and another one, $AF_d(\vartheta, \varphi)$, that takes into account the direction of arrival. In this case the dipole is calculable and the following $AF_d(\vartheta, \varphi)$ function is reported

$$AF_d(\vartheta, \varphi) = AF_d + 20 \log_{10} \frac{\cos(\pi/2 \, \cos \vartheta)}{\sin \vartheta} \qquad (3.2.9)$$

whereas for the bilog antenna numeric simulations were done by the authors.

The result was expressed as error in antenna factor when replacing the bilog with the dipole, but accounting for differences in free-space AFs separately (using a correction factor K), so that the resulting antenna factor error ΔAF takes into account only the errors in directivity as they reflect in the antenna factor function.

$$\Delta AF = AF_b(\vartheta, \varphi) - AF_d(\vartheta, \varphi) - K \qquad (3.2.10)$$

The results are shown in Figure 3.2.3 in terms of ΔAF span at variable angle of incidence for the three measurement distances of 3, 10 and 30 m. It is evident that for the 30 m distance the error is negligible and that reducing the distance to 10 m still keeps the error within about ± 1 dB, very small on average; the 3 m distance on the contrary makes the error increase to unacceptable values, if also combined with the other sources of uncertainty.

3.2.5.3 Antenna radiation pattern [155]

Antennas suggested by EN 50121-2 and -3-1 have almost standardized radiation patterns, for which no appreciable changes are expected for antennas of different manufacturers with slightly different sizes. For biconical antennas, the largest ones have better antenna factors but are more exposed to coupling to ground. Logperiodic antennas have a length of about 0.7 m, possibly going to 0.8 m if the frequency range is extended beyond 1 GHz by adding one or two of the shortest elements at antenna tip.

All dipole-type antennas see multiple lobes appearing in the radiation pattern when the frequency is short enough for the first resonances to occur; for a basic example see sec. 1.3.3 on the dipole antenna used beyond the first resonance.

A dipole antenna will show multi-lobing if antenna length is above approximately $1.4 - 1.5\lambda$ (see Figure 1.3.3). This multi-lobing may be less apparent with a biconical antenna of same length, because of its broadband behavior; however, also biconical pattern deteriorates and multi-lobing appears at a slightly larger frequency, when antenna length is around 1.6λ, that roughly corresponds to $320 - 350$ MHz, well beyond the maximum frequency of use for biconical antennas, usually limited to $200 - 230$ MHz.

(a)

(b)

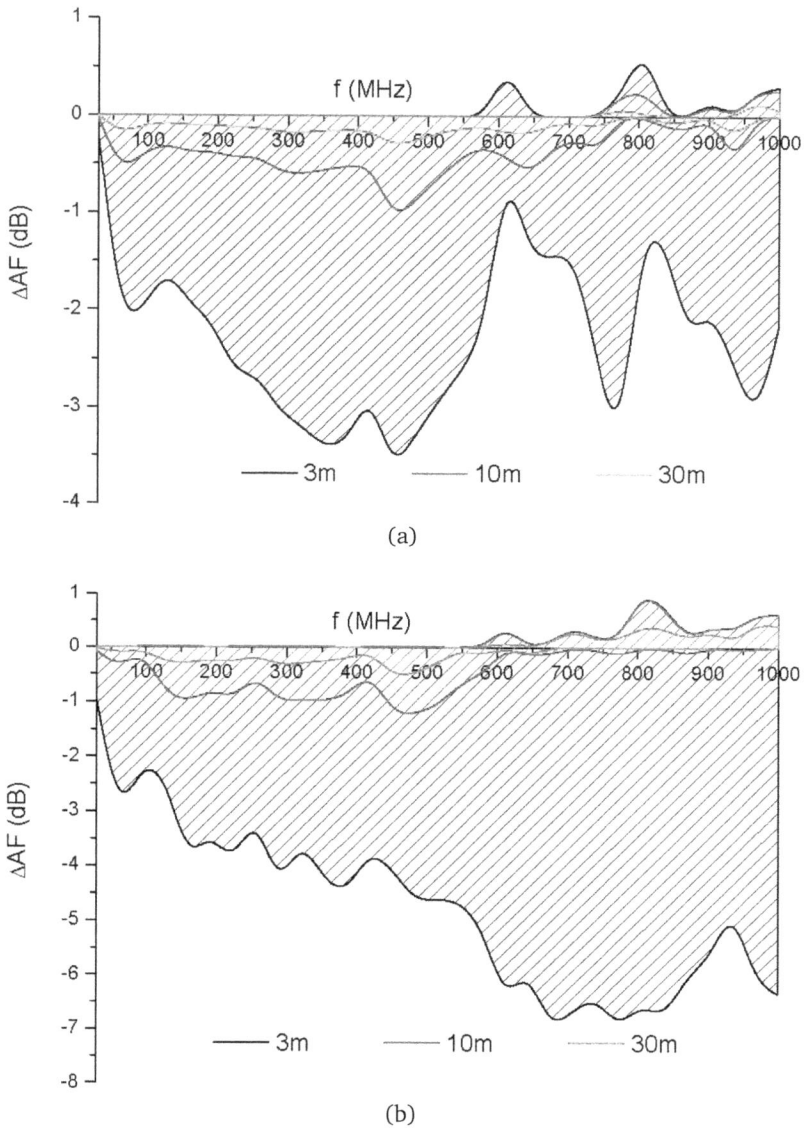

Figure 3.2.3 – Span of the antenna factor error ΔAF for the bilog with respect to the dipole in (a) horizontal and (b) vertical orientation for three different measurement distances of 3, 10 and 30 m [17].

Logperiodic antennas, when the frequency increases and the dipole elements are brought into resonance one by one, exhibit periodic fluctuations of the antenna factor and correspondingly in the otherwise smooth doughnut-like radiation pattern of a single dipole; the adjacent dipoles couple to the active one, not only increasing directivity, but largely reducing such dips and swells of radiation pattern.

3.2.5.4 Antenna calibration

Uncertainty in antenna calibration is directly related for electric field antennas to the uncertainty of the adopted calibration method with respect to others and the intrinsic uncertainty of the calibration procedure and adopted references. Without going into the details of calibration, methods (all accepted and all well documented in recognized standards) may differ by as much as $4-5$ dB. Such variability is exemplified by the variability of site imperfection of OATS (Open Area Test Site[7]), normally ranging between ± 1.5 dB to ± 3.5 dB, without excluding sporadic larger deviations, maybe over a limited number of frequencies, as resulting from various round robin tests. OATS is one of the methods used for antenna calibration, that thus transfers the site imperfection uncertainty into the uncertainty of the antenna factor.

This is larger than the instrumental uncertainty of the measurement chain and references used by the calibration laboratory and is related to the physical principles, assumptions and approximations the chosen method is based upon.

3.2.5.5 Probes calibration

Probes have smaller dimensions than antennas and may be calibrated inside a reference magnetic or electric field source, where also higher field intensity is available:

- for magnetic field calibration is achieved by coil arrangements:

    - Helmholtz coils may reach a very accurate and uniform field distribution, easily about or below 0.1 % [44] (see Figure 3.2.4): the relevant elements of the uncertainty budget are the dimensional stability of the Helmholtz pair, including the separation distance, and the measurement of the input current and its stability during measurements; when characterizing the setup other quantities are relevant to the estimated uncertainty, such as the positioning error and the dimensional stability of the search coil calibrated by construction, the influence of background noise and the accurate correlation between the input current and the total Ampere-turns of the bulk current along the circumference with a Rogowski coil [26];

---

[7] Open Area Test Site is the site meeting the specifications for the measurement of electric emissions from equipment under test, in particular ensuring a known reflection and scattering of the electromagnetic waves hitting the nearly perfectly reflective and flat conductive floor. Emission standards such as EN 55011 and 55022 specify how to use the OATS for measurements; electromagnetic characteristics and uncertainty are given in CISPR 16-1-4 [39].

- the Helmholtz pair features a 1-D field aligned with coil axis and thus requires that the probe under calibration is rotated to align its axes one at a time, with a consequential non-negligible positioning error; for a 3-D magnetic field orientation during calibration leaving the probe in the same position, the so-called 3-D Helmholtz coils;

- for electric field, depending on frequency, a parallel plate capacitor, an open stripline or a TEM/GTEM cell are used:

  - the open stripline design equations appeared in the old immunity standard to radiated electromagnetic field IEC 801-3 and it appears now in CISPR 20[8], App. E; calibration procedure by means of an additional square plate positioned parallel to the internal plane and used as a voltage divider appears in the App. F of the same standard; a more complex procedure appears in the EN 61000-4-20[9], where calibration EUT of different sizes, containing a known comb generator, are required.

Almost all calibration methods for magnetic and electric field probes calibrate one axis at a time. One relevant characteristic when using the probe is the uniform response and calibration of its sensors along the three axes and the amount of coupling between the axes (i.e. verifying how a field component lying along one axis is partially read also by the other sensors).

3.3 Design of setup

It is evident that the uncertainty that characterizes the final result may be reduced and kept under control by several techniques, both operating on measured data during post-processing (e.g. using statistical techniques as in sec. 2.2.4) or optimizing setup and instrument settings (e.g. antenna choice and installation, distance from external sources, cable noise pickup, adequate sweep time, detector choice, correct earthing, etc.), together with test conditions (e.g. driving style, synchronization, number of tests and verification of repeatability).

3.3.1 Post-processing of measured data

Depending on the aim of the specific measurement (e.g. evaluation of emissions or of background noise, estimate of channel power, etc.) and the nature of the involved physical mechanisms (e.g. transient or steady, narrowband or broadband) different techniques may be adopted, always within the scope of acceptability set forth by the applicable standards. Processing available on-board spectrum analyzer and on an external computer shall be used wisely avoiding undesirable data manipulation; examples of available trace processing functions are shown in sec. 3.1.2:

---

[8] CISPR 20, *Sound and television broadcast receivers and associated equipment – Immunity characteristics – Limits and methods of measurement*, 2013-10.

[9] EN 61000-4-20, *Electromagnetic compatibility (EMC) – Part 4-20: Testing and measurement techniques - Emission and immunity testing in transverse electromagnetic (TEM) waveguides*, 2003-04.

(a)

(b)

(c)

(d)

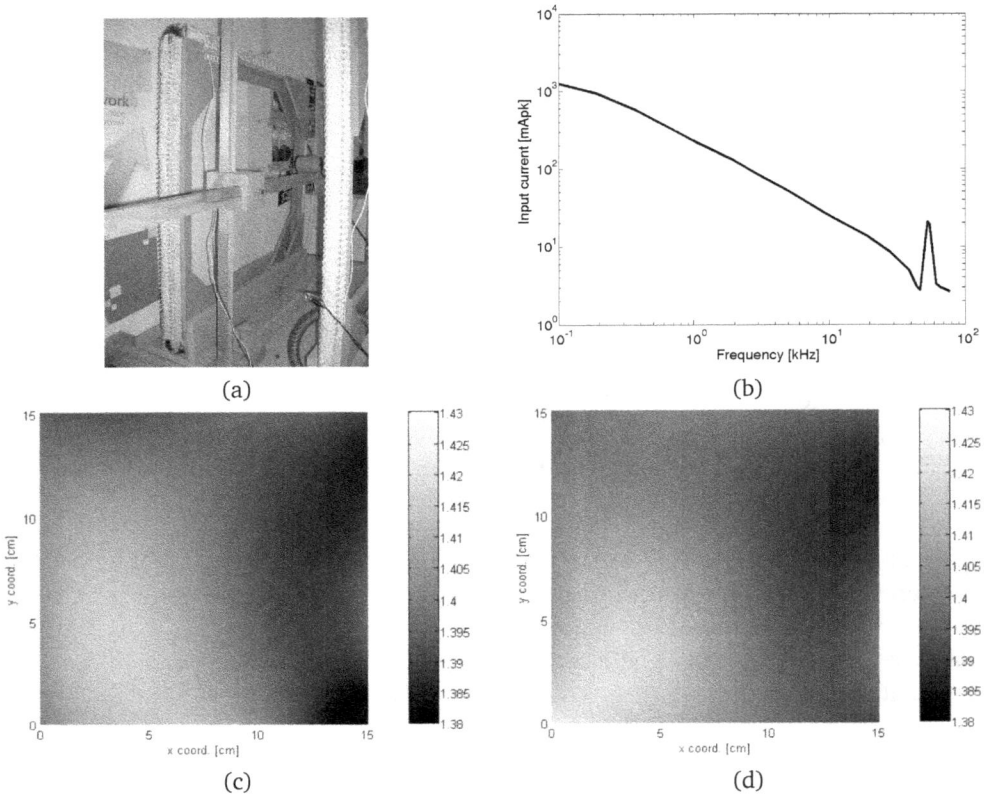

Figure 3.2.4 – Helmholtz pair for probe calibration [44]: (a) picture of the Helmholz pair, search coil with positioning system and Rogowski coil for bulk current measurement; (b) frequency response between magnetic induction in the center and input current; (c) and (d) magnetic field distribution in a cross sectional quarter of positive x and y coordinates, for $z = 0$ cm (center) and $z = 2$ cm (2 cm offset). The two field distributions were obtained for the tight separation of 60 cm ensuring 0.35 % spread over the volume of a 11 cm sphere (the diameter of the largest triaxial coil in [97]).

- max hold is a built-in function for spectrum analyzers that retains the maximum since the last reset for each pixel (or frequency bin) over multiple sweeps; it is thus useful to get the worst-case picture of background noise and to capture altogether intermittent sources, such a radio ham and non-commercial radio sources; the other side of the coin is that the possibility of statistical characterization of the set of traces is lost;

- average with sample dispersion estimate: whereas average is readily available in spectrum analyzers as incoherent averaging of successive spectra, dispersion shall be calculated offline during post-processing;

- post-processing may use robust average to discard weird values and aberrations, e.g. with evident transients and glitches, as suggested by EN 50121-2, provided that such "transients" can be unambiguously recognized and identified;

- if the spectrum analyzer does not have the "channel power" function (see sec. 3.1.11), (3.1.15) may be implemented in post-processing.

3.3.2 Size of EUT

From the standpoint of emissions measurement as commonly applied to electrical and electronic equipment (EUT), the size of the EUT is relevant to the correct measurement with acceptable uncertainty boundaries. We have considered the problem of large EUTs whose emissions may arrive at the measuring antenna with different and variable angles of incidence, e.g. depending on frequency and the various internal sources of emissions inside the EUT. In general, however, the application of open-field (or open-area) test site measurement techniques require that the coupling between EUT and antenna is avoided, limiting the size of the EUT for a given measurement distance d separating EUT and antenna and minimum wavelength $\lambda_{\min}$ [103], so that

$$D_{\max}^2 < \frac{d\,\lambda_{\min}}{2} \tag{3.3.1}$$

For measurements up to $1\,\mathrm{GHz}$ and at $10\,\mathrm{m}$ distance the maximum size of the EUT is only $1.23\,\mathrm{m}$. It is evident that for railway applications the typical EUTs such as rolling stock are much larger, even limiting to the portion of it where the emissions-relevant equipment (e.g. traction converters, auxiliary converters and traction motors) is located. This implies in general an underlying approximation for railway measurements with respect to OFTS (or OATS) measurements. The requirement may be relaxed once the lack of relevant emissions is acknowledged above about $100\,\mathrm{MHz}$, resulting thus in a ten times larger limit on EUT size, that fits for example the rolling stock size.

3.3.3 Antenna use

Antennas are quite well exemplified by the EN 50121 standards, namely loop, biconical and logperiodic: other antennas may be used provided that they feature an as good sensitivity (especially in the MHz range and up to a hundred MHz) and known issues of coupling and related countermeasures. Their characteristics were considered in sec. 3.2; now we focus on their correct choice and use.

3.3.3.1 Loop antenna

The loop antenna is a shielded loop and is quite insensitive to coupling to nearby objects, other parts of the setup and the operator. The active loop type is particularly sensitive and suitable for the frequency interval up to $30\,\mathrm{MHz}$; the other side of the coin is an excessive susceptibility of active antenna types to intense out-of-band magnetic field (such as that due to traction supply fundamental and main harmonics), responsible for bringing often the antenna into saturation. When the source of low frequency magnetic field is not the line catenary subject to tests, this is a symptom of unwanted coupling also of higher frequency components from a different source, despite the visible effect is that of saturation due to low order components: the antenna shall be kept far away from cables carrying traction and supply current.

The active type features a stable output impedance for which, in conjunction with the fairly low frequency range, no matching attenuators are required. However, on this aspect datasheets often characterize VSWR only with a "less than" statement with quite large value, ensuring that the statement is in no way false, but maybe not so accurate. For the EMCO 6502 active loop the output impedance is very close to the nominal $50\,\Omega$, so that a much better VSWR is expected than the usually assumed one of 2.5; an S_{11} measurement over $[10\,\text{kHz}, 30\,\text{MHz}]$ says that the return loss is nearly constant up to about some MHz and then increases linearly with frequency, from $-30\,\text{dB}$ at $5\,\text{MHz}$ to $-20\,\text{dB}$ at $27\,\text{MHz}$, with a reflection coefficient that thus increases from 0.03 to 0.1 and the VSWR is correspondingly going from 1.06 to 1.22. The linear increase may be explained by the parasitic capacitance of the output stage.

3.3.3.2 Biconical antenna

The biconical antenna is a broad dipole with enhanced operating frequency range, but substantially the same radiation pattern and directivity (see sec. 1.3.3). So, the biconical antenna is immune from the problem of directivity correction for a broad range of angles of incidence, with its half-power beamwidth larger than $100\,°$.

The major problems of correct use of the biconical antenna are related to its input impedance and coupling with the surroundings:

- the input VSWR may be very large, e.g. up to $7:1$ at the lowest frequency around $20-40\,\text{MHz}$, where by the way the largest emissions of the CISPR frequency interval C (see Table 3.1.2) are expected: this is confirmed by the considerations accompanying Fig. 3 and 4 in [79]; a matching attenuator at the antenna end of the cable is required, the other attenuator possibly provided by the spectrum analyzer itself (see sec. 1.5.3);

- the fairly large extremities of the antenna not only achieve a good impedance matching with free space impedance and improve the frequency response, but are also responsible for a remarkable coupling with the ground and too close conductive objects; in particular, in the vertical polarization orientation one end is very close to the ground (at about $1.5\,\text{m}$ at the lowest position of $2.5\,\text{m}$), creating an unbalance in the capacitive coupling terms and thus affecting the common-to-differential mode rejection of the internal balun [99]: the effect is that of creating common mode resonances between the antenna, the cable shield and the ground, influencing the measured differential signal; as for uncertainty and frequency response, they were considered in sec. 3.2.2.2 and 3.2.5.

Combined bilog antennas have often a smaller biconical section, with less coupling problems, but lower sensitivity.

3.3.3.3 Logperiodic antenna

Logperiodic antenna is easier to handle because the antenna factor and the VSWR are more stable with frequency, the size is smaller than the biconical and also the risk of stray coupling is averted.

Although no relevant emissions are expected in the antenna frequency range for typical on-site measurement configurations for rolling stock and line emissions, the logperiodic as a directive antenna is in any case exposed to the problem of oblique incidence (see sec. 3.2.5.2): if not aiming at the source of emissions or if emissions are expected from different directions with different angles of incidence, then the readout is not accurate for the variation of the directivity function. This is what is corrected with the procedure for substation emissions measurement in the 2006 version of EN 50121-2 (see sec. 6.3.5.1), when the angle with the maximum reading at a given frequency is annotated and then the whole frequency range is swept with that orientation.

A known source of uncertainty is related to the measuring distance varying with frequency, because the active dipole element changes and the electric center of the antenna moves: the EN 50121-2 requires that the mechanical center of the logperiodic is used and it is recognized that the distance error spans about $\pm 40\,\mathrm{cm}$ at most for the longest antennas that in terms of field intensity translates into $\pm 0.35\,\mathrm{dB}$.

The advantage of using a bilog antenna is not really such, because in any case the $[200 - 1000\,\mathrm{MHz}]$ interval needs to be split to keep the sweep time conveniently fast.

3.3.4   Earthing and connections

Earthing of instrumentation shall be ensured for the vast majority of equipment, unless battery supplied and isolated equipment is used (handheld spectrum analyzers). When supplying from hazardous voltage, earthing is mandatory for safety, besides interference exigencies: earthing is particularly relevant when supplying by means of a motor generator, that may often lack of these safety precautions.

Long power supply cables may be a source of interference, if common mode currents take place, also as a consequence of various earthing arrangements. Common mode chokes may be installed on the cable that is laid down straight away from the antenna location with the motor generator at a convenient distance (e.g. $50\,\mathrm{m}$).

Of course, with battery supplied instruments, it is always possible to perform sweeps for comparison with the motor generator on and off and decide on the relevance of its emissions and any other effect.

3.3.5   Synchronization and rolling stock operation

Several factors related to logistics, organization, and human factors are very important both for the successful test execution and as an entry of the uncertainty budget, besides instrumental uncertainty. Human factors are quite relevant whenever the operator may influence the measurement result or at least her/his behavior may affect data quality and uncertainty, always quite difficult to quantify precisely: we see in sec. 4.4.5.2 and 5.4.4.3 that the driving style and the synchronization of sweeps with acceleration and braking events determines the shape of the spectrum coming out from a single sweep and the intensity of captured transients while sweeping. Post processing of the set of traces for identical (or very similar) test conditions improves the repeatability of the results, removing singular points and aberrations, and thus reducing the impact of synchronization errors and operator's influence. However, it shall be taken into account that several runs are needed to collect a meaningful set of traces covering satisfactorily the frequency range when post-processed.

In any case the agreement with the driver shall be on the reference speed to keep, on the acceptable amount of fluctuations of the speed value, on the intensity of acceleration and braking and the steadiness of the applied effort. To this aim it is observed that often the driver modulates the effort changing it while moving the throttle forward and backward, but causing several transients (especially in the traction converters) and unsteady conditions. A few trial runs are advisable to verify that driver and people on-board understood the program and the desirable driving conditions: not only it may be difficult to understand desired conditions whose relevance cannot be perceived from a pure driving and movement perspective, but also language problems may be an obstacle.

Different choices are possible to carry out all the necessary measurements for rolling stock and measurement conditions: it is possible to change antennas while keeping the rolling stock always in the same operating (and driving) condition, or have the driver to change cyclically between e.g. acceleration and braking, also to keep her/his attention alive and avoid anticipation of maneuver and other psychological side effects.

Synchronization of sweeps may be achieved by putting visible marks along the track, at which the driver begins the required maneuver and the sweep may be triggered. These trial runs are also useful to verify such synchronization and the time duration of sweeps that shall fit train passage.

3.4 Measurement uncertainty

3.4.1 Overall uncertainty requirements

The EN 50121-2 doesn't require explicitly calibrated instrumentation, and nonetheless certified and traceable calibration: sec. 5.1.5 of the EN 50121-2 (2006) speaks about "calibrated antenna factors", and the same is in the 2015 version at sec. 5.1.1.3. However, it is known that there exist many methods of antenna calibration [96], based on different principles and techniques, and that they bear different uncertainties, so that comparison between two different calibrations of the same antenna bring to several dB of uncertainty. The statement is neither accompanied by an indication of the required calibration method, target uncertainty or similar, and it is thus to be intended as underlying the need of verification of the correct operation of the antenna, so that the declared antenna factor curves are reliable.

The $\pm 4\,\mathrm{dB}$ statement regarding "difference from EN 55016-1-1 [38] equipment" appearing in sec. 5.1.3 of the EN 50121-2 (2006) is a statement of compliance of performance, not regarding calibration, otherwise it should be again accompanied by a quantification of uncertainty with the related coverage factor and confidence level. We will see, however, that this value is not so different from the $k=2$ expanded uncertainty estimate for a good-quality and well documented open area test site.

In the EN 50121-3-1 (2006), below Table B.1, the caveat is against the use of low-cost spectrum analyzers that require calibration to ascertain that they work correctly.

3.4.2   Variability of operating conditions and synchronization

For the wide range of operating conditions and elements of test setup and instrumentation that have influence on the measurement results, it is evident that a sound verification of repeatability and Type A uncertainty is much more relevant than the exhibition of certificates of only a part of the whole system under test, that is for example antennas and spectrum analyzer (featuring a low uncertainty contribution if properly connected and configured). In other words, certification of equipment is easy to handle, but does not ensure neither the correctness nor the overall uncertainty of measurement results: ineffective driving style and operating conditions, synchronization, poor traction line response, external disturbance and unwanted coupling are all elements that may be kept under control with an adequate initial verification of the test site, a thorough knowledge of phenomena and expected behavior, and consistent post-processing techniques with an adequate number of repeated measurements in each test condition.

3.4.3   Uncertainty of radiated emissions measurement

The expected uncertainty of radiated emission measurements accounting for site imperfections, antenna characteristics, connecting cable and measuring receiver (or spectrum analyzer) may be estimated using CISPR 16-4-2 [43], sec. 7.1.3, which lists the input quantities of estimation of uncertainty of radiated disturbance measurements in the frequency range 30 to 1000 MHz (see Table 3.4.1):

- receiver reading: mainly due to measurement instability and scale interpolation errors, but also to the combined effect of gain and compensation of internal blocks; for the former the random nature in principle allows to achieve lower uncertainty by averaging successive readings, whereas the latter represents a residual systematic effect (see sec. 3.1.8.4 and [99] for practical examples);

- attenuation of the connection between antenna and receiver: for cable parameters the standard offers the possibility of using manufacturer's data (in this case taking a rectangular distribution with amplitude equal to the declared tolerance); if the cable, possibly including attenuators, is calibrated with a statement of uncertainty, a normal distribution may be assumed; measurement of return and insertion loss with VNA gives the lowest uncertainty, if compared to an amplitude-only measurement using spectrum analyzer (see sec. 1.4.2);

- antenna factor: the variability of antenna factor for ground effects was considered in sec. 3.2.2.2, showing that it may represent a relevant term of uncertainty for biconical antennas, depending on height above ground; conversely, the problem of direction of arrival or, in other words, the antenna factor for oblique illumination, is more relevant for directive antennas, such as logperiodic, and is evaluated in sec. 3.2.5.2;

- receiver related input quantities:

  - *receiver sine-wave voltage accuracy*: normally declared hastily as complying to the CISPR 16-1-1 requirement of $\pm 2\,$dB to apply as a tolerance, assuming rectangular distribution; for more accurate and better supported statements, a Gaussian distribution may be assumed;

  - *receiver pulse amplitude response*: normally compliance to CISPR 16-1-1 is declared for the various detectors, within the required tolerance of $\pm 1.5\,$dB, leading again to assume a rectangular distribution; whereas for electric arc emissions it is a relevant parameter, for the other slower transients of rolling stock emissions it is expected to be of scarce application; it may be in any case estimated using more accurate time domain setups;

  - *receiver pulse response variation with repetition frequency*: in many cases for spectrum analyzers this is not a well documented parameter, sometimes declared with the same accuracy of pulse amplitude response above; but again, this is a parameter that is more relevant for assessment of compliance of equipment and products;

  - *receiver noise floor*: it is normally well below the limits once correction factors of the measurement chain are applied; however, if for some measurements the antenna factor is large and also the adopted resolution bandwidth is large, a reduced dynamic range with an exaggeratedly high displayed noise floor may be experienced (we have seen that different choices of resolution bandwidth and detector affect the amount of displayed noise);

- *mismatch between antenna port and receiver*: as seen for biconical antenna in sec. 3.2.2.2 and 3.3.3.2, the input impedance may largely vary with frequency especially before the first resonance; CISPR 16-4-2, sec. A.2, comment A7, declares a VSWR of 2:1 as acceptably large, whereas it is known that values as large as 5 or 6:1 may be easily encountered, in the first frequency points around 30 MHz; using a matching attenuator greatly improves the impedance matching at antenna connector (see sec. 1.5.3);

- *antenna factor frequency interpolation*: the problem of reproducibility of antenna factors using different calibration techniques was already considered; the error here is related to the interpolation of values reported in the antenna calibration certificate; sometimes certificates show only a curve and taking numeric values by hand is exposed to an even larger error; in general, interpolation may be more accurate if a `spline` or `pchip`[10] polynomial is used, rather than simple linear interpolation; CISPR 16-4-2 proposes a $\pm 0.3\,$dB rectangular distribution;

- *antenna factor variation with height*: this is detailed in CISPR 16-4-2, sec. D.3, comment D2, focusing on height scanning (and not on the simpler method of fixed height used by EN 50121); considering the problem of coupling with the ground image, especially for biconical antennas (see sec. 3.2.2.2), maximizing

---

[10] Matlab functions taken as example of high-order interpolation functions.

the height above ground, e.g. positioning at 3.5 m instead of the minimum of 2.5 m, largely reduces the problem, although it may be troublesome for antenna stability and the amount of the necessary fixing elements and standoffs;

- antenna directivity: this factor is not exactly the antenna directivity properly said, that is already included in the antenna factor, but the error related to the combination of direct and ground reflected rays, considered at sec. 3.2.5.2; it may be said that EN 50121 does not allow the tilting of the antenna suggested by CISPR 16-4-2 to partially compensate for the directivity difference, but at the same time the ground is not reflective as at an OATS or semi-anechoic chamber;

- antenna phase centre location: such correction is negligible for biconical antennas, whereas for logperiodic antennas a maximum change of $\pm 0.35/0.4$ m is expected for the position of the active element with respect to the physical center of the antenna; assuming a $1/r$ dependency of field intensity with distance, the CISPR 16-4-2 determines the applicable uncertainty factor;

- antenna cross-polarization response: negligible for the biconical antenna, it is taken as a worst-case value of ± 0.9 dB with rectangular distribution for the logperiodic, under the assumption of 20 dB cross-polar suppression and equal intensity for vectors in the horizontal and vertical polarization planes;

- antenna balance: applicable to biconical, it is an extremely variable factor that not only is influenced by the quality of the feeding balun, but also by the distance with respect to ground in vertical polarization, see sec. 3.3.3.2, and the coupling with connecting cable, when this is parallel to antenna elements, again particularly relevant when in vertical polarization, see sec. 3.2.5.1;

- normalized site attenuation of the test site: site imperfection for OATS is required by CISPR 16-1-4 [39] to comply with a maximum ± 4 dB difference between the measured and the theoretical site attenuation; in reality, CISPR 16-4-2 at comment D7 observes that since the uncertainty of the measured site attenuation is heavily affected by the uncertainty of the used antennas, and this is also in direct relationship with the different methods used to calibrate antennas, a real OATS is expected to behave better than the overall ± 4 dB limit; OATS for emissions measurement are controlled sites, designed and built for the purpose, whereas real railway sites featuring ballast, concrete structures and soil are quite far from such performance, and a larger variability may be expected, affecting reproducibility and to author's knowledge have never been evaluated before;

- distance between EUT and measurement antenna: when measuring at 10 m, errors in quantifying the distance from the track axis may be considered not relevant, being limited to maximum 0.1 m; what is creating the largest uncertainty is that the perimeter of the EUT recalled by CISPR 16-4-2 is to be taken on the vehicle enclosure and that the location of on-board sources may span the vehicle width;

- height of table supporting the EUT and effect of setup table material supporting the EUT (not applicable to our configurations);

- **effect of ambient noise at an OATS:** the determination of background noise is considered in the next chapters for each type of test, requiring in general a sufficient number of recordings to reach statistical consistency and to monitor the variability of external sources.

A few publications may be used as a useful reference for the uncertainty budget, beyond the good example in CISPR 16-4-2 [43]: the calculation in [79] is shown in Table 3.4.2 for completeness; similar values may be found in [96], where, however, the unfavorable VSWR of biconical antenna is pointed out, which, conversely, seems underestimated in [79] and Table 3.4.2.

For the other measurements covered in chapters 7 to 9, considerations similar to those regarding the EN 50121 tests may be made: calibration of the magnetometer or probe is only one of the factors, whereas the selection of measurement points, the identification of intense field gradient that increase the sensitivity to positioning errors and movements, and the identification of external sources and other influencing factors are much more relevant affecting both correctness and uncertainty of measurements.

Contributions to uncertainty	PDF	biconical [dB]	logperiodic [dB]
Receiving part			
Antenna-factor (AF)	norm. $(k=2)$	± 2.0	± 2.0
Cable-loss measurement	norm. $(k=2)$	± 0.2	± 0.2
Antenna corrections:			
- AF variation with height	rect.	± 1.0	± 0.3
- AF frequency interp.	rect.	± 0.3	± 0.3
- directivity difference (3, 10 and 30 m)	rect.	± 0.0 (H); ± 0.5, ± 0.25, ± 0.1 (V)	± 1.0, ± 0.2, ± 0.1 (H,V)
- phase center variation (3, 10 and 30 m)	rect.	± 0.0	± 1.0, ± 0.3, ± 0.1 (H,V)
- cross-polarization	rect.	± 0.0	± 0.9
- balance	rect.	± 0.3 (H), ± 0.9 (V)	± 0.0
Mismatch antenna-receiver: $|\Gamma_{RCV}| = 0.333$	U-shape	$+0.9/-1.0$	$+0.9/-1.0$
Receiver specification:			
- voltage accuracy	norm. $(k=2)$	± 1.0	± 1.0
- absolute pulse response	rect.	± 1.5	± 1.5
- pulse-repetition response	rect.	± 1.5	± 1.5
- noise floor proximity	rect.	$+0.5/0.0$	$+1.1/0.0$
Open area test site			
Site imperfection	triang.	± 4.0	± 4.0
Separation dist. (3, 10 and 30 m)	rect.	± 0.3, ± 0.1, ± 0.0	± 0.3, ± 0.1, ± 0.0
Total expanded unc. $(k=2)$		± 4.9 (H) ± 5.0 (V)	± 5.2, ± 4.9 (H,V)

Table 3.4.1 – Typical uncertainty in RE measurements for Horizontal and Vertical polarizations and 3, 10, 30 m of measurement distance (values separated by comma for distance when different, only one given if distance is non-influential; indicated by "H" or "V" for polarization), as reported in CISPR 16-4-2 [43].

Contributions to uncertainty	PDF	biconical [dB]	logperiodic [dB]
Receiving part			
Antenna-factor (AF) calibration	norm. ($k=2$)	± 2.5	± 2.5
Cable-loss measurement	norm. ($k=2$)	± 0.3	± 0.3
Antenna corrections:			
- AF variation with height	rect.	± 1.5	± 0.5
- AF frequency interp.	norm. ($k=2$)	± 0.3	± 0.3
- directivity difference	rect.	$+0.2/-0.0$	$+0.5/-0.0$
- phase center variation	rect.	0.0	0.2
Mismatch antenna-receiver:	U-shape	$+1.7/-2.1$	$+0.9/-1.0$
$\|\Gamma_{RCV}\|=0.333$		$\|\Gamma_{ANT}\|=0.643$	$\|\Gamma_{ANT}\|=0.333$
Receiver specification:			
- voltage accuracy	rect.	± 1.0	± 1.0
- absolute pulse response	rect.	± 0.0	± 0.0
- pulse-repetition response	rect.	± 0.0	± 0.0
Open area test site			
Site imperfection	triang.	$+3.2/-3.0$	$+3.2/-3.0$
Overall system repeatability	Type A, std	0.5	0.5
Total expanded uncertainty ($k=2$)		± 4.9	± 4.0

Table 3.4.2 – Uncertainty in RE measurements at the distance of 10 m for a specific OATS among the 12 sites [79]; it is observed that some terms that appear in the CISPR 16-4-2 uncertainty budget (see Table 3.4.1) were not included.

Calibration is a relevant part of the budget when performing measurements at laboratories, where the other remaining factors are well evaluated and controlled. For on-site measurements verification procedures should always be used, performing for example simple measurements and checks whose verification shall be clear and indisputable, covering possible installation mistakes and wrong settings. The other mentioned factors affecting repeatability and reproducibility may be controlled by statistical post-processing, critical analysis of abnormal variations, and experience, conveyed into detailed and really applicable procedures and checklists.

Part II

Electromagnetic field emissions

4

Electromagnetic radiated emissions from rolling stock

4.1 Problem description

The subject of the test may be a locomotive or a vehicle or an entire train consist; in the latter case, of course, the test configuration depends on the type of cars (or coaches), if they are motorized or trailed, if they contain relevant electrical/electronic equipment (such as an air conditioning converter), if compliance by similarity to other coaches or units may be invoked.

The vehicle cannot be considered a point source as long as the wavelength and the measuring distance are both comparable with its dimensions [96]: even considering only emissions originating from the vehicle body and neglecting the contribution of the pantograph and catenary, electromagnetic field terms caused by different equipment and circuits overlap with significant phase displacement at the highest frequency. It is quite possible that while the vehicle is passing in front of the antenna significant variations of the measured intensity are observed for both time and frequency axes if a conveniently fast scan and sampling rate are used.

As long as the vehicle moves, the distance between the hypothetical source and the measuring antenna changes: when the vehicle approaches the antenna, the distance

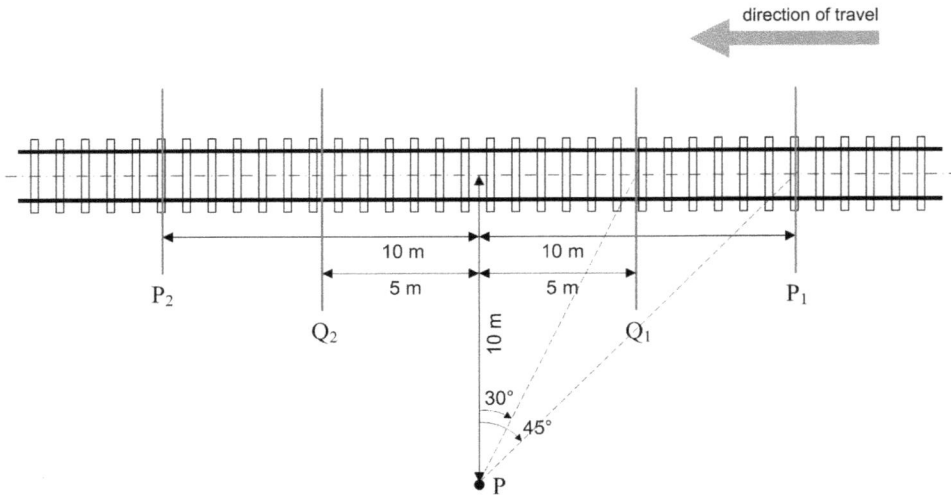

Figure 4.1.1 – Scheme of the measuring area with vehicle passage points.

reduces to the minimum distance r^* that is the distance between the antenna and the track axis, prescribed by the EN 50121-3-1 standard [28]. The reduction of antenna-vehicle distance causes a slow amplitude modulation of the measured components: when the spectrum is captured over a time interval that is relatively long (as for a complete scan using a frequency-domain equipment), this phenomenon is particularly evident. Taking the position in front of the antenna as the reference position, the electric field under far-field assumption reduces with the inverse of the distance r. Fixing an acceptable reduction of 1 or 3 dB (that with acceptable approximations correspond to 10 and 30%), the corresponding distance may be easily determined:

$$E(r) = \frac{E_1}{r} \qquad E^* = E(r^*) \tag{4.1.1}$$

$$\frac{\delta E}{E} = \frac{E(r^*) - E(r)}{E(r^*)} = \frac{\frac{1}{r^*} - \frac{1}{r}}{\frac{1}{r^*}} = 1 - \frac{r^*}{r} \tag{4.1.2}$$

Thus the reduction of electric field intensity is directly proportional to distance: considering the criteria of -1 dB and -3 dB maximum reduction, the distance increase shall be limited to 12 and 41%, corresponding to angles of 30° and 45°, as shown in Figure 4.1.1. The so-determined intercepts define track points P_1, P_2 and Q_1, Q_2 that delimit the measuring area for train passage: the sweep is started when the reference point vehicle passes past point "1" and the sweeping parameters adjusted so that the sweep has finished before the vehicle leaves the measuring area after point "2".

As will be underlined later on, triggering the sweep to vehicle passage is prone to some approximation and as a consequence to a synchronization error: the reason is that the sweep over the frequency axis is combined with the time evolution of emissions as the vehicle changes its position (slight effect on line resonances) and its distance from the measuring antenna (effect on the received amplitude).

Repeated measurements in the same test and operating conditions are highly advisable to improve statistical consistency and to reduce the influence of various sources of errors, such as the synchronization error. Due to the concomitant influence of driving transients (due to the driving style considered for its impact on repeatability in sec. 5.4.4), subsequent sweeps are usually processed using max hold, so that the amplitude change due to offset between sweep instant, vehicle position and instantaneous operating condition is much reduced, picking up the maximum at each frequency bin, resulting from several slightly time and position shifted sweeps.

4.2 Reference standards and limits

Limits are defined in EN 50121-3-1 differently for the stationary test and the slow moving test: for the former rolling stock is classified as city (tram, trolley buses, etc.) or other rail vehicles, while for the latter distinction is made for the catenary voltage ($600-750\,\mathrm{Vdc}$ for tram and metro applications, $1.5-3\,\mathrm{kVdc}$ for metro and railway applications, $15-25\,\mathrm{kVac}$ for railway applications) and the traction system type, as it is for the EN 50121-2 and traction line emissions. It is remembered that stationary test limits are in Quasi-Peak and those for slow moving test are in Peak mode.

The limits shown in Figure 4.2.1 and 4.2.2 are those reported in EN 50121-3-1 [28], Figure 1 and 2, respectively. The limits of the new 2015 version of the standard are identical, except that the $[9-150\,\mathrm{kHz}]$ interval has been removed and left in Annex C [30] as informative.

Compliance to the limits is reached if all the recorded spectra have values that after post-processing and application of any correction factor are below the limits. The EN 50121-3-1 standard mentions a $6\,\mathrm{dB}$ criterion for the margin with respect to limits when evaluating background noise, but does not give any tolerance or safety margin criterion, nor a statistical approach based on uncertainty, for the assessment of compliance of rolling stock emissions.

4.3 Measurement procedures and techniques

This section is very similar to the next corresponding section for traction line emissions measurements appearing in Chapter 5.

4.3.1 Preliminary checks and conditions

Besides the usual care in taking RF measurements regarding connectors, cables, earthing of equipment, verification of undesirable antenna coupling, etc., the applicable standard EN 50121-3-1 [28, 30] prescribes additional conditions to be checked for environment and line.

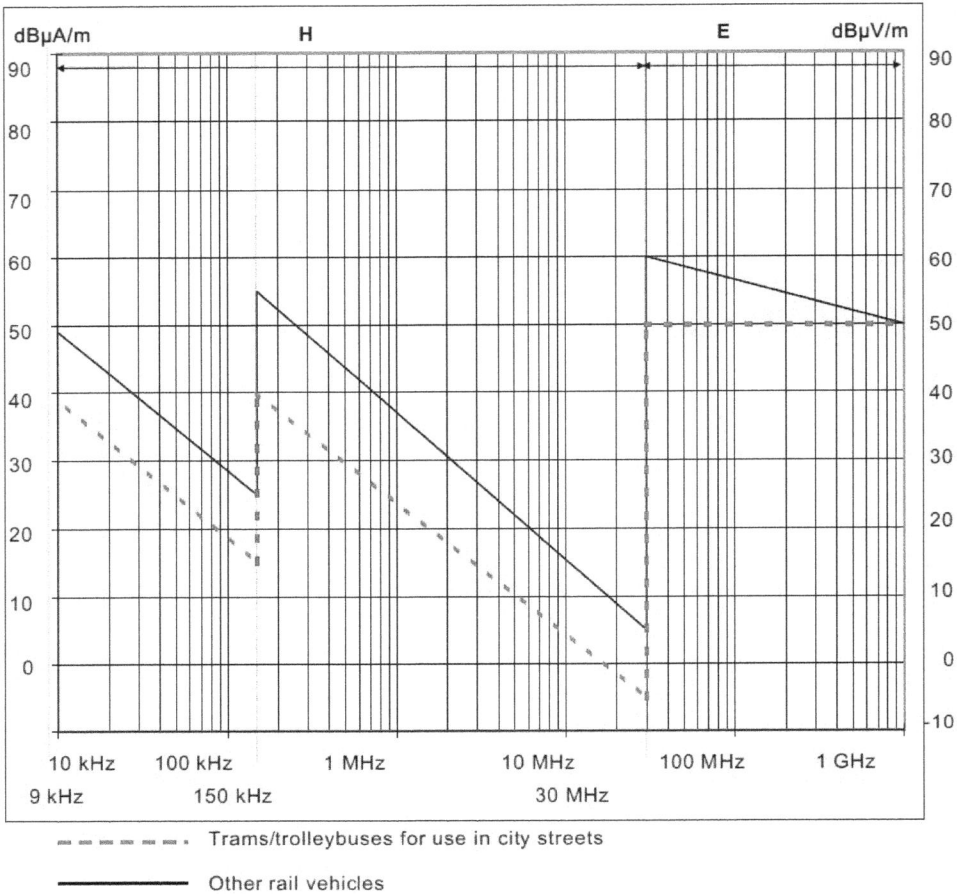

Figure 4.2.1 – Limits for rolling stock emissions as per EN 50121-3-1 [28], Stationary condition, Quasi-Peak mode, 10 m distance.

4.3.1.1   Weather conditions

EN 50121-3-1 standard does not specify weather conditions for the measurements, so that reference is made to what the EN 50121-2 requires at sec. 5.4.1 of 2006 version and A.1.9 of 2015 version.

To minimize the possible effect of weather on measured values, measurements should be carried out in dry conditions (after 24 hours during which not more than 0.1 mm rain has fallen), with a temperature of at least 5 °C, and a wind velocity of less than 10 m/s. Humidity should be low enough to prevent condensation on power supply conductors.

Since it is necessary to plan tests beforehand, they may be made in weather conditions, which do not meet the target conditions. In these circumstances, the actual weather conditions shall be recorded, unless they are evidently biasing the results for which tests shouldn't be performed.

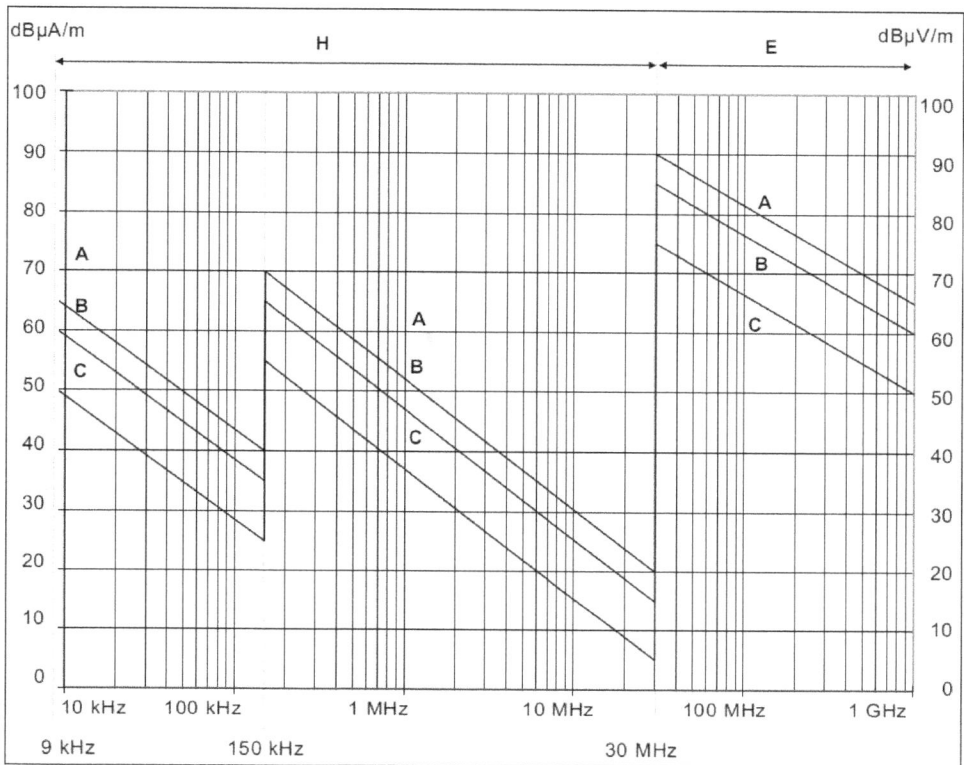

Figure 4.2.2 – Limits for rolling stock emissions as per EN 50121-3-1 [28], Dynamic condition, Peak mode, 10 m distance.

The requirement on humidity and the quantity of water on parts and soil may be explained not only by the condensation on the catenary reported at sec. 5.4.1 of EN 50121-2, but also considering the effect on field propagation especially at high frequency: soil conductivity has impact on field attenuation and distribution between horizontally and vertically polarized components; also the wet vehicle body, especially when made of non-conductive material, may slightly attenuate emissions. However, this is generally true for electric field at very high frequency, so that the entire magnetic field measurement interval up to 30 MHz is not appreciably influenced, as well as the first portion of the electric field interval above 30 MHz, and it is exactly in these intervals that significant and recognizable emissions from rolling stock are expected.

Temperature requirement corresponds more or less to an acceptability condition for personnel and equipment: in several countries several months of the year are characterized by lower temperatures, nonetheless affecting seriously measurement results, provided that equipment is stored in a warmer place, it is switched on well before the measurement and it is protected from cold exploiting self heating.

Wind speed requirement is usually met being a $10 \, \text{m/s}$ (or $36 \, \text{km/h}$) speed already relevant for antennas: loading and securing the tripod with sand bags is always advisable, but may be a valid countermeasure only for reasonably fast wind; by experience wind above $60 \, \text{km/h}$ is very hard to manage.

4.3.1.2   Traffic, line conditions and extraneous sources

Traffic and line conditions shall be annotated, with care for "physically-remote but electrically-near" trains or vehicles (EN 50121-2 [27], sec. 5.4.3.); in the new EN 50121-3-1 (2015) [30] this condition is transformed into the statement that vehicles in the same supply section may affect the measurement results.

When measuring emissions it is important to identify contributions of other parts of the railway system (e.g. substations and signalling) and of the external environment (e.g. power lines, industrial sites and apparatus, radio and television transmitters): whereas high-frequency RF sources may be identified as occurring in known bands, HF transmitters, mains signalling and similar phenomena give place to emissions that may be easily confused with emissions of rolling stock converters.

4.3.1.3   Measurement position with respect to infrastructure

The measurement location shall be positioned midway suspension masts (or any similar suspension mechanism, that can influence electromagnetic emissions) and not at a discontinuity of the supply conductor (this causes transients due to the interaction with the current collection mechanism), as specified in EN 50121-2 (2007), sec. 5.1.9, and EN 50121-2 (2015), end of sec. 5.1.2. In case of third rail systems the standard recommends a distance of 100 m from "gaps in the rail, to avoid inclusion of the transient fields associated with the make and break of collector contact". EN 50121-3-1, sec. 6.3.1, specifies consistently but with different wording, that is requiring a continuity of at least 200 m.

Even if not explicitly accounted for in the EN 50121-3-1, power lines may have a negative impact on the measurements: "if overhead power lines are nearby, other than those which are part of the railway network, they should be no closer than 100 m to the test site" (EN 50121-2 (2007), sec. 5.1.10, and EN 50121-2 (2015), end of sec. 5.1.2.).

4.3.2   Test operating conditions

The purpose of the test is evaluating rolling stock emissions that originate mostly in the pantograph/catenary system, rather than the vehicle body, the standard underlines. A certain degree of similarity with the traction line emission test (as per EN 50121-2 [27, 29]) is expected: the test shall be based on vehicle (or train) runs in specific operating conditions, such as stationary, accelerating, braking and the measurements shall be synchronized with vehicle operation, with frequency scans triggered with respect to the passage in front of the antenna.

The list of test phases is:

1. *Stationary test*, during which the auxiliary converters shall operate and the traction converters shall be under voltage, but not operating; the loading or voltage condition that produces the maximum of emissions is not necessarily that of maximum load and maximum voltage, so that some preliminary analysis is necessary and an explanation will be added to the test report. Stationary conditions are useful to discriminate between the various sources of emissions on-board that may be switched on and off with the vehicle standing still in front of the antenna and allowing some flexibility in scanning, detection and processing, although diversified stationary conditions and on-off sequences are not required for testing. For line emissions measurement stationary tests are optional and may be used for comparison with results of dynamic tests: to this aim they shall be carried out in the same measurement conditions, that is in particular using a Peak detector, rather than the Quasi-Peak detector required for rolling stock testing.

2. The *dynamic test* condition is called *"slow moving test"* (and there is no other faster moving test) with a recommended speed range of (20 ± 5) km/h for urban vehicles and (50 ± 10) km/h for main line vehicles, as stated in the EN 50121-3-1 [28], sec. 6.3.2 (the speed shall be low enough to avoid arcing at or bouncing of the sliding contact and high enough to allow for electric braking):

   (a) the standard reads "when passing the antenna the vehicle shall *accelerate* at 1/3 of its maximum tractive effort within the given speed range"; there are a few details to clarify in order to carry out the test satisfactorily:

      i. the maximum tractive effort shall be determined using the vehicle characteristics, but keeping in mind that enough ballasting is required to avoid too a fast acceleration to the maximum allowed speed for the test site (thus maximum speed is 25 km/h and 60 km/h);

      ii. the sentence in the standard seems to suggest that the vehicle shall accelerate when passing in front of the antenna, but two elements shall be remarked and corrected: first, the vehicle is not a zero-dimension object, so that a reference point shall be considered and this is often the center of the vehicle where the pantograph is located; in this case the driver located at the front of the vehicle shall be signaled in some way (e.g. a light, a person standing and waving a flag, etc.) when the maneuver shall begin; second, based on the reasoning regarding the traveled distance in front of the antenna and the tolerable change of measuring distance (see sec. 4.1), sweep triggering will occur at a convenient distance before the antenna location (see Figure 4.1.1);

   (b) the standard reads "when passing the antenna the vehicle shall *decelerate* at 1/3 of its maximum tractive effort"; the considerations for acceleration are equally applicable to the deceleration test;

   (c) in the case of vehicles with only mechanical braking, or with braking circuit that uses the same "circuit" (complex of converters, other equipment an connections) as in acceleration, both tests may be replaced by a stationary test where the vehicle accelerates at 1/3 of tractive effort against mechanical brakes (provided that this kind of test is possible at zero speed and it is not impeded by e.g. safety circuits and alarms);

3. Background noise measurement is needed to identify external sources (EN 50121-2, sec. 5.1.12), but no instructions are given to determine suitable margins with respect to the limits, as indicated by CISPR 16 standards. The aim is recognizing external sources and, in particular, to check at which frequency the background noise level is too large, above a margin of 6 dB from the limits, so to exclude such frequencies from the evaluation: it is thus apparent that for a meaningful comparison the same measurement method used for emissions shall be used; however, both Peak and Quasi-Peak measurements are required and limit specified correspondingly, so that both conditions in stationary and dynamic test mode can be verified. Needless to say that, anticipating the stationary test in QP mode and observing that the limit is not violated exempts from the execution of background noise QP sweeps, with a significant time saving; conversely, since dynamic tests measured in Peak mode are more likely close to the limit curve, it is worth and advisable to perform background noise measurements at least in Peak mode: several repeated recordings may be passed to post-processing, or "max hold" may be set directly on the spectrum analyzer or receiver; several separate sweeps allow statistical evaluation of background noise that a straight "max hold" reading cannot.

4.3.3 Antenna orientation and position

Antenna orientation is such to capture relevant emissions from the system, that the standards in a simplification assume mostly from the catenary system and in any case lying in the plane that contain the line cross section, orthogonal to the direction of travel.

For clarity, the used antennas are loop antennas up to 30 MHz and biconical, log-periodic antennas or a combination for the 30 – 1000 MHz frequency range.

Positioning with respect to the track along which rolling stock is running is strictly regulated by the EN 50121-2 standard:

- the antenna distance from the axis of the closest track is 10 m; even if not explicitly said for rolling stock testing in EN 50121-3-1, for different distances the corrective coefficient given in sec. 5.1.6. of the EN 50121-2 shall be applied, being the electromagnetic phenomena of the same nature and thus keeping valid the rationale to determine the said corrective coefficient; the use of different distances is permitted only if strictly necessary and shall be justified;

- the heights are measured with respect to horizontal plane passing through the rails head, that is the "top of rail"; in special cases the height may be measured from the ground level, see EN 50121-2 (2007), sec. 5.1.8;

- the orientation of the antennas shall be:

  - loop antenna shall be oriented to capture the horizontal polarization of the magnetic field only (i.e. the loop antenna stands vertically, with the plane containing the loop perpendicular to the ground and parallel to the line of the track;

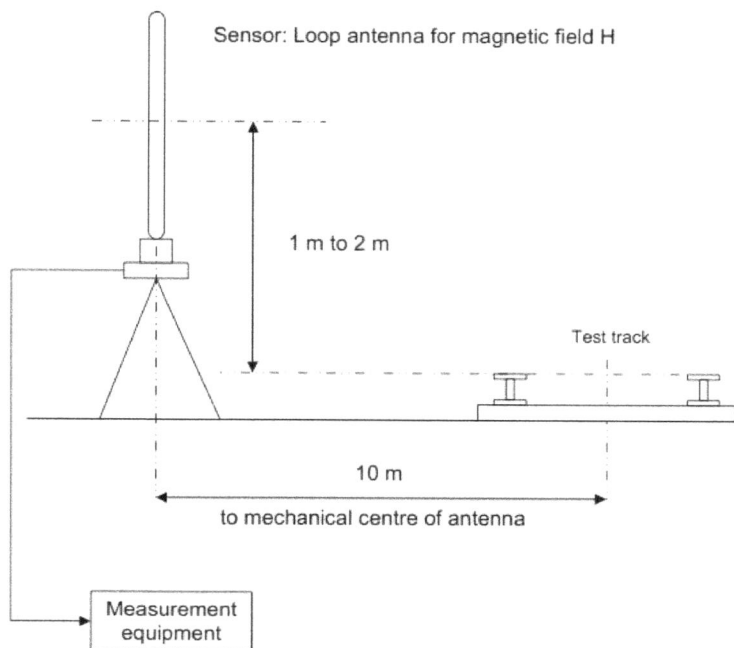

Figure 4.3.1 – Position and distance of loop antenna for measurement of radiated emissions as per EN 50121-3-1 (identical to EN 50121-2).

Interval	A	B	C	D
Frequency range	$9 - 150\,\text{kHz}$	$0.15 - 30\,\text{MHz}$	$30 - 200\,\text{MHz}$	$30 - 200\,\text{MHz}$
Resolution Bandwidth	$200\,\text{Hz}$	$9\,\text{kHz}$	$120\,\text{kHz}$	$120\,\text{kHz}$

Table 4.3.1 – Resolution bandwidth (RBW) prescribed by EN 50121-2 as per CISPR specifications related to Quasi-Peak detector and by extension Peak detector.

- the biconical antenna, when measuring the vertical component of the electric field, is positioned perpendicular to the ground; when measuring the horizontal component it is turned by 90° and its axis parallel to the line of the track;
- the log-periodic antenna aims always at the track, with its dipole elements perpendicular or parallel to the ground to measure the vertical or the horizontal components of the electric field.

4.3.4 Measurement settings

The frequency range is subdivided into frequency intervals for which different Resolution Bandwidths (RBW) are to be used, as for intervals A, B, C and D, as shown in Table 4.3.1.

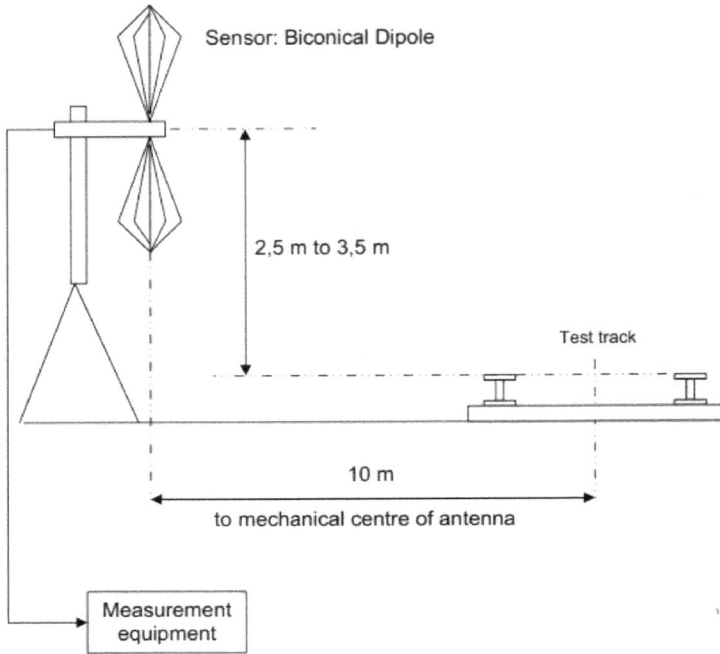

Figure 4.3.2 – Positioning and distance of biconical antenna for measurement of radiated emissions as per EN 50121-3-1 (identical to EN 50121-2).

Sub-intervals are then recommended by EN 50121-3-1 and are reported later in Table 4.4.1, when considering sweep time specifications: sub-intervals are partially over-lapped in order to reconstruct the whole spectrum of emissions and are sufficiently reduced to allow the measurement of a moving source; smaller sub-intervals are not forbidden and allow a shorter measurement area.

Recognizing a moving source, measurements are all taken in Peak mode: they can be performed by either sweep mode (advisable) or by selecting a set of test frequencies, to speed up the measurement, but with a minimum number of three frequencies per decade, as specified in EN 50121-2 [27], sec. B.11.

Slightly different RBW values may be used, as available on the instrumentation and to increase measurement speed (sweep time reduction). Normally, when this is done, during post-processing correction is applied to the measured values for the ratio of the chosen RBW with respect to those prescribed in Table 4.3.1. This correction is not always justified and depends on the nature of measured data; to avoid improper data manipulation, RBW values shall be always the nearest to the prescribed ones (for most spectrum analyzers $300\,\mathrm{Hz}$ for $200\,\mathrm{Hz}$, $10\,\mathrm{kHz}$ for $9\,\mathrm{kHz}$ and $100\,\mathrm{kHz}$ for $120\,\mathrm{kHz}$), thus keeping the correction factor small. Under broadband input signal assumption (e.g. white noise), the correction factor is the ratio of bandwidth for power measurements and its square root for amplitude measurements: $k_P = 10\ \log_{10}(\mathrm{RBW}_1/\mathrm{RBW}_2)$ for power, $k_A = 20\ \log_{10}(\mathrm{RBW}_1/\mathrm{RBW}_2)$ for amplitude, where RBW_1 and RBW_2 stand e.g. for prescribed and available resolution bandwidth values and the resulting correction factor shall be summed to the reading performed with the available RBW value.

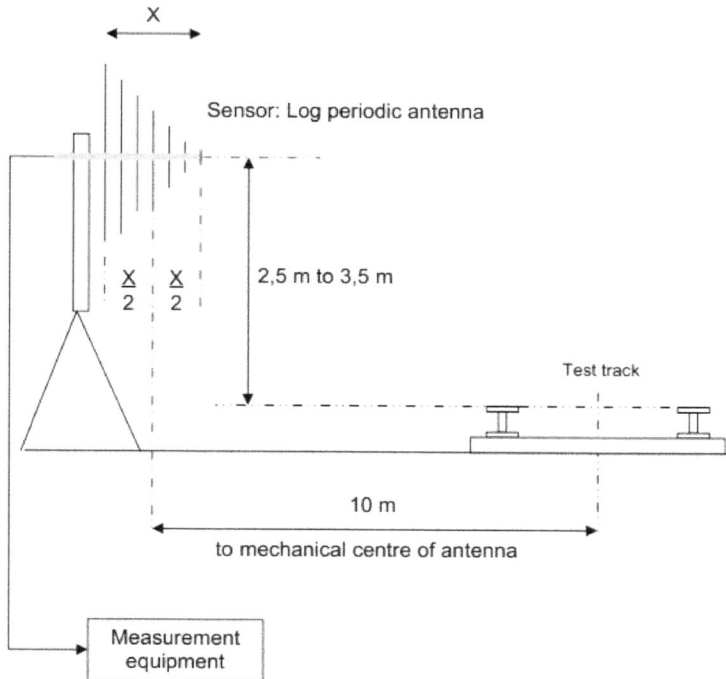

Figure 4.3.3 – Positioning and distance of log-periodic antenna for measurement of radiated emissions as per EN 50121-3-1 (identical to EN 50121-2).

Sweep time shall be fast enough to capture transient emissions, but slow enough not to worsen measurement accuracy. Spectrum Analyzers normally set automatically a sweep time value barely sufficient for an acceptable accuracy (setting of sweep time to "auto"): the EN 50121-2 recommendation at sec. 5.2.2 gives the minimum values for the present application; a longer value is strongly advised to improve accuracy, of course reaching a trade-off. In CISPR 16 standards and [99] general values are given, establishing a relationship between RBW, VBW and sweep time.

Because of the smoothing effect on the measured traces and the slight adverse effect on sweep time, video bandwidth shall be excluded: to this aim, similarly to other EMC standards, a setting in 3:1 proportion with RBW is sufficient.

4.3.5 Measurement execution

Recalling the vehicle operating conditions considered in sec. 4.3.2, measurements will be performed in stationary, acceleration and braking modes. The exact sequence is not relevant. Measurements will be preceded, and possibly also followed, by background noise measurements.

Train operating conditions, as also considered in sec. 5.3.2, are: BKG (for background noise), STA (for stationary), ACC (for acceleration) and BRK (for braking).

To validate the measurement results and the compliance with limits, background noise measurements are needed, so that external sources can be identified. To this aim background noise may be evaluated in several conditions (as anticipated in sec. 4.3.2 for stationary test):

- vehicle switched off and pantograph lowered (so taking into account of the effect of the vehicle as a conductive obstructing and reflecting object);

- the effect of raised pantograph, that defines a voltage constraint for the otherwise floating catenary (in the case of third-rail supply, this condition is much less relevant);

- on-board auxiliaries switched on with lowered pantograph, accounting for direct emissions from vehicle body.

Several recordings are needed in order to reach statistical significance and a complete picture of external sources and the amount of emissions caused by the various configurations and auxiliaries contribution. How to reach the statistical significance depends on the confidence level and on the statistical behavior of the various sources: white noise can be easily managed based on sec. 2.6.2 and 3.1.8; often external sources are radio transmissions and narrowband emissions from other electrical or electronic systems, with a pattern of variable operating conditions or featuring on and off periods. In general the quantification for an accurate correction of amplitude reading (see sec. 3.1.8) is almost impossible and the background noise measurement aims at excluding specific frequency intervals from successive emissions evaluation. All these measurements may be identified as BKG and, as said, may be differently post-processed and evaluated, depending on the observed behavior of spectral components. Recordings of background noise shall be made prior to the beginning and at the end of the tests.

Since line emissions are mostly caused by transient phenomena, especially during acceleration and braking of trains, a minimum number of measurements shall be performed and recorded for each condition and each frequency range and antenna polarization. The minimum number of recordings for each combination of these conditions may be in general set to five, unless experimental evidence demonstrates that the measured levels are much lower than the limits and thus able to cope with a wider confidence interval due to the larger sample dispersion.

Looking at Figure 4.1.1, the scan shall begin when the train enters the measurement position "1" (either P_1 or Q_1) and finishes before the train leaves position "2" (either P_2 or Q_2). Usually some preliminary scans are necessary to determine their time duration with respect to train speed and the time interval between the passage in front of the two reference points "1" and "2". The error due to synchronization of train passage and sweep start is further considered at the end of the next chapter in sec. 5.4.4.

To summarize the train operating conditions are $N_{t.o.c.} = 3$ (STA for stationary, ACC for acceleration and BRK for braking), there are $N_{f.i.} = 4$ frequency intervals (unless they are split into sub-intervals as it is always the case for dynamic tests) and a total of $N_{ant.} = 5$ antenna orientations (Loop, Biconical Vertical, Biconical Horizontal, Log-Periodic Vertical, Log-Periodic Horizontal). When using split frequency intervals to capture train passage in front of the antenna, $N_{f.i.} = 4$ increases significantly: interval

B is usually split into $5-6\,\mathrm{MHz}$ sub-intervals for the most common speeds and acceleration/braking efforts; similarly interval D needs to be split into two sub-intervals. It is remembered that interval splitting is done when in dynamic conditions to cope with the varying train position while sweeping (see sec. 4.4.3).

Because of the peculiar nature of some emissions (in particular low-frequency magnetic emissions in Interval A, that propagate at a much longer distance due to the lower line attenuation), additional measurements may be performed when the train is arriving, but is still before the first landmark of the measurement area and when the train has left the measurement area and is beyond the second landmark. Practically speaking, the span to cover may be as extended up to some hundreds meters centered around the projection of the measurement point P. If this distance were determined based on what is observed in sec. A.8 of the standard regarding repeatability (the chance that remote vehicles will produce significant emissions at the test point is considered relevant for other vehicles up to 20 km for catenary systems and 2 km for third-rail systems), then the area would be stretched unacceptably far. This measurement condition is an additional one and requires that sample readings are made when the vehicle is approaching or leaving the measurement area: of course the operating condition will be that of nearly constant speed in a crossed combination, that is before the measurement area when going to brake, and after the measurement area when going to accelerate. A few samples may be recorded to identify if resonance conditions occur in the catenary system: measurements are performed with the loop antenna over the first two intervals A and B (covering interval B only partially up to about 1 MHz).

4.4   Further considerations

4.4.1   Definition of the measurement area in front of the antenna

Emissions are caused by the current components in the catenary system in front of the measurement point: the measured intensity of the direct emissions is a function of the distance r; to avoid a change of the measured intensity while the vehicle is moving and the distance r between it and the antenna is changing, a maximum span distance shall be identified. Looking at Figure 4.1.1, for vehicle positions both at the left and at the right of the antenna projection P onto the track, the antenna-vehicle distance is increasing and the measured intensity is correspondingly reducing: under the assumption of a $1/r$ dependency of electric field components, the defined positions at the left and right for $1\,\mathrm{dB}$ and $3\,\mathrm{dB}$ tolerance criteria are those with a multiplying factor of the orthogonal measurement distance of 1.12 and 1.41, for $10\,\mathrm{m}$ and $20\,\mathrm{m}$ wide measuring area, respectively.

As said, for testing rolling stock emissions, the scan shall start when the train enters position "1" (either P_1, P_2 and Q_1, Q_2 depending on the adopted criterion) and finish before the train leaves and clears the track at position "2" (vehicle movement is set from right to left).

4.4.2   Intermittent emissions from a moving source

Test conditions and methods were conceived in the EN 50121-3-1 (and EN 50121-2) standards taking into account two major aspects: electromagnetic emissions are transient and intermittent with the vehicle representing a moving source; the catenary system is responsible for the largest part of emissions. Both assumptions are now reviewed and discussed.

Emissions from on-board converters are repetitive and follow a switching pattern, so that they can be adequately measured with a long enough dwell time and the correct detector (as for the so-called "clicks"); of course this is not in agreement with the exigencies imposed by a moving source.

The standard is based on the assumption that the source of emissions is the catenary, defining thus two important mechanisms: the magnetic field polarization is mainly vertical and the catenary current and related emission precede the vehicle.

The former is true for the magnetic field vector, whereas for the electric field at higher frequency the polarization is both horizontal and vertical, depending on modes: the vertical polarization is typical of many electric field emissions, vertical with respect to the soil that is the potential reference; the horizontal polarization, parallel to the soil, is justified considering that the electric field is in relationship with longitudinal voltage drop along the catenary, especially when line losses are relevant.

The latter is a well-known phenomenon for which low frequency current components undergo much smaller attenuation and propagate at a much longer distance, preceding and following the vehicle passage in front of the antenna by several hundreds meters. However, direct emissions from the vehicle are also relevant, when the many on-board cables are considered, to which internal sources are directly connected: in this case emissions have different polarizations and may be considered in the reactive field region in the first meters of distance. This is not so relevant when measuring at 10 m distance, rather when space constraints come into play and the distance between the vehicle and the antenna is reduced (measuring distance is discussed in sec. 5.4.2).

Based on practical observations and by simple source models relevant emissions may be ascribed also to the pantograph and to the complex of on-board equipment, in particular converters, and their connections: the total ampere-turns of a filter inductor are much larger than the single turn of the catenary loop; modern EMUs and distributed traction trains, featuring also connection to different supply systems through different pantographs require a large number of cables, running both vertically (from pantograph to underfloor) and horizontally between units, and thus responsible of relevant emissions in both polarizations.

4.4.3   Sweep time

The EN 50121-3-1 (2006), Annex B, specified resolution bandwidth values as in Table 4.4.1 and focused on the first frequency interval A: "the measuring apparatus shall be in accordance with the EN 55016-1-1 requirements described in Subclause 4.2: "Peak measuring receivers for the frequency range 9 kHz to 1 GHz." However, for the 9 kHz to 150 kHz range (band A), the 200 Hz bandwidth may give the following

Interval	Subrange	Span	RBW	Sweep time
A1	$9-59\,\text{kHz}$	$50\,\text{kHz}$	$1\,\text{kHz}$	$300\,\text{ms}$
A2	$50-150\,\text{kHz}$	$150\,\text{kHz}$	$1\,\text{kHz}$	$300\,\text{ms}$
B1	$0.15-1.15\,\text{MHz}$	$1\,\text{MHz}$	$9/10\,\text{kHz}$	$37\,\text{ms}$
B2	$1-11\,\text{MHz}$	$10\,\text{MHz}$	$9/10\,\text{kHz}$	$370\,\text{ms}$
B3	$10-20\,\text{MHz}$	$10\,\text{MHz}$	$9/10\,\text{kHz}$	$370\,\text{ms}$
B4	$20-30\,\text{MHz}$	$10\,\text{MHz}$	$9/10\,\text{kHz}$	$370\,\text{ms}$
C	$30-230\,\text{MHz}$	$200\,\text{MHz}$	$100/120\,\text{kHz}$	$42\,\text{ms}$
D1	$200-500\,\text{MHz}$	$300\,\text{MHz}$	$100/120\,\text{kHz}$	$63\,\text{ms}$
D2	$500-1000\,\text{MHz}$	$500\,\text{MHz}$	$100/120\,\text{kHz}$	$100\,\text{ms}$

Table 4.4.1 – Frequency interval and sweep times indicated as guideline for test by EN 50121-3-1, Annex B.

problems: it is not always available in standard spectrum analyzers; the scan duration is excessive for moving sources; this would make it necessary to multiply the number of sub-ranges which is contrary to the objective of the method. For these reasons, the bandwidth for band A may be higher and 1 kHz is a convenient value. Proper corrections shall be carried out on the measurement results assuming that the noise is a broad band white noise." The 2015 version recalls the EN 55016-1-1 (2010), Clause 5, and presents directly the information of Table 4.4.1, having removed interval A and with no considerations on the 1 kHz RBW value.

Table 4.4.1 reports sweep times referring to a given span interval, without specifying the number of points, or in other words the frequency step. Thus a convenient way of expressing the sweep time for further comparison is first of all per unity of span: with analog spectrum analyzers specifications in MHz/ms were commonplace, as it can still be found in MIL STD 461 [110]. Then, assuming a frequency step slightly smaller than the RBW value (step $= 0.5\text{RBW}$ as advisable and required by some standards), it is possible to estimate the dwell time for each point and compare it to the time required for the IF filter with RBW bandwidth to settle having the transient response vanished). In B1 $\text{RBW} = 10\,\text{kHz}$, span $= 1\,\text{MHz}$ and thus the number of points with 50% step is $N = 200\,\text{pts}$; with a prescribed sweep time of $37\,\text{ms}$, the dwell time per point $\text{DT}/\text{pt} = 37/200\,\text{ms} = 185\,\mu\text{s}$. The settling time of the IF filter for a given accuracy, while neglecting the remaining transient response, can be quantified accurately only by the manufacturer; however, it is easy to show that the estimated $185\,\mu\text{s}$ dwell time covers some time constants ensuring a satisfactory reading of IF filter output.

4.4.4 Background noise

Regarding the evaluation of background noise the standard suggests (at its sec. 6.3.1) "to perform ambient noise measurement also with the vehicle completely powered down in front of the antenna." In reality manufacturers may have different approaches, especially when measurements are taken on new vehicles, for which more investigations are appropriate. Sec. 4.3.5 already considered from an operative viewpoint the various rolling stock configurations and conditions that progressively add contributions The effect of raised pantograph on the catenary system and preexisting disturbance may be considered, as well as the masking effect of the vehicle mass with respect to external ambient noise with directive characteristics (e.g. radio cellular and broadcasting sources). Then different on-board systems may be switched on and off, in order to assess individual contributions: dc bus voltage and battery chargers, air conditioning, compressors, on-board electronics, etc.. When switching on on-board auxiliaries on battery supply with lowered pantograph, it is possible to evaluate the emissions that are directly radiated by the vehicle body without contribution of the catenary (see sec. 4.4.2).

4.4.5   Repeatability

An overview of the most relevant repeatability issues for rolling stock and line emissions tests will be also given in sec. 5.4.4, focusing on line emissions tests.

When measuring vehicle emissions a short track section is used so that it is not expected that line resonances may vary significantly as the vehicle moves. Conversely, instrumental uncertainty, as for all electromagnetic field measurements, driving style and operating conditions of on-board apparatus represent the most relevant elements as for repeatability.

4.4.5.1   Instrumental uncertainty

Instrumental uncertainty is well documented and as commonplace for this kind of measurements supported by certificates and possibly a Type B approach to its evaluation for the entire measurement chain. For the commonly used equipment the combined uncertainty for antenna factor, distance error, cable compensation and spectrum analyzer amplitude reading may range between 4.5 and $5.5\,\mathrm{dB}$ for a confidence level of $95\,\%$, so with a coverage factor of $k = 2$ under Gaussian distribution assumption (see sec. 3.4).

What instrumental uncertainty fails to cover is the remaining indeterminacy related to environment and setup conditions (driving conditions, operating conditions, synchronization of sweeps, intermittent and variable external sources), for which a Type A approach is highly advisable.

4.4.5.2   Sweeping, driving style and synchronization

Driving style and synchronization between measurement area landmarks (P and Q points in Figure 4.1.1) and spectrum analyzer sweeps are probably the most relevant elements that determine the overall uncertainty and affect repeatability.

Emissions are intermittent and characterized by several transients (see also sec. 4.4.2): the driving style also has a significant impact on the spectrum shape of recorded emissions; lack of synchronization between frequency scanning and vehicle passage affects the location on the frequency axis of spectrum peaks captured while sweep and transient occur together; this gives the false impression of a source of emission or a line resonance with variable frequency and this is particularly evident for short braking transients creating humps in the scanned spectrum. Synchronization is achieved by the spectrum analyzer operator by starting sweeps with respect to vehicle passage at landmarks: different starting times of frequency sweeps may occur because of uncertainty in detecting vehicle position and pressing the corresponding button on the spectrum analyzer or controlling computer.

Whereas with a max hold interpretation of the collected results changing slightly the synchronization is beneficial for a complete picture of line emissions, retrieving at the end the envelope of all peaks, if averaging or other statistical estimators are used, such shifts of peaks have the dramatic impact of a parametric disturbance.

On the driver's side, similarly, the driver might get used to the requested maneuver and may anticipate the acceleration or braking with respect to the agreed landmark, or the applied effort may slightly change from the prescribed 1/3, or in some cases it may be modulated rather than steady (following quite a common driving attitude of applying acceleration and braking in split intervals, not in one single stretch).

Since such elements cannot be fully controlled, they shall be addressed:

- before the measurements, clarifying and agreeing how landmarks are approached, what is the intention of the test, how get visual feedback from the personnel standing wayside (radio communications to be kept to a minimum and only when the vehicle is far away and no measurements are performed);

- during the measurements, when the spectrum analyzer operator or the person coordinating the measurements and giving feedback during train passage may perceive deviations from the agreed test procedure and have also the confirmation of unusual variability of the acquired trace once the train has passed.

A certain amount of experience is thus always necessary to perform such measurements at best.

4.4.5.3   Operating conditions

For operating conditions, only the manufacturer knows if slight changes of tractive effort or speed may dramatically affect the operating mode of e.g. traction converters, or, similarly, if load cycles of auxiliaries are such that their contribution may be varying from sweep to sweep.

In this case it is the manufacturer's interest to investigate in the most complete way even unusual conditions of its rolling stock, in order to collect pictures and signatures if various operating modes that may sooner or later occur during site tests and other kind of tests during test and commissioning phases. In general, when such conditions and operating mode changes are not perfectly known and the possibility of controlling and influencing the is limited (for example, no diagnostic or low level access is possible to the converter modulators or the designer is not available for support), the variability that is observed is all included in the repeatability estimate using a Type A approach.

4.4.5.4   Examples

It is thus interesting to evaluate repeatability for real measurements analyzing the diuspersion and possibly the statistical distribution of readings at each frequency bin and deriving a statement of uncertainty for a 95 % confidence level, so to compare it with the expected instrumental uncertainty.

Figure 4.4.1 shows various measurement conditions:

- in Figure 4.4.1(a) various forms of background noise and stationary noise are shown where harmonic emissions due to auxiliaries are clearly visible in black, but also pre-existing components when auxiliaries are still switched off, as well as a line resonance around 47 kHz;

- dispersion of traces during acceleration and braking is shown in Figure 4.4.1(b), where the black traces of acceleration condition are quite consistent with maximum standard deviation σ of 7 dB at few low frequency points, but on average limited to about 3 dB; traces collected for braking conditions, on the contrary, are quite dispersed showing consistently about 10 dB of standard deviation;

- the above commented "line resonance" may be in reality due to the combination of line and on-board cables and reactive components, as it is demonstrated by Figure 4.4.1(c) where the same traction line used for the previous measurements was used with a different locomotive: the line resonance is shifted to about 70 kHz with a slightly larger factor of merit.

It is also observed that the resulting dispersion is not really as such, if traces are collected and then max hold profile is determined during post-processing: the dispersion in particular of traces during braking is the result of artificially altering synchronization between sweeps and train passage, purposely applied to satisfactorily cover the whole frequency range, in view of the application of the max hold function. Max hold profiles obtained by processing batches of a statistically significant number of traces (e.g. minimum five) are very similar and show a limited spread, usually in the range of $3 - 5$ dB.

It is quite commonly stated that locomotive emissions are relevant only in the low and medium frequency range, up to about a ten of MHz. Although the FM radio frequency interval is commonly recognized as an interval to be excluded from any evaluation, at higher frequency the presence of radio sources that cause the exclusion of the frequency interval depends on the specific country and things have changed with the passing years. Observing Figure 4.4.2, it appears that one of the black traces corresponding to a measurement of an acceleration passage is above the limits between 178 and 195 MHz: the trace superposed to the thick gray background noise has evidently a remarkable correlation, but this does not explain why the other traces in acceleration and braking conditions are much lower; additionally this acceleration trace is also showing some peculiar emissions at about 33, 38 and 43 MHz, demonstrating that there are vehicle emissions. The reason of such a weird behavior was the combination of three phenomena: background noise is directive because of near sources around the FM band and at about 176 MHz and above it; the vehicle in reality has some narrow-band emissions in the lowest frequency range, whose captured amplitude, however,

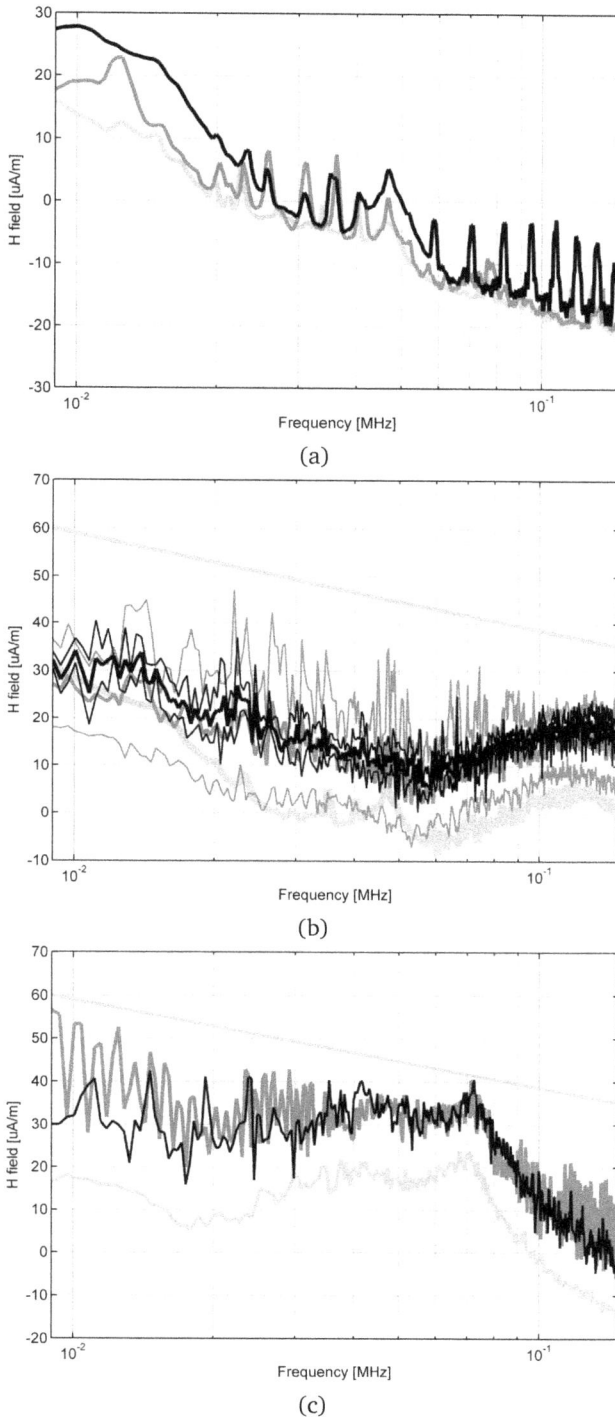

Figure 4.4.1 – Measured traces: (a) background noise with no vehicle in the measurement area (light gray), pantograph down and auxiliaries off (gray), auxiliaries on (black); (b) dispersion of braking (gray) and acceleration (black) traces, with $\pm 1\sigma$ boundaries around mean value (pantograph up and auxiliaries on, thick light gray); (c) resonance of catenary system due to on-board cables interaction; prevalence of braking emissions (gray). Limit curves are the straight gray lines.

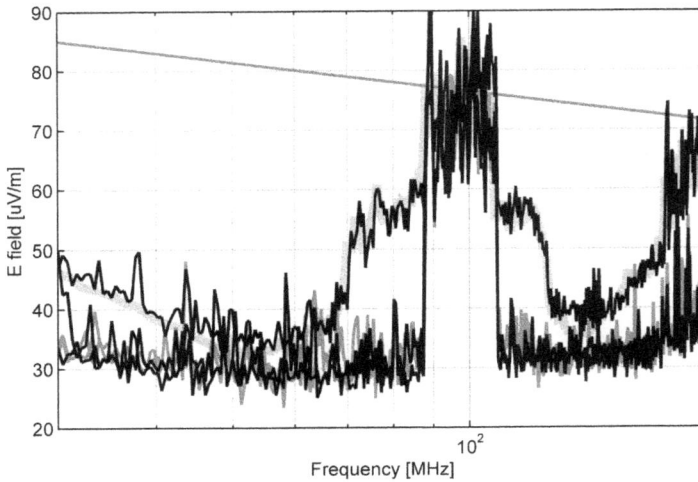

Figure 4.4.2 – Apparent emissions above limits (straight gray line) for acceleration.

highly depends on the synchronization with the sweep (also the other lower traces have such components); last, the troublesome trace was recorded in an attempt of enhancing the components at lower frequency, delaying the sweep, so that the vehicle was already clearing the measurement area while the sweep was passing over the remaining part of the frequency range, thus not shielding the antenna against the said directive sources. Of course, this sweep was a bad attempt to capture transient phenomena, because no attention was paid to the duration of the sweep with respect to the occupation of the measurement area: such measurement had to be split into two sub-intervals, maybe also skipping the FM band that is generally recognized as "useless".

4.4.6   Antenna position and maximum emissions

Regarding the positioning of the antenna the standard says "the antenna shall be in front of the middle of each vehicle unless an alternative location is expected to produce higher emission levels": understanding if there are positions with higher emissions is of course a very complex task, because a uniform behavior of constant increase or decrease for different positions is an oversimplifying assumption and measurements shall be taken for all positions over the entire frequency range. This is of course a relevant problem for measurements in stationary mode that may require to shift the antenna location parallel to the vehicle axis, towards the head or the tail. For dynamic tests, emissions are captured over the observation width while the spectrum analyzer (or receiver) is sweeping and the vehicle is moving in front of the antenna: slightly anticipating or delaying the instant of time at which the scan is started with respect to train position is beneficial to avoid ambiguous synchronizations of phenomena that make correspond some peaks of emission on the frequency axis to specific positions of the vehicle on the track; the frequency at which the vehicle passes in front of the antenna is correspondingly shifted.

4.4.7 Operating conditions and maximum emissions

The standard observes that for the stationary test with the auxiliary converters in operation, "it is not inevitably under maximum load conditions that the maximum emission level is produced": maximum of emissions depends on the emission mechanisms, distinguishing between emissions due to flowing current (usually maximized at large loads) and due to voltage gradients (usually higher at light or no load); a change of converter modulation scheme or triggering of internal resonances may have as well a relevant effect; the identification of the operating conditions is thus left to the manufacturer (see sec. 4.4.5.3).

In the most recent version EN 50121-3-1 (2015) [30], sec. 6.3.2, we read "the tests shall cover the operation of all systems on-board the rolling stock which may produce radiated emissions."

To this aim, avoiding the huge number of scans that an accurate analysis of emissions variability would require for all systems and equipment on-board including each various operating points, the simpler criterion of switching on sequentially the most significant loads while stationary and on battery supply, and then repeating the same with the raised pantograph is acceptable and gives quite a complete picture with the most significant electromagnetic signatures. For the most relevant sources changing the operating point, such as modulating the load in two or three steps between no load and full load may contribute even more information.

Regarding the operating conditions of traction converters (and braking converters), the standard itself fixes the prescribed loading at 1/3 of the maximum tractive effort (see point 2.(b) in sec. 4.3.2).

4.4.8 Diesel locos and coaches

Assuming that diesel electric locomotives are not relevant related to electromagnetic emissions for the absence of an electric traction line (e.g. catenary) is oversimplifying: tests are performed similarly to full electric units without the problem of catenary resonances and distance from suspension masts or line discontinuities. As prescribed by the standard, the applicable limits are those shown in Figure 4.2.1 and 4.2.2, using the "tram/trolleybus" and the "C" curves, respectively.

Trailed coaches may be considered passive from an EMC standpoint unless they are equipped with electrical and electronic units, that are EMC relevant, e.g. switching electric current, using microprocessors and digital systems, etc. In this case coaches are tested for stationary conditions only and, similarly, the "tram/trolleybus" curve of Figure 4.2.1 applies. Regarding coaches, when the standard says "test for identical coaches or wagons are performed only once", it may be interpreted as the confirmation that this is a type test and that similar units shall not be repeatedly tested, even if there are several in a train consist.

5

Electromagnetic radiated emissions from traction line

5.1 Problem description

The traction line, especially when of the overhead catenary type, is an aerial for all conducted disturbances caused by substations and rolling stock. At low frequency the behavior is that of a loop antenna where positive catenary current returns through the return circuit (e.g. rails and other conductors); the return circuit is imperfect, dispersive and leaky, because of rail non-linearity, additional hysteresis losses, and significant leakage to earth. This model is acceptable for the traction current components at supply frequency and harmonics up to the first resonance of the supply system, that is as said of the catenary or third rail type.

Catenary type systems have the largest loop inductance due to the distance of about $6\,m$ between the overhead contact wire and the running rails; loop capacitance is mainly due to the catenary to ground self capacitance (ranging between 10 to $20\,pF/m$ for single and double track systems), adding minor but significant contributions due to suspension mast and rolling stock.

Suspension masts introduce additional lumped capacitance, loading the catenary with a periodic pattern, being masts separated by about $40-60\,m$. Vehicles and trains capacitively load the catenary by means of their metallic roof, much closer to the catenary than track and soil, so that an increase of p.u.l. line capacitance by about a factor of five in the line section occupied by the train may be expected.

Significant loading is due to substations and traction supply apparatus, so that depending on the type of supply (dc or ac) and the voltage level the first resonance frequency of the traction supply circuit may vary between 10 to 20 kHz. This resonance is that of the entire traction supply circuit and influences how conducted disturbance enters the overhead line through the pantograph. If the measured magnetic field intensity is considered, the resonance is a combination of the electric resonance of the traction circuit and the phenomenon of standing waves for the overhead line only, that thus moves to a higher frequency, neglecting lumped loading at substations.

Line losses are significant due mainly to skin effect: the increase of longitudinal resistance with the square root of frequency begins at a fairly low frequency considering the typical diameter of catenary ropes and third rails.

Resistive losses are much more relevant than radiation losses: the attenuation of high-frequency components is increasingly larger and the effective length of the aerial is progressively reducing, having indicated with effective length the distance from the feeding point at which wave power is reduced by a given amount, e.g. by an order of magnitude, at 10 % of the original injected power.

Line resonances are a complex phenomenon: as long as attenuation is not too large and the wavelength is long compared to the transverse cross section, line behavior may be modeled by known expressions for propagation along a center-fed transmission line. For a long line the approximation that the line is infinite and that the considered section is terminated on its characteristic impedance (and is thus absorbing without reflections) is probably oversimplifying: end terminations are of the low impedance type if considering substations and other supply apparatus, whereas if the catenary is interrupted at the end of the line or at a neutral section the terminating impedance may be considered nearly infinite. For test tracks, stations tracks, and in any case in the presence of significant discontinuities in the supply line, a transmission line model with one end open- or short-circuited is preferable. This notwithstanding, all efforts shall be done to avoid ending sections and such termination conditions that have the side effect of making the distribution of current asymmetric and increasing the effect of some components due to multiple resonances.

Practically speaking, for traction supply sections the length is determined by the separation between substations or neutral sections: assuming a reference length value L of $1-2$ km and a velocity factor of 0.5 (or slightly less, accounting for some additional capacitive loading), the first resonance frequency is located between 30 and 100 kHz. Some variability is expected in relationship to the capacitive loading of masts and additional conductors of the return circuit, the number of the tracks and the presence of vehicles and trains. Higher order resonances are of course possible, but feature increasing losses, flattening resonance peaks significantly, still altering phase relationships; for frequencies around 10 MHz the wavelength is short enough to be comparable with the cross section dimensions.

In the low frequency range below the first resonance the traction line is a low impedance loop with prevailing magnetic field emissions: the spatial distribution of the field is that of a loop with a return current spreading among rails, other conductors and soil, and with sharing between them variable with frequency. So, a prevalence of vertical magnetic component is expected for measurement points located on the loop mid plane, that is a few meters about the top of rail. This is considered in more detail in Chapter 8, dealing with low frequency magnetic emissions.

The source of emissions is the rolling stock, the substation giving a minor contribution. Conducted emissions through the current collection point (pantograph to the catenary or sliding shoe to the third rail) propagate along the line, undergoing attenuation and phase rotation and progressively radiating in the surrounding environment. This Chapter considers emissions from the whole system, focusing on the emissions from the traction line originating from rolling stock: the rolling stock is thus part of the test and it is required that specific operations take place. In this case the difference with the EN 5121-3-1 standard considered in the previous chapter is in the required operating conditions and the span of the track included in the measurements (restricted to the rolling stock acceleration and braking in front of the antenna in the EN 50121-3-1, and with several hundreds meters before and after the measurement point including also cruising at high speed for the EN 50121-2).

5.2   Reference standards and limits

5.2.1   Removal of interval A from limits

The most relevant change passing from the 2006 to the 2015 version of the EN 50121-2 is the removal of the lowest portion of the frequency interval between 9 and 150 kHz from the mandatory normative part of the standard, moving it into an informative Annex C. The technical justifications given in the "Foreword" section are:

- "there are very few outside world victims", that however does not rule out completely possible interference, considering also that low-frequency emissions propagate at a much longer distance;

- the measurement is not useful to assess internal compatibility, because "the radiated emission measured at 10 m is not representative", probably because possible victims are closer than 10 m, but this does not in reality justify the complete removal of the test, whereas a reduction of the distance or the introduction of a compensation factor would be suitable; additionally, EMC with other railway apparatus for this frequency range is covered by other standards such as EN 50238, that means that coupling and interference cannot be excluded and are not so unlike, but necessitate specific procedures at least for track circuits and axle counters covered by EN 50238 standards;

- other systems, such as loops and tags, are, however, left unaddressed and the standard at Note 2 of sec. 6.1 clearly says that Annex C (reporting as indicative values what in the 2006 version of the standard were limits) is informative and no guarantee is given for an undisturbed operation of external services with working frequencies below 150 kHz; it is not clear if the lack of guarantee is based on the experience matured in the last ten years;

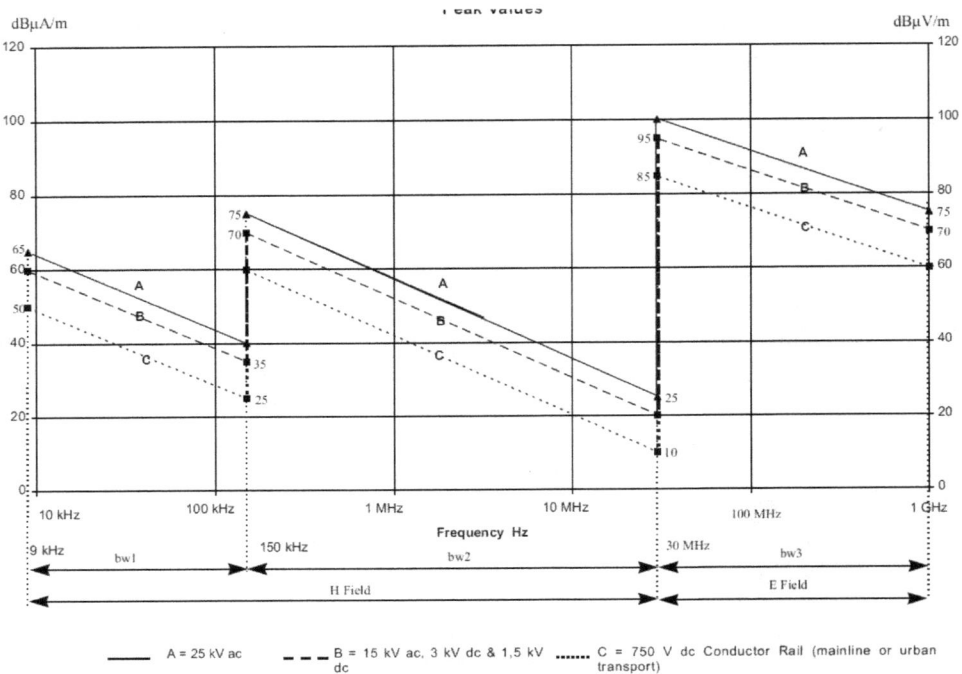

Figure 5.2.1 – Limits for traction line emissions as per EN 50121-2, Peak mode, 10 m distance.

- low reproducibility is invoked as a final justification: in other parts of the standard the problem of repeatability is considered without quantification; whereas repeatability is mostly influenced by the stability of operating conditions, site geometry and electromagnetic characteristics, regarding reproducibility influence at large extent is expected by how standard requirements and procedure are able to limit the many degrees of freedom, fixing measurement conditions and settings, with the least ambiguity.

5.2.2 Limits

Limits are defined in EN 50121-2 depending on the catenary voltage ($600 - 750$ Vdc for tram and metro applications, $1.5 - 3$ kVdc for metro and railway applications, $15 - 25$ kVac for railway applications) and the traction system type. Due to the dynamic characteristics of measurements and the use of Peak detector all limits are expressed for Peak mode, both for magnetic and electric field emissions.

The limits shown in Figure 5.2.1 are those reported in EN 50121-2 [27], Figure 1. The limits of the new 2015 version of the standard are identical, except that the $[9 - 150\,\text{kHz}]$ interval has been removed and left in Annex C [29] as informative. For non-electrified lines, the limits are as those given for 750 V dc systems (EN 50121-2 (2006), sec. 4.1.

Compliance to the limits is reached if all the recorded spectra have values that after post-processing and application of any correction factor are below the limits. The EN

50121-2 standard does not indicate any tolerance or safety margin, nor a statistical approach based on uncertainty (either estimated from the characteristics of the used instruments and test setup or by statistical analysis of measurements themselves) for the assessment of compliance.

5.3 Measurement procedures and techniques

This section is similar to the corresponding section for rolling stock emissions measurements appearing in Chapter 4.

5.3.1 Preliminary checks and conditions

Besides the usual care in taking RF measurements regarding connectors, cables, earthing of equipment, verification of undesirable antenna coupling, etc., the applicable standard EN 50121-2 [27, 29] prescribes additional conditions to be checked for environment and line.

5.3.1.1 Weather conditions

This section reports the same information from EN 50121-2 that was already introduced in sec. 4.3.1.1 is identical to the corresponding sec. 4.3.1.1 of the previous chapter, where specifications appearing in sec. 5.4.1 of 2006 version and A.1.9 of 2015 version of EN 50121-2 were reported.

To minimize the possible effect of weather on measured values, measurements should be carried out in dry conditions (after 24 hours during which not more than $0.1\,\mathrm{mm}$ rain has fallen), with a temperature of at least $5\,°\mathrm{C}$, and a wind velocity of less than $10\,\mathrm{m/s}$. Humidity should be low enough to prevent condensation on power supply conductors.

Since it is necessary to plan tests beforehand, they may be made in weather conditions, which do not meet the target conditions. In these circumstances, the actual weather conditions shall be recorded, unless they are evidently biasing the results for which tests shouldn't be performed.

The requirement on humidity and the quantity of water on parts and soil may be explained considering the effect on field propagation especially at high frequency: soil conductivity has impact on field attenuation and distribution between horizontally and vertically polarized components; also the wet vehicle body, especially when made of non-conductive material, may slightly attenuate emissions. However, this is generally true for electric field at very high frequency, so that the entire magnetic field measurement interval up to $30\,\mathrm{MHz}$ is not appreciably influenced, as well as the first portion of the electric field interval above $30\,\mathrm{MHz}$, and it is exactly in these intervals that significant and recognizable emissions from rolling stock are expected.

Temperature requirement corresponds more or less to an acceptability condition for personnel and equipment: in several countries several months of the year are characterized by lower temperatures, nonetheless affecting seriously measurement results, provided that equipment is stored in a warmer place, it is switched on well before the measurement and it is protected from cold exploiting self heating.

Wind speed requirement is usually met being a $10\,\text{m/s}$ (or $36\,\text{km/h}$) speed already relevant for antennas: loading and securing the tripod with sand bags is always advisable, but may be a valid countermeasure only for reasonably fast wind; by experience wind above $60\,\text{km/h}$ is very hard to manage.

5.3.1.2 Traffic and line conditions

Traffic and line conditions shall be annotated, with care for "physically-remote but electrically-near" trains or vehicles (EN 50121-2 [27], sec. 5.4.3); in reality, with the new EN 50121-2 (2015) [29], sec. 5.4.3, this factor has been regarded as insignificant for the purpose of tests. Whether this change is due to the removal of the limit over the frequency interval A for very low frequency emissions below 150 kHz or to some other factor is not clarified.

The measurement location shall be positioned midway suspension masts (or any similar suspension mechanism, that can influence electromagnetic emissions) and not at a discontinuity of the supply conductor (this causes transients due to the interaction with the current collection mechanism), as specified in EN 50121-2 [27], sec. 5.1.9. In case of third rail systems the standard recommends a distance of 100 m from "gaps in the rail, to avoid inclusion of the transient fields associated with the make and break of collector contact".

"If overhead power lines are nearby, other than those which are part of the railway network, they should be no closer than 100 m to the test site" (EN 50121-2 [27], sec. 5.1.10).

5.3.2 Test operating conditions

The purpose of this test is the evaluation of line (or system) emissions, that are in turn caused by rolling stock and depend on rolling stock condition. Thus the test shall be based on vehicle (or train) runs: whereas for substation measurements (see Chapter 6) the standard specifies a traction line load that necessarily requires multiple trains and conditions near to the full exploitation of the line, for line emissions the standard talks about the dynamic behavior of a single vehicle without clearly stating how measurements are synchronized with the vehicle. When testing rolling stock (see Chapter 4) scans are triggered with the passage nearly in front of the antenna (see Figure 4.1.1 and the measurement area); line emission tests differentiate for measurements being extended also to vehicle positions relatively far from the antenna (as anticipated at the end of sec. 5.1) and the dynamic conditions are different from those prescribed by EN 50121-3-1 and generally faster.

The list of test phases is:

1. as stated in the EN 50121-2 [27], sec. 5.4.2, "two dynamic test conditions" are specified for the traction mode:

    (a) "measurement at a speed of more than 90 % of the maximum service speed, (to ensure that the dynamics of current collection are involved in the noise level) and at the maximum power which can be delivered at that speed;

    (b) at the maximum rated power and at a selected speed, (particularly if the lower frequencies are of concern)." This condition on maximum traction effort may be difficult to fulfill if the train used for the test is not enough loaded to ensure adhesion and stability; when testing on a new line during for example trial runs or endurance tests, ballast will be necessary to reach the required load.

2. "If the vehicle is capable of electric braking, tests are required at a brake power of at least 80 % of the rated maximum brake power." This condition may again be difficult to fulfill if the trains used for the test are not enough loaded to ensure adhesion and stability; furthermore, if the train is too light intense braking decelerates too fast and the stopping distance is very short, in less time than that of a complete scan. This condition is particularly important when performing rolling stock tests, as seen in Chapter 4. It is also briefly commented that braking is required, but acceleration is not mentioned, leaving the doubt if recordings are required or allowed when the train is accelerating and how to deal with small accelerations that are normally applied also during cruising and when passing curves, crossings and neutral sections.

3. Stationary conditions are useful to discriminate between the various sources of emissions, but are not required for testing. They are for rolling stock testing as per EN 50121-3-1: for line emission tests to be usefully compared with the results of dynamic tests they shall be carried out in the same measurement conditions, that is in particular using a Peak detector, rather than the Quasi-Peak detector setting, we are used to for rolling stock testing.

4. Background noise measurements are needed to identify external sources (EN 50121-2, sec. 5.1.12), but no criteria were given in the 2006 version to determine suitable margins with respect to the limits, as indicated for example by CISPR 16 standards; conversely, the 2015 version indicates a 6 dB margin with respect to limits and frequencies at which background emissions are closer to the limit or above it shall be excluded from the successive evaluation. It is thus apparent that for a meaningful comparison the same measurement method used for emissions shall be used, that is the Peak mode: several repeated recordings may be passed to post-processing, or "max hold" may be set directly on the spectrum analyzer or receiver. The standard includes also further measurements at a significant distance from the line, but this only helps in identifying the frequencies where external sources exert their influence, the amplitude being different changing position. Background noise shall be recorded when there is no traffic on the line, or at least traffic may be excluded for a long distance, such as 2 km (although the EN 50121-2 is stricter and prescribes 20 km for catenary supply system, whereas 2 km is for third rail lines).

Because of the peculiar nature of some emissions (in particular low frequency magnetic emissions in the Interval A) interacting with traction circuit resonances, additional measurements of magnetic emissions for Interval A (and some in Interval B for better confidence) are to be performed when the train is at a moderate distance from the measurement location (that is between 100 and 500 m approximately)[1]. The first paragraph of sec. 5.1.9 of EN 50121-2 states "if resonance exists, this should be noted in the test report". It is however extremely difficult to recognize a resonance and distinguish it from a peak of emissions, such as the switching frequency of some on-board converter: usually resonances have a lower factor of merit and most of all they move with train position, so that beginning the scan at different train positions may result in a peak shift (see the comment at the end of sec. 4.1).

5.3.3 Antenna orientation and position

Antenna orientation is such to capture relevant emissions from the system, that the standards in a simplification assume mostly from the catenary system and in any case lying in the plane that contain the line cross section, orthogonal to the direction of travel.

For clarity, the used antennas are usually loop antennas up to 30 MHz and biconical, log-periodic antennas or a combination for the $30-1000$ MHz frequency range.

The positioning with respect to the line under measurement is strictly regulated by the EN 50121-2 standard (that remains the sole definitive reference):

- the antenna distance from the axis of the closest track is 10 m (for different distance a corrective coefficient shall be applied, see sec. 5.1.6 of the EN 50121-2; the use of different distances is permitted only if strictly necessary and shall be justified);

- the heights are measured with respect to horizontal plane passing through the rails head, that is the "top of rail" (in special cases the height may be measured from the ground level, see EN 50121-2, sec. 5.1.8);

- the orientation of the antennas shall be:

  - loop antenna shall be oriented to capture the horizontal polarization of the magnetic field only (i.e. the loop antenna stands vertically, with the plane containing the loop perpendicular to the ground and parallel to the line of the track;

  - the biconical antenna, when measuring the vertical component of the electric field, is positioned perpendicular to the ground; when measuring the horizontal component it is turned by 90° and its axis parallel to the line of the track;

---

[1] A similar consideration appears in Chapter 4, end of sec. 4.3.5, justifying the need for a few sample measurements aiming at identifying line resonances; here these measurements distributed along a longer line section at left and right of the measurement position are part of the scope of line emissions assessment. In any case the identification of traction line resonances is due to fulfill the requirement appearing in sec. 5.1.9 and B.10 of EN 50122-2 (2006) (corresponding to sec. 5.1.2 and A.8 of the 2015 version) and sec. 6.3.1 in the EN 50121-3-1 (2006 and 2015).

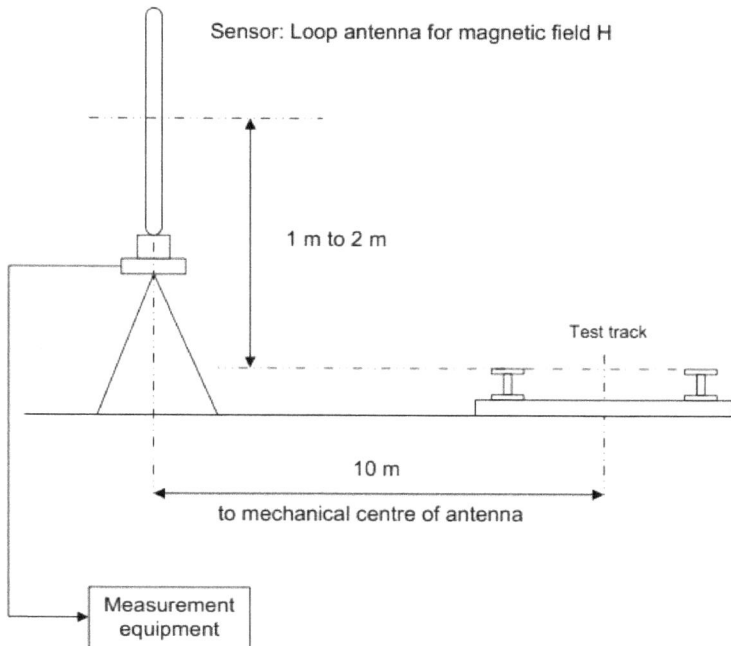

Figure 5.3.1 – Position and distance of loop antenna for measurement of radiated emissions as per EN 50121-2 (identical to EN 50121-3-1).

- the log-periodic antenna aims always at the track, with its dipole elements perpendicular or parallel to the ground to measure the vertical or the horizontal components of the electric field.

Vehicle emissions propagate along the catenary and especially at low frequency their intensity may be relevant even when the vehicle is at a significant distance; measurements are taken both when the train is at some distance along the track and when the train passes in front of the measurement position and antennas. Emissions are thus caused by the current components in the catenary system in front of the measurement point and by direct emission of the vehicle itself: the measured intensity of the direct emissions is a function of the distance r, that changes while the vehicle is moving; under the assumption of a $1/r$ dependency of electric field components, two positions P_1, P_2 and Q_1, Q_2 for $1\,\text{dB}$ and $3\,\text{dB}$ criteria were determined in sec. 4.1 (see Figure 4.1.1) and may be used as a reference, although not required by the standard.

5.3.4 Measurement settings

The frequency range is subdivided into frequency intervals for which different Resolution Bandwidths (RBW) are to be used, as for intervals A, B, C and D, as shown in Table 5.3.1.

Recognizing a moving source, measurements are all taken in Peak mode: they can be performed by either sweep mode (advisable) or by selecting a set of test frequencies,

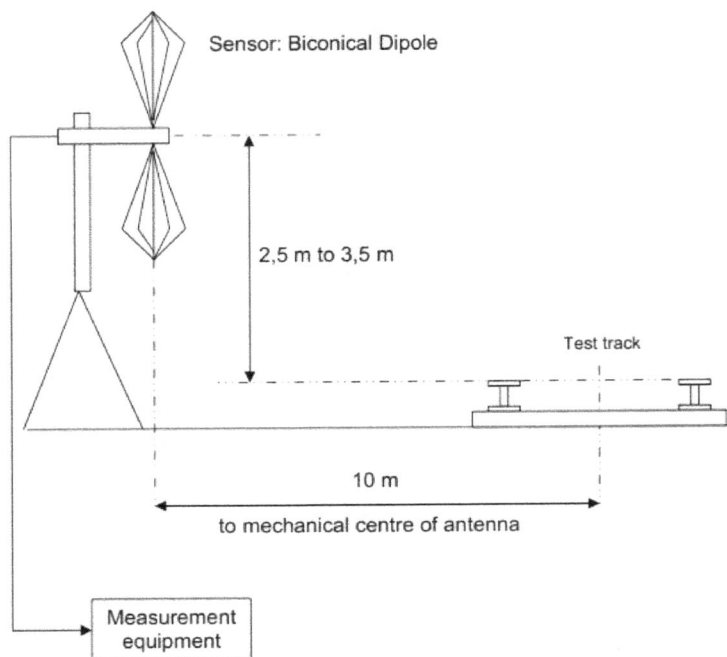

Figure 5.3.2 – Positioning and distance of biconical antenna for measurement of radiated emissions as per EN 50121-2 (identical to EN 50121-3-1).

Interval	A	B	C	D
Frequency range	$9-150\,\text{kHz}$	$0.15-30\,\text{MHz}$	$30-200\,\text{MHz}$	$30-200\,\text{MHz}$
Resolution Bandwidth	$200\,\text{Hz}$	$9\,\text{kHz}$	$120\,\text{kHz}$	$120\,\text{kHz}$

Table 5.3.1 – Resolution bandwidth (RBW) prescribed by EN 50121-2 as per CISPR specifications related to Quasi-Peak detector and by extension Peak detector.

to speed up the measurement, but with a minimum number of three frequencies per decade, as specified in EN 50121-2, sec. B.11.

Slightly different RBW values may be used, as available on the instrumentation and to increase measurement speed (sweep time reduction). Normally, when this is done, during post-processing correction is applied to the measured values for the ratio of the chosen RBW with respect to those prescribed in Table 5.3.1. This correction is not always justified and depends on the nature of measured data; to avoid improper data manipulation, RBW values shall be always the nearest to the prescribed ones (for most spectrum analyzers $300\,\text{Hz}$ for $200\,\text{Hz}$, $10\,\text{kHz}$ for $9\,\text{kHz}$ and $100\,\text{kHz}$ for $120\,\text{kHz}$), thus keeping the correction factor small. Under broadband input signal assumption (e.g. white noise), the correction factor is the ratio of bandwidth for power measurements and its square root for amplitude measurements: $k_P = 10\,\log_{10}(\text{RBW}_1/\text{RBW}_2)$ for power, $k_A = 20\,\log_{10}(\text{RBW}_1/\text{RBW}_2)$ for amplitude, where RBW_1 and RBW_2 stand

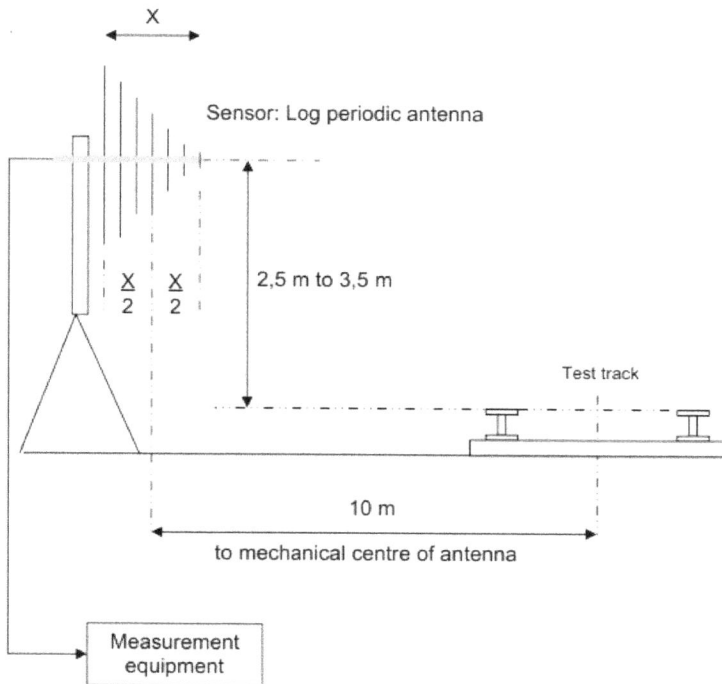

Figure 5.3.3 – Positioning and distance of log-periodic antenna for measurement of radiated emissions as per EN 50121-2 (identical to EN 50121-3-1).

e.g. for prescribed and available resolution bandwidth values and the resulting correction factor shall be summed to the reading performed with the available RBW value.

Sweep time shall be fast enough to capture transient emissions, but slow enough not to improve measurement accuracy. Spectrum Analyzers normally set automatically a sweep time value barely sufficient for an acceptable accuracy: the EN 50121-2 [27], sec. 5.2.2, gives a generic recommendation in this sense; probably using twice the minimum sweep time value set by the spectrum analyzer is advisable to improve accuracy [99]. The 2015 version of the EN 50121-2 [29], however, indicates a much slower dwell time of 50 ms for each frequency point, that is in contrast with the exigency of measuring fast moving objects and is discussed later on in sec. 5.4.5). In CISPR 16-1-1 [38] and [99] general values are given, establishing a relationship between RBW and sweep time, that give sound and accurate readouts, without however verifying the compatibility with the constraint represented by the fast moving source (see sec. 5.4.5 for a general discussion). As already commented in sec. 4.3.4, the effect of video bandwidth shall be removed in terms of smoothing traces and slight slowing down of the sweep time, so that a 3:1 setting with respect to RBW is advisable.

5.3.5 Measurement execution

Train operating conditions for dynamic tests were described in sec. 5.3.2: they are identified as SPD (traveling at 90 % nominal speed) and POW (maximum rated power at a selected speed)[2], but include also phases with acceleration (that may be indicated as ACC). An additional operating condition is braking (electric braking at 80 % of the maximum brake power), identified as BRK.

To validate the measurement results and the compliance with limits, background noise measurements are needed, so that external sources can be identified. To this aim several conditions may be considered (similarly to the stationary test performed in EN 50121-3-1), including background noise.

Several recordings are needed in order to reach statistical significance and a complete picture of external sources and disturbance. These measurements may be identified as BKG. Recordings of background noise shall be made prior to the beginning and at the end of the tests.

Since line emissions are mostly caused by transient phenomena, especially at high speed and at sudden changes of speed, that is acceleration and braking of trains, a minimum number of measurements shall be performed and recorded for each condition and each frequency range and antenna polarization. The advisable minimum number of recordings for each combination of these conditions is five, unless experimental evidence demonstrates that the measured levels are much lower than the limits and allow to cope with a wider confidence interval with larger sample dispersion.

To summarize, the train operating conditions are SPD for traveling at nearly maximum speed, POW for maximum power at selected speed, ACC for acceleration and BRK for braking, there are four frequency intervals (or three in the 2015 version), possibly split into sub-intervals, and a total of five antenna orientations (Loop, Biconical Vert., Biconical Horiz., Log-Periodic Vert., Log-Periodic Horiz.).

Because of the peculiar nature of some emissions (in particular low-frequency magnetic emissions in Interval A, that propagate at a much longer distance due to the lower line attenuation), measurements shall be performed for train positions far from the measurement position and keeping the sweep on running until and after the measurement position is reached (the measurement position corresponds to the measurement area used for rolling stock emissions, see Figure 4.1.1). Practically speaking, the span to cover may be as extended to include a distance of about 1000 m around the measurement area. However, this distance should be evaluated in direct relationship to what is observed at sec. A.8 of the standard regarding repeatability, saying that "the chance that remote vehicles will produce significant emissions at the test point" is considered relevant for other vehicles up to 20 km for catenary systems and 2 km for third-rail systems. By experience it is known that such safety distances used to setup the background noise measurement are really too cautious and are not then used as a reference distance for measuring train emissions farther away, as it would have been useful. This measurement condition is identified as FAR and applies to vehicle operating conditions (speed, acceleration and braking) altogether, if they cannot be distinguished, annotated and put in relationship with each sweep. The relevance of FAR measurements is particularly evident for measurements performed with the loop

---

[2] The distinction between SPD and POW conditions and the possibility of verifying that a statistically consistent number of runs in such conditions have been performed are critical for the spectrum analyzer operator wayside, for which communication with personnel on-board is necessary or recording of train conditions during tests.

antenna over the first two intervals A and B (with emphasis on interval A, while interval B may be partially included for completeness, possibly limited between 150 kHz and 1 MHz).

5.4 Further considerations

5.4.1 Selection and exclusion of transients

The EN 50121-2 standard at sec. 5.3 addresses "transients due to switching (...), such as those caused by operation of power circuit breakers": these shall be disregarded when "selecting the maximum signal level found for the test". This expression is confusing for several reasons: first, from an operative viewpoint, if the frequency sweep is made as requested with a max-hold setting then it will be impossible to extract or separate the contribution of such transients; second, there is indication on how to identify such transients, provided that circuit breaker tripping is perfectly monitored and perceived, while other transients that are not exemplified might go unnoticed; third, if the concept of transient is widened to embrace transient emissions, then any change of status or operating condition of power converters and apparatus is a transient, and discarding such emissions heavily influences and biases the results.

Identifying and discarding aberrations of data, possibly due to abnormal transients, may be approached by using the so called robust estimates, using e.g. outliers removal, inter-quartile range, etc.

5.4.2 Measurement distance

The prescribed distance between the track axis and the electric center of the antennas is 10 m: this is the result of a compromise between opposite exigencies, well explained in the EN 50121-2 standard [27] at sec. B.7. Reducing the measuring distance increases the signal strength, with obvious benefits for the signal to noise ratio and in general to keep several sources of uncertainty under control. However, considering the normally adopted 1, 3, 10, 30 m sequence of distance values, we may observe that:

- 1 m is immediately ruled out as impossible;

- 3 m, even if barely in a safe portion of space with respect to train passage and dynamic gauge, would leave a large influence of train body on the received signal strength when in front of the antenna; the reasons are not only a generic perturbation of the field, but also the obstruction of the line of sight between the catenary and the antenna, as well as the fact that the antenna would not aim at a portion of space towards the pantograph and the catenary, from which the most significant portion of emissions is expected;

- 30 m would be very useful for a uniform measurement of emissions from the line section, including vehicle body, with very minor influences due to moving parts; it would be also very attractive to increase the time available for scanning during train passage, thus reducing the number of needed train runs; however, the

Frequency range	n
[150 kHz, 400 kHz]	1.8
[0.4 MHz, 1.6 MHz]	1.65
[1.6 MHz, 110 MHz]	1.2
[110 MHz, 1000 MHz]	1.0

Table 5.4.1 – Corrective conversion factor n as a function of frequency (EN 50121-2 (2006), sec. 5.1.6, or EN 50121-2 (2015), Table 1).

measured signal strength would be lower with local background noise becoming more relevant; it is conversely underlined that such distance is the preferred one for measurements on elevated railways, as indicated in the fourth bullet of sec. 5.1.2, EN 50121-2 (2015);

- 10 m resulted thus in a good compromise among the various exigencies of safety, influence of vehicle body, minimum signal strength and portion of line intercepted during one scan; minor variations of the reference distance due to contingency and specific site characteristics may be accommodated by applying a corrective coefficient.

When the used measurement distance is different from the 10 m prescribed one by an acceptable amount not to compromise measurement results, correction may be applied accounting for the attenuation with distance depending on frequency, using the following expression:

$$X(10\,\text{m}) = X(d) + n\,20\log_{10}(d/10) \tag{5.4.1}$$

where given the measured $X(d)$ at distance d, it may be converted to the equivalent 10 m value to compare with limits. In this expression the letter "X" was used rather than the "E" used in the standard that falsely suggests that the correction can be applied only to electric field. The corrective coefficient n (that takes into account the attenuation with distance) is indeed defined also down to 150 kHz, where emissions are measured for magnetic field only (see Table 5.4.1). Moreover, the EN 50121-2 (2006), sec. B.8, equivalent to EN 50121-2 (2015), sec. A.6, clearly extends the formula to the correction of magnetic field measurements when talking of the necessity of converting in V/m of equivalent electric field the magnetic field levels expressed in A/m, multiplying by the free space impedance 120π: in reality, magnetic field values may be directly used without any conversion, that is neither useful to compare to the limits that are again already expressed in A/m.

5.4.3 Distance from other vehicles

Differently from the EN 50121-3-1 (2006) [28] for vehicle testing that prescribes a minimum separation of the vehicle under test from other vehicles of 2 km for urban vehicles and 20 km for main line vehicles, in the EN 50121-2 for line emission testing "the presence of physically-remote but electrically-near vehicles out of the test zone is regarded as insignificant when considering radio noise". It may be assumed that the rationale is that whatever the trains on the line they all contribute to line emissions that are the subject of the measurement; however, it turns out that the testing conditions are badly specified, being unknown the number of vehicles contributing to the measured values. Additionally, the term "radio noise" is in reality vague, suggesting either that portion of the frequency interval where the emissions are "radiated" (that is mainly electric and in quasi far-field conditions) or the entire range of emissions, that is by the way already the scope of the standard. However, sec. B.10 of EN 50121-2 (2006), equivalent to sec. A.8 of EN 50121-2 (2015), dealing with repeatability considers "the chance that remote vehicles will produce significant emissions at the test point" as relevant, requiring for catenary systems a distance of other vehicles of 20 km and for third-rail systems 2 km.

5.4.4 Repeatability

Repeatability is not addressed as a purely metrological issue to be evaluated in proper terms as part of the more general uncertainty problem [15, 43]. Sources of variability between repeated measurements are many, both spatial and temporal, testifying the non-stationarity of the process: shifts of emission peaks due to line resonances as the vehicle moves, change of operating conditions and driving style occurring in fraction of seconds, low frequency instability due to the transient response of the on-board filters to driving changes, variability of secondary noise phenomena, also external, overlapping to the main emissions.

The sources of uncertainty affecting repeatability are many and with different characteristics and effects, so that it is extremely difficult to derive a rigorous and complete approach from a metrological viewpoint [15]; many elements of uncertainty and repeatability are common to the similar section 4.4.5 in the previous chapter. The EN 50121-2 (2006) [27], sec. B.6, talking of the accuracy of measurement equipment comments on an expected repeatability of ±10 dB, that is quite a large value stunting expectations of accurate measurements[3]: such repeatability value corresponds more or less to the large dispersion of traces recorded during braking shown in Figure 4.4.1; as it is seen in the following, taking care of the most relevant factors improves significantly repeatability, as it may be directly appreciated by a Type A approach for the evaluation of the uncertainty of a sample of recordings in similar (or ideally identical) conditions.

5.4.4.1 Instrumental uncertainty

Instrumental uncertainty is well documented and as commonplace for this kind of measurements supported by certificates and possibly a Type B approach to its evaluation for the entire measurement chain; for the adopted equipment the combined uncertainty for the antenna factor, distance error, cable compensation and spectrum

---

[3] The 2015 version of EN 50121-2 [29] has this part removed with its sec. A.8 very generic on this and simply recalling the need of keeping other trains far away.

analyzer amplitude reading may be between 4.5 and 5.5 dB for a confidence level of 95 %, so with a coverage factor of $k = 2$ under Gaussian distribution assumption (see sec. 3.4).

What instrumental uncertainty fails to cover is the remaining indeterminacy related to environment and setup conditions (line resonances, transients due to considerable speed, driving conditions, operating conditions, synchronization of sweeps, intermittent and variable external sources).

5.4.4.2 Line resonances

Uncertainty of position influences directly the measured intensity and possible traction line resonances. As described at the beginning in sec. 5.1, line resonances may occur in the tens or hundreds kHz and enhance line emissions, resulting in humps and peaks with low quality factor that increase the amplitude of some components of the rolling stock discrete spectrum of emissions (please, see Figure 4.4.1 for an example).

5.4.4.3 Driving style and synchronization

As already explained for vehicle emissions in sec. 4.4.5.2, lack of synchronization between frequency scanning and vehicle passage affects the location of spectrum peaks captured while sweep and transient occur together. Line emissions are less strictly related to vehicle passage since recordings are performed in more variable vehicle operating conditions and positions with respect to the measuring point.

The driving style has in any case a significant impact on the spectrum shape of recorded emissions: the consideration on the use of max hold approach or statistical methods is again valid; the former is much more useful in this case lacking a strong synchronization with vehicle operation.

Similarly, clarifying and agreeing the driving style and how acceleration and braking is applied is beneficial for repeatability, of course complying also to all other rules of comfort and safety and to any restriction that might be in place in case of test and commissioning activities and during trial runs. A communication link is in any case mostly useful to require slight adjustments of driving conditions and asking confirmation of exceptional events, that may occur when the vehicle is far from the measuring point and thus not under direct supervision of the operator.

5.4.4.4 Examples

The simple straightforward calculation of sample variance might not be appropriate when the variability is due to a frequency shift: in other disciplines, such as model validation, this is a parametric uncertainty that is not properly evaluated if a simple amplitude error criterion is applied (as it is for the sample dispersion). However, sample dispersion is the commonly accepted method for the estimation of uncertainty with a Type A approach [15].

Some examples are reported in the following, where problems of artificial frequency shift and spectrum humps distributed over the frequency axis due to lack of synchronization are visible (see Figure 5.4.1). To stress the phenomenon a case study is selected where trains are tested already in commercial service, thus following a predetermined pattern of braking to station and acceleration from station stopping always in a predetermined platform zone: at the beginning of the platform (called position 1) the train is approaching with conditions corresponding to SPD and then BRK; similarly, at the other end (called position 2) when the train is leaving the platform the corresponding conditions are ACC and SPD. In either case one of the two dynamic transient conditions (ACC and BRK) is occurring farther away from the antenna measurement point, whereas the other one is occurring approximately in front of it: in each case thus attenuation is expected of measurements performed farther away and a shift of spectrum humps due to different catenary resonances and the mentioned lack of accurate synchronization with the sweep. It is observed that position 1 is closer to the catenary feeding cables and thus the background noise is larger.

Observing the single traces (or small groups of traces), local increase of the spectrum profile in various positions along the frequency axis: the broad humps especially in BRK traces in the half figure above and in ACC traces in the lower half are due to transients captured by the sweep; the fact that they are not fixed in frequency confirms that they are due to the reciprocal relationship between the transient and the sweep and possible misalignment of synchronization of a few seconds.

The profiles are not smooth but jagged by the presence of harmonic components (comb shape), clearly caused by on-board converters; the spacing between harmonic components indicate the switching frequency of the source, that assuming that only even or odd harmonics appear is about $600\,\mathrm{Hz}$, being the separation between adjacent lines in the order of $1.2 - 1.3\,\mathrm{kHz}$.

The narrower peak located at $25\,\mathrm{kHz}$ is caused by train components exciting a catenary resonance; it is also visible in Figure 5.4.3(b) at position 2. When measurements are performed at position 2 the train is closer to the antenna during accelerations for which the profile values are much larger than in position 1; however, the nearly direct emission when passing in front of the antenna creates more dispersion, and the spread of ACC curves is slightly larger than the said $20\,\mathrm{dB}$.

In Figure 5.4.2 and 5.4.3 both the max hold profiles and the distributions with $\pm 1\sigma$ boundary curves are reported: in many cases (in terms of train operating conditions and portions of the frequency intervals) the lowest frequency interval between 9 and $150\,\mathrm{kHz}$ features the largest dispersion; the maximum dispersion is around $20\,\mathrm{dB}$ but lower on average: position 1 is slightly less disperse (about $7\,\mathrm{dB}$ on average), whereas position 2 keeps the $\pm 1\sigma$ profiles running parallel, but slightly convergent, for the entire $[9 - 150\,\mathrm{kHz}]$ interval. For the $[0.15 - 30\,\mathrm{MHz}]$ interval the distribution of traces is less dispersed; accelerations captured nearly in front of the antenna allow to spot out a local on-board resonance between 5 and $15\,\mathrm{MHz}$ (see Figure 5.4.3).

By observing that the max hold profile lies almost always very near to the $+1\sigma$ profile, we may conclude that the distribution of trace values has negligible tails, as it is for a truncated Gaussian: the limited number of samples (nine in some conditions, but

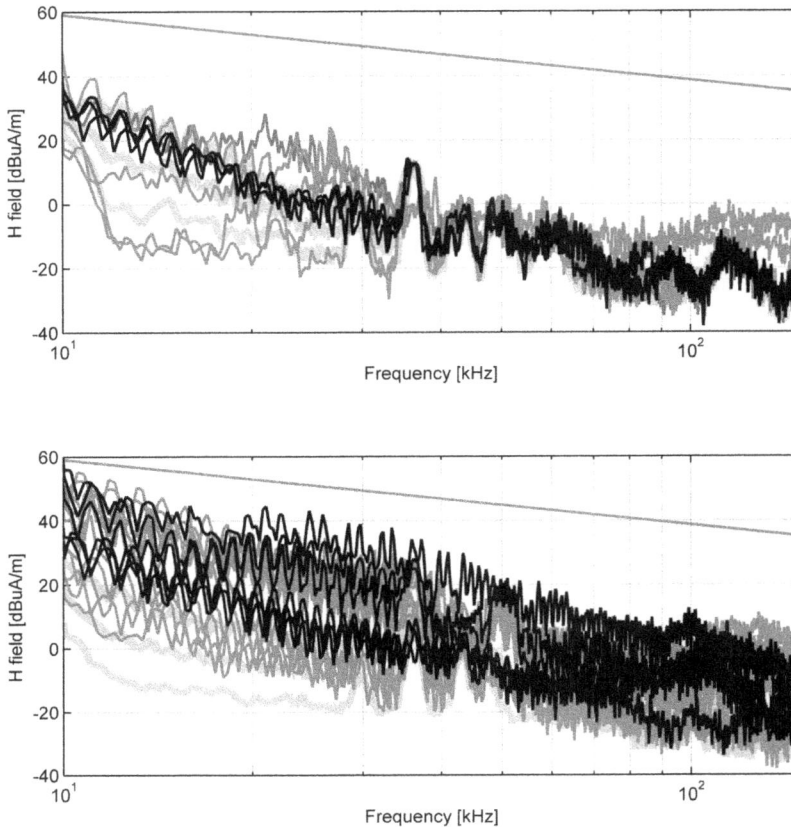

Figure 5.4.1 – Dispersion of background noise (light gray), braking (gray) and acceleration (black) traces measured at two positions on a platform, position 1 above and position 2 below: background noise is plotted as a center trace for the mean value and two traces for 1σ boundaries; individual traces for acceleration and braking are plotted overlapped, graphically showing trace dispersion and frequency shifts.

about six on average) does not allow to draw conclusions on the distribution, using for example histogram fitting.

5.4.5 Sweep time and speed

The problem of measuring a moving object by sweeping frequency, and thus substantially assuming steady condition, leads to the definition of acceptable sweeping speeds. Sweep time (called "scan rate") is given in Table 2 of EN 50121-2 for a maximum movement of $5\,\text{m}$ of the train at various speeds during a single sweep; for speeds above $100\,\text{km/h}$ the required sweep time gets quite fast, hard to achieve for conventional SA operating in sweep mode; in this case FFT mode shall be used as generically indicated by the standard itself (see sec. 5.2.1 of version 2015 of EN 50121-2 [29]). Also the FFT mode for spectrum analyzers is subject to a minimum equivalent seep time (that we called observation time for FFT calculation), due not only to exigencies

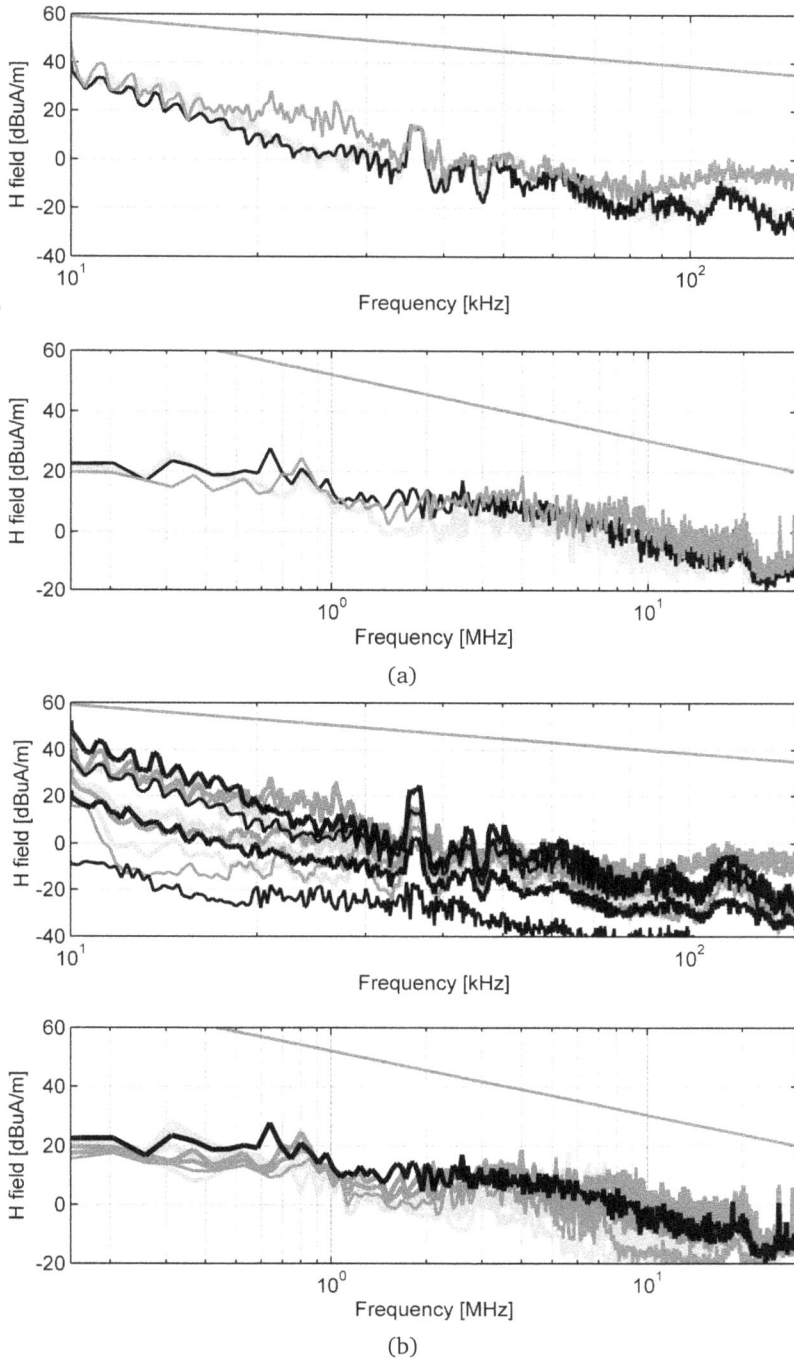

Figure 5.4.2 – Dispersion of background (thick light gray), braking (gray) and acceleration (black) curves for [9 − 150 kHz] and [0.15 − 30 MHz] frequency intervals at position 1: (a) max hold profiles and (b) ±1σ boundaries around mean value and max hold profile.

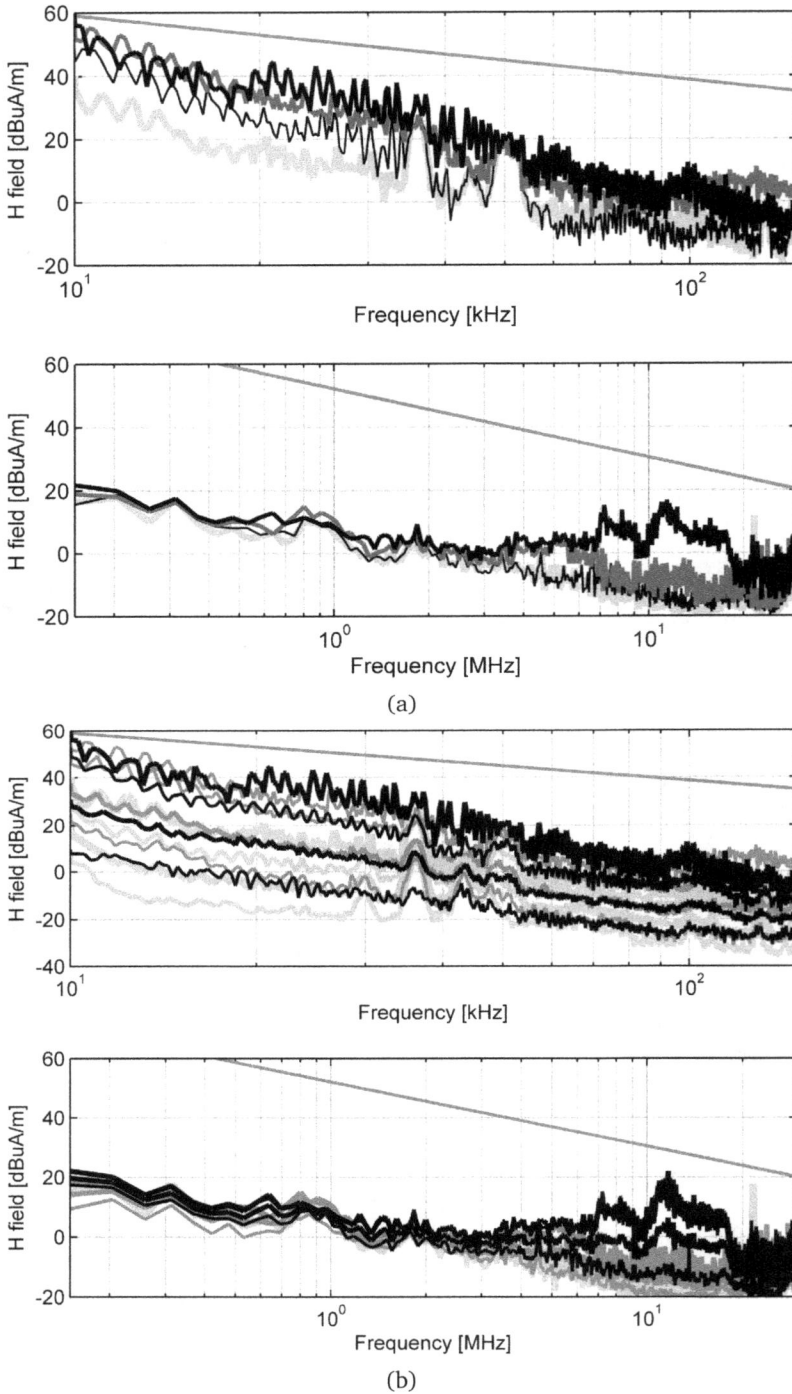

Figure 5.4.3 – Dispersion of background (thick light gray), braking (gray), acceleration (thick black) and coasting (thin black, added to (a)) curves for [9 − 150 kHz] and [0.15 − 30 MHz] frequency intervals at position 2: (a) max hold profiles and (b) ±1σ boundaries around mean value and max hold profile.

Speed [km/h]	Speed [m/s]	Sweep time [ms]
60	16.67	300
100	27.78	180
200	55.56	90
300	83.33	60
320	88.89	56

Table 5.4.2 – Sweep times as per EN 50121-2, Table 2, for "observation width" of 5 m (train movement while scanning).

of number of points and frequency resolution, but also to the required computation time between successive frames (see sec. 3.1.10).

The observation of having a nearly constant vehicle-antenna distance during one sweep is correct, as it was preliminarily considered for the example of direct emissions from the vehicle in sec. 4.1; however, it is underlined that this same observation is not re-ported in the EN 50121-3-1 standard, for which this time the measured emissions are exactly those of the vehicle!

Regarding the meaning of the "sweep time" header in Table 5.4.2, in the standard it is indicated as "time in s for an observation width of 5 m (scan rate)": one may wonder if it is really identifying the sweep time for the entire span (as reported in the table) or the dwell time for a single frequency. The sentence "this ensures that the frequency results are measured at least every 5 m of train movement" clarifies that not a single point, but all the frequency points, is to be measured in the indicated time.

The same EN 50121-2 standard [27], sec. 5.1, indicating prescriptions for the measure-ment of a moving source reads "the duration at selected frequency shall be sufficient to obtain an accurate reading. This is a function of the measuring set and the recom-mended value is 50 ms." So, the dwell time, rather than the sweep time for the entire span, is set to 50 ms. The reason was to be able to capture the maximum of emissions at each swept frequency point, where emissions are likely to occur with largest inten-sity at phase reversals of the supply frequency for ac systems: if 50 ms is long enough to capture more than two periods of 50 Hz fundamental, it embraces only half a period for 16.7 Hz systems. In any case the prescribed 50 ms dwell time is much longer than the sweep time values reported in Table 5.4.2.

Finally, observing that the source of emissions is the vehicle and that the sweep time values in Table 5.4.2 are justified considering vehicle speed, the comparison with the corresponding sweep time prescriptions in Table 4.4.1 (EN 50121-3-1, Annex B) is justified. This comparison, however, soon points out that for high nominal speeds the already conservative sweep time values given in Table 4.4.1 shall be reduced, increasing correspondingly the number of sub-intervals and test runs: at 200 km/h the required 90 ms are fulfilled if the 370 ms corresponding to 10 MHz interval are split into four chunks, thus reducing to 2.5 MHz. It is evident that the effort to carry out the measurement campaign becomes relevant and that other techniques shall be investigated: FFT mode or time-domain acquisitions followed by time-to-frequency transform are viable solutions.

5.4.6 Other than open area site measurements

It is immediately clarified that with the expression "open area site measurements" it is not intended the use of an "OATS", or "Open Area Test Site", in the CISPR sense, characterized by a normalized site attenuation whose uncertainty is usually low and known. The expression indicates rather measurements taken open air without significant reflections from e.g. tunnel walls.

5.4.6.1 Underground measurements

Especially when crossing city centers or for integrated stations with integrated city and suburban lines, traction lines are often of the underground type. Strictly speaking the new version of the EN 50121-2 [29], sec. 5.1.1.4, reads "no measurements are necessary for total underground railway systems with no surface operation (no victim outside this railway system can be affected). However, if the system is not fully underground measurements either shall be taken in the few line sections that are at grade (extending then the results to the untested underground sections) or shall be conceived for a satisfactory execution coping with space constraints of tunnels and underground stations. It is evident that both for safety reasons and for a dramatic lack of space, tunnel measurements are not an option: even at line-side niches and recovery sites the distance from the track axis and the space around the measuring antennas is not sufficient; thus, station platforms shall be used using at best the available space.

The EN 50121-2 standard does not explicitly cover it, but such measurements are not forbidden if the fulfillment of standard requirements is carefully verified and demonstrated:

- the distance between the antenna and the track axis may be shorten with respect to the prescribed 10 m and corrective coefficient applied, provided that the distance reduction is not dramatic and correction is kept within a few dB;

- with measurements performed at station platform, ceiling shall be high enough to accommodate for the maximum antenna height, including the antenna physical size and some margin to avoid unwanted coupling; considering the biconical antenna in vertical polarization at the minimum height of 2.5 m, 5 m of ceiling are barely necessary;

- reflections from tunnel walls, floor, ceiling and station walls shall be due taken into account:

  - floor behavior may be assumed similar to that of soil in open area site conditions, even if at a different height with respect to the top of rail (about 80 cm above it, rather than 20 cm or more below it); the EN 50121-2 (2015) does not clearly treat the case of a ground level higher than the top of rail; conversely, at sec. 5.1.7 of the 2006 version, it is said that when the ground is lower than 0.5 m with respect to the top of rail measurements can be done, but this condition shall be annotated;

- ceiling creates further reflections, especially considering that the catenary
  will be very close to it, so that an increase of emission level should be
  observed, except at some specific frequencies where there is a partial compensation and cancellation of reflected terms;

- for station walls, positions where walls are close behind the antenna shall be
  avoided, preferring those where large passages for passengers are located;
  such passages usually should be large and tall enough to allow antenna
  positioning avoiding major coupling issues;

- it shall be remembered that two elements of the traction supply system may
  negatively impact on the measurements:

  - the supply points of catenary, or third rail, are often located at stations that
    host integrated substations, so that feeding cables orthogonal to the trac-
    tion line may represent additional sources as well as capacitively loading
    the catenary; neutral (or isolating) sections may also create abnormal ve-
    hicle emissions due to pantograph bounces, line resonances and electrical
    oscillations;

  - the feeding cables themselves carry the full dc line current and several low
    frequency harmonics that easily bring into saturation active loop antennas;
    this is of course not a permanent condition that may be easily detected
    while setting up the measurement, but occurs when the line absorption is
    at maximum, that is normally with approaching and leaving trains, thus
    wrongly suggesting a particularly intense emissions from the same trains.

5.4.6.2 Measurement position within the right-of-way

In several cases there might be the impossibility of staying off from the outer track for
the required 10 m measurement distance: lines sections of viaduct at several meters of
height with respect to ground and lines surrounded by walls and metallic fences due
to safety reasons. Such situations to author's knowledge have never been considered
nor quantified in terms of impact on measurement results; in any case an accurate
evaluation of the impact would be a very expensive and time consuming activity, with
the need of providing simulated conditions in an otherwise ideal open area site. The
issues and the critical conditions that may be identified are several:

- the prescribed 10 m distance cannot be easily or at all fulfilled;

- it is not possible to find a measurement location outside tracks with only available space on other tracks, remaining inside the right of way; this is of course
  risky and not satisfactory from a measurement and metrological perspective:

  - staying on other tracks implies that circulation of trains shall be prohibited
    and that safety measures shall be enforced to avoid risk for personnel and
    equipment, including for close-by traffic on the track under measurement;

– the presence of a third rail also near the area used for the measurements is another safety issue, especially for accidental contact by personnel directly or by means of cables or equipment; additionally, even in the case of a catenary, when raising antennas during installation and measurements, especially those with sharp corners may be subject to induction for example when the catenary voltage is particularly high, e.g. at 15 kV or 25 kV;

– the presence of the catenary above the measuring antennas have influence both in terms of slightly affecting propagation of emissions from the other track and loading the antennas; but most of all it is a source of additional emissions, both occurring at the same frequency interval under measurement and at supply frequency and harmonics, again with the risk of saturation of active antennas;

• the most convenient locations along the line from a practical point of view are at line ends, because of the least impact on line operation, even if measurements may be performed during off periods; however, this choice violates the requirement of a homogeneous line which is continuous both at left and right of the measurement location: the consequence is that resonances and stationary waves may take place at some frequencies, that may become critical if excited by components of vehicle emissions.

5.4.6.3   Measurement position for elevated track

For measurements at elevated tracks the option is to remain within the right of way (if the lateral clearance is enough to accommodate for the required 10 m, or slightly shorter, distance) or to perform measurements from the ground level (aiming at the track with antennas slanted with respect to the horizontal plane). Such measurements shall reckon with a larger height from the ground reference of the line under test and necessitate special test conditions regarding height of antennas and distance from the track. The reason for selecting elevated line section is simply that the whole line is elevated or the remaining line portions are in tunnel.

In the vast majority of cases antennas are located at the ground level, with height measured from ground rather than the top of rail, and aiming slanted at the track above. It is obvious that the usually adopted 10 m distance is not satisfactory and that shall be lengthened to aim at the elevated track with a suitable angle, as small as possible. As already commented considering several measuring distance in sec. 5.4.2, longer distances such as the 30 m value (suggested also by the EN 50121-2, sec. 5.1.2, as preferred for elevated systems) lead to lower measured values of emissions and a more significant influence of background noise, that shall be thus evaluated thoroughly and carefully.

Additionally, low aiming angles are effective with catenary systems to capture line emissions, but a third rail would be in any case hidden by parapets and walls.

5.4.7 Normal commercial conditions

In some cases, due to delays and contingency, tests are postponed and must then be performed on the complete system, already in commercial service after the trial runs phase. This of course limits the degrees of freedom, the possibility of selecting locations and of controlling train operation: patterns of acceleration, cruise speed and braking are those dictated by time table and stations, unless night tests are performed; the latter are an option, but often line availability is quite limited between the time the line is cleared and available for tests (it may be 1 am or later, depending on the time to recover the last train and to get the permit from the Operation and Control Center) and the first train the day after (leaving the Depot in the early morning).

Executing measurements at platforms ensures that weather conditions are less troublesome, because light rain can be tolerated. On the contrary, the far from ideal conditions of many stations and platforms are the real problem: metallic canopies, too close columns and masts, low ceiling.

In addition, the sequence of train operation and the intensity of acceleration and braking when leaving and approaching platform are dictated by comfort constraints. There is no position at platform to capture adequately both acceleration and braking: if the train arrives from the right it will always brake before the antenna and accelerate past the antenna, staying stationary approximately in front of the antenna; it is thus advisable to select two positions at the beginning and end of platforms, combining measurement results.

Even without a dense timetable, e.g. five-minute headway, the overall duration of measurements is acceptable.

6

Electromagnetic radiated emissions from substation

6.1 Problem description

The substation is a complex combination of electrical equipment and machinery (e.g. transformers, rectifiers and protections) interconnected by cables and line segments, including the feeders to the traction line. Defining and modeling its emissions is not straightforward as significant low-frequency harmonics (also brought in by the high-voltage in-feed) overlap to commutation byproducts (for dc rectifiers and in general when converters are present, usually located between a hundreds kHz up to several MHz); broadband radiofrequency emissions are also possible, e.g. related to converters and in general to the overhead lines picking up external disturbance, originating from powerful radio sources, such as TV and radio broadcast.

6.2 Reference standards and limits

The applicable standard is the EN 50121-2 [27, 29]. For the 2006 version [27] the Annex A reports the measurement method; in the 2015 version [29] the two parts re-

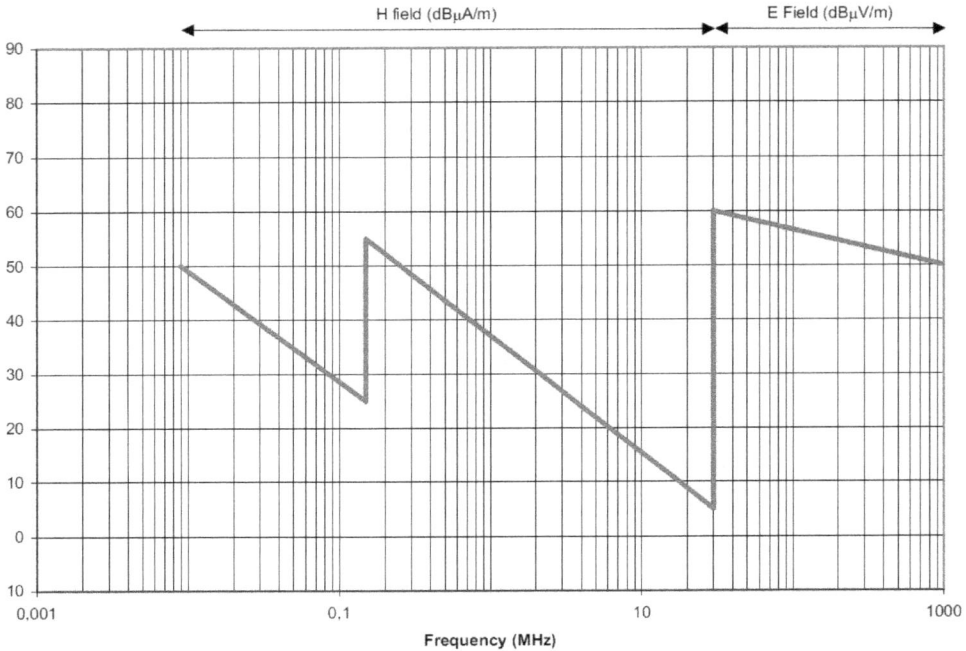

Figure 6.2.1 – Limits for traction line emissions as per EN 50121-2 (2006) [27], Peak mode, 10 m distance. In EN 50121-2 (2015) [29] the [9 − 150 kHz] frequency interval has been removed and related values moved to Annex C as informative.

garding line emissions and substation emissions have been harmonized, reporting the relevant information in the standard body and leaving all the remaining information in annex A that has been changed to "informative". In this way the detailed procedure of the 2006 version is lost, leaving only general prescriptions that are equally applicable to line and substation measurements.

Another relevant change passing from the 2006 to the 2015 version of the EN 50121-2 is the removal of the lowest portion of the frequency interval between 9 and 150 kHz from the mandatory normative part of the standard, moving it into the informative Annex C: the reported values are the same, but are not regarded as limits any longer.

Limits appearing in EN 50121-2 [27], Figure 2, for substation emissions are in terms of Quasi-Peak detection: in Figure 6.2.1 the more complete limits are shown, including the now informative [9 − 150 kHz] interval.

Compliance to the limits is reached if all the recorded spectra have values that after post-processing and application of any correction factor are below the limits. The EN 50121-2 standard does not indicate any tolerance or safety margin, not a statistical approach based on uncertainty.

6.3 Measurement procedures and techniques

6.3.1 Preliminary checks and conditions

Besides the usual care in taking RF measurements regarding connectors, cables, earthing of equipment, verification of undesirable antenna coupling, etc., the applicable standards EN 50121-2 [27, 29] prescribe additional conditions to be checked for environment and line.

6.3.1.1 Weather conditions

This section is identical to the corresponding sec. 5.3.1.1 of the previous chapter; for completeness this information is reported in both chapters.

To minimize the possible effect of weather on measured values, measurements should be carried out in dry conditions (after 24 hours during which not more than $0.1\,\text{mm}$ rain has fallen), with a temperature of at least $5\,°\text{C}$, and a wind speed of less than $10\,\text{m/s}$. Humidity should be low enough to prevent condensation on power supply conductors.

Since it is necessary to plan tests beforehand, they may be made in weather conditions, which do not meet the target conditions. In these circumstances, the actual weather conditions shall be recorded, unless they are evidently biasing the results for which tests shouldn't be performed.

The requirement on humidity and the quantity of water on parts and soil may be explained considering the effect on field propagation especially at high frequency: soil conductivity has impact on field attenuation and distribution between horizontally and vertically polarized components; also the wet vehicle body, especially when made of non-conductive material, may slightly attenuate emissions. However, this is generally true for electric field at very high frequency, so that the entire magnetic field measurement interval up to $30\,\text{MHz}$ is not appreciably influenced, as well as the first portion of the electric field interval above $30\,\text{MHz}$, and it is exactly in these intervals that significant and recognizable emissions from rolling stock are expected.

Temperature requirement corresponds more or less to an acceptability condition for personnel and equipment: in several countries several months of the year are characterized by lower temperatures, nonetheless affecting seriously measurement results, provided that equipment is stored in a warmer place, it is switched on well before the measurement and it is protected from cold exploiting self heating.

Wind requirement is usually met being a $10\,\text{m/s}$ (or $36\,\text{km/h}$) speed already relevant for antennas: loading and securing the tripod with sand bags is always advisable, but may be a valid countermeasure only for reasonably fast wind; by experience wind above $60\,\text{km/h}$ is very hard to manage.

6.3.1.2 Substation loading

As per EN 50121-2 (2006), sec. A.2, "the load can change widely in short times. Since emission can be related to load, the actual loading of the substation shall be noted during emission tests". It is specified in the following that the minimum substation load shall be 15 or 30 %, depending on test. Starting from the observation that substation load can change widely in short times and that when emissions measurements are carried out the system may be still in trial run, it is not clear what the term "load" refers to: it is sensible to interpret it as nominal loading due to running trains, that doesn't imply the substation is loaded at the prescribed level all the time. To obviate, either nominal substation load is increased, to reduce the time intervals at which the effective loading is less than the prescribed 15 or 30 % or the measurement method is conceived to account for fluctuations of the emission intensity during the sweep. The former is achieved by increasing the number of trains and the individual power demand (longer train consists, ballasting with water or sand, etc.), observing that, in any case, due to substation oversizing during design (to accommodate for the shortest headway and future expansion of the system), loading levels of 15 or 30 % imply already a significant number of circulating trains. The latter may be addressed for example by using faster Peak sweeps (and also FFT mode), rather than the prescribed Quasi-Peak sweep, as observed in the EN 50121-2 (2015), sec. 5.1.3, suggesting a max-hold Peak measurement to be supplemented by Quasi-Peak readouts only for those frequency intervals at which limits are exceeded.

6.3.2   Test operating conditions

As anticipated, the detailed prescriptions regarding substation loading of the 2006 version have been removed in the new 2015 version.

Loading conditions are specified for each type of measurement in EN 50121-2 (2006), sec. 3.7. In synthesis, for all measurements the loading of the substation shall be at least either 15 % or 30 % of the rated load. This requirement is not easy to fulfill if tests are carried out during e.g. trial runs. Usually systems are oversized with substation design based on the results of electromechanical simulations run at the maximum contractual headway during peak time and applying some additional margin for future expansion: one or few trains available during trial runs or endurance tests cannot absorb that large fraction of the total rated power, resulting thus in a forced choice: either derogating to the required substation load or perform tests when the system is in commercial service (see sec. 5.4.7).

Background noise measurements are needed to identify external sources and to establish the necessary margin with respect to the limits (see sec. 5.1.12 of the EN 50121-2 (2006) [27]). Background noise shall be recorded when there is no traffic on the line, or at least traffic for which we may exclude supply by the substation under measurement: in many cases this can be achieved only when the line is completely empty, that during the forced pace of trial runs and test and commissioning period may be quite a hard condition to meet. The measurement method might be at the discretion of the Test Responsible, for example using Peak mode for fast scan and several repeated recordings for post-processing, or directly "max hold" condition on the spectrum analyzer.

No specific indication is given in the standards for the handling of background noise results, as for the statistical methods to adopt and how to translate the results in margins with respect to limits for successive emissions measurements. What was said in sec. 5.3.2, item 4, is equally applicable.

As for line emissions measurements, the presence of far trains may be relevant, so that the line shall be empty with no trains up to 2 km/20 km, as observed in sec. 5.3.2.

6.3.3 Antenna position

The entire set of antennas (loop antenna up to 30 MHz, biconical and log-periodic antennas, either separated or combined, above it) is used. Positioning with respect to substation is indicated in the EN 50121-2 standard, Annex A: "emission shall be measured at a distance of 10 m from the outer fence of the substation, at the midpoints of the three sides, excluding the side which faces the railway, unless this side is more than 30 m from the centre of the nearest electrified railway track. In this case all four sides shall be measured. If the length of the side of the substation is more than 30 m, measurements shall be taken additionally at the corners." If no fence exists, the measurements shall be taken at 10 m from the apparatus or from the outer surface of the enclosure if it is enclosed. Three measurement points are thus required as a minimum (provided that they are accessible and that enough space is available to position the antennas and to avoid unwanted coupling and influence on antenna parameters. Except for small substations, e.g. for tram and metro systems, railway substation side is likely to exceed 30 m, so that additional points are required.

Heights are measured with respect to the ground level. Antenna orientation will vary depending on the specific measurement: in two cases antennas shall be oriented first to capture the maximum emission (see sec. 6.3.5.1 of 2006 version at steps 1 and 3).

6.3.4 Measurement settings

The frequency range is subdivided into frequency intervals for which different Resolution Bandwidths (RBW) are to specified; the same settings reported in Table 5.3.1 apply also to substation emissions measurements.

It is prescribed that measurements are all taken in Quasi-Peak (QP) mode, except for the observation put forth by the EN 50121-2 standard itself and commented below in sec. 6.4, regarding the interval split and the application of QP detector the least number of times. QP detector time constants shall be as specified in CISPR 16-1-1 [40], as implemented in compliant receivers and spectrum analyzers (it is underlined that in spectrum analyzers when removing the QP setting also the CISPR-compliant RBW values are usually no longer available and the closest standard values shall be used, such as e.g. 300 Hz for 200 Hz, 10 kHz for 9 kHz and 100 kHz for 120 kHz). For the calculation and application of resolution bandwidth correction factor, please see sec. 4.3.4.

Measurements shall be performed by either sweep mode or by selecting some test frequencies. When using QP settings, scan times as specified by CISPR are correspondingly set: some margins for settings are still possible when adjusting dwell time at each frequency. Please see sec. 3.1.1.9 for an estimate of speed-up when passing from Quasi-Peak to Peak detection setting (commented also in sec. 6.4).

Antennas are facing the substation and measurements are taken for all the trains positions along the traction line; the Quasi-Peak mode requires such a slow frequency scan that, correctly, various different traffic conditions are measured and collected while scanning only a few points. So, no attention shall be given to synchronize the measurements with specific train behavior, provided that the traffic is normal and the loading conditions specified in sec. 3.7 of the standard are met.

6.3.5   Measurement execution

The measurement procedure follows the requirements of the reference standard EN 50121-2 (2006), sec. A.3, clarifying steps and operations. The procedure is as most complete as possible, including considerations on acquisition of background noise traces and verification of correlation with system operating conditions. The procedure has changed in the 2015 version of the EN 50121-2 [29], that makes now no distinction between measurements of emissions from the line and from the substation as far as antenna positioning: as we see below, antennas are not rotated any longer, but left fixed as for ordinary line emissions measurements (see sec. 5.3.3 and 5.3.5).

6.3.5.1   Measurement procedure

Measurement positions are identified as described above in sec. 6.3.3. For each of them the following steps are applied as per sec. A.3 of the 2006 version:

1. the maximum emission level around 1 MHz (exact frequency to select on site to avoid the overlapping of other radio transmissions), measured by loop antenna in vertical plane polarization, annotating the angular orientation of the antenna for which the maximum occurs; substation loading shall be at least 30% of rated load; the base of the loop antenna shall be between 1 m and 1.5 m above ground;

2. keeping the loop antenna in the same position and orientation of step 1, measure emissions over the frequency range 9 kHz to 30 MHz; again substation loading shall be at least 30% of the rated load; not all the maximum values at each frequency occur at the same height above ground and angular orientation (this is what the standard says in the note that reads "it is accepted that the fixed antenna position may result in values being less than the absolute maximum at some frequencies;"

3. the maximum radio emission over the frequency range 30 MHz to 300 MHz is measured by vertical dipole or vertical biconical antenna (the standard says "typically", so other antennas may be used, provided that a demonstration f equivalence can be given); substation loading shall be at least 15% of the rated load; the antenna center shall be 3 m above ground;

4. the maximum radio emission at a frequency around 350 MHz (again, selected on site to avoid the overlapping of other radio transmissions) is measured with a vertically polarized log-periodic antenna, annotating the angular orientation of the antenna; substation loading shall be at least 15% of the rated load; the antenna center shall be 3 m above ground;

5. keeping the log-periodic antenna in the same position and orientation of step 4, measure emissions over the frequency range 300 MHz to 1 GHz; again substation loading shall be at least 15 % of the rated load.

As anticipated, the procedure has changed in the 2015 version of the EN 50121-2 [29] and now antennas are oriented as for line measurements (fixed height, fixed orientation), keeping valid the criterion for positioning appearing in sec. 6.3.3. Without any further specification, substation emissions as per 2015 version may be measured in sweep mode.

6.3.5.2 Background noise

To validate measurement results and compliance with limits, background noise measurements aim at identifying external sources and their variability: several recordings are in general needed in order to reach statistical significance and a complete picture of pre-existing external sources and disturbance. Recordings of background noise shall be made prior to the beginning of the test and at the end of the tests, if possible.

6.4 Practical considerations

The time required for measurements is quite long as Quasi-Peak scanning of the frequency intervals at steps 2, 3 and 5 in sec. 6.3.5.1 above require several hours. Over such a long time interval it is quite difficult to keep loading conditions constant and to assume that background noise and external sources keep unaltered. Even taking background noise specifically for the frequency range of each step of the procedure one at a time, the single QP scan time may take several hours to finish. Faster measurements can be achieved by changing the detector, even if it shall be underlined that QP setting is required by the standard.

Quasi-Peak detector is required for substation emissions, on the ground that they are a stationary measurement; the possibility of time-varying conditions and non stationary behavior is not considered: however, traffic and loading conditions may change widely within seconds and thus the level of emissions, related to load conditions. To cope with this intrinsic variability, a faster method than QP detector shall be adopted: as noted in EN 50121-2 [27], sec. 5.1.3, a max-hold setting may be used (implicitly assuming a Peak detector) and only if limits are exceeded a Quasi Peak reading shall be used; it is intended thus to extract the frequency intervals where limit is violated (or the necessary margins are not ensured) and to pass them to Quasi-Peak detector measurements. This statement needs some clarifications and further considerations:

- by experience, the probability of exceeding limits using Peak detector and max-hold setting is relevant and this may occur at several sub-intervals of the swept frequency interval;

- from a practical viewpoint, in some cases (in particular for Spectrum Analyzers with respect to EMI Receivers) it may be difficult to quickly retrieve and combine the desired frequency intervals for measurement;

- additionally, the standard does not say anything about Peak detector and sweep settings: it is known that in case of noise and noise-like signals the readout of Peak detector is larger when adequately long dwell times are used: see end of sec. 3.1.1.9 for an estimate of the correction factor under broadband assumption; conversely, for narrow-band components correction is no longer valid and the difference between Peak and Quasi-Peak detector readouts is much smaller.

7

Radio interference and emissions from sliding contact

Interference to radio is only partially due to the intermittent emissions caused by electric arc at the current collection point: emissions from on-board equipment, and in particular static converters, may also be relevant up to hundreds of MHz, besides the fact that the vehicle or train body may affect the propagation of radio signals, causing e.g. scattering, attenuation, etc. and represent a source of variability for some measurements.

In the following we focus on arc emissions and the impact on victim systems nearby: the problem is divided into the characterization of electric arc emissions, their impact on radio transmission, and the effect of the vehicle on radio transmission and quality.

7.1   Sliding contact emissions

Current collection relies on sliding contact at the vehicle-catenary interface that is represented by the sliding contact: depending on the type of system (railway, metro, tram, etc.) current collection is implemented by mean of one or two pantographs, or several sliding shoes, featuring different current intensities, contact surfaces, materials. Addi-

(a)

(b)

Figure 7.1.1 – Examples of electric sparking for current collection at pantograph.

tionally, also a variety of shapes of the catenary system (overhead contact wire, rigid catenary, third rail) and of vehicle speeds increase the complexity of the problem of describing electric arc dynamics and electromagnetic emissions.

There exists a series of electric arcs to transfer the current from the contact wire to the pantograph, or shoe, as shown in Figure 7.1.1. At the current collection point there are always several arcs at any instant of time, whose characteristics (average current, life time, mobility, length) depend on the total collected current, each carrying ordinarily a few tens of amperes; the formation and extinction of the electric arcs generates RF noise. Two radiating mechanisms may be considered: emissions from the electric arc itself as a short Hertzian element and emissions from the overhead contact wire (or third rail), excited by the electric arc current.

7.1.1 Electric arc characteristics

The characteristics of the electric arc are considered, focusing both on arc geometry (arc length and arc orientation), but also on its dynamic behavior (time duration and arc stability).

7.1.1.1 Arc length

The arc length is statically determined by the inter-electrode gap. For a sliding contact the arc root will move slightly to follow the previous arc root: the reason is that the extinguishing root is surrounded by ionized air creating a lower impedance path, that is easier to follow for the arc current; the arc length will be, however, slightly longer than the gap length between pantograph and catenary. So, arc length increases for moving electrodes, changing also its orientation to follow the moving spot.

At a given current the arc length can increase up to some limit value after which the arc blows out; a stable arc may be defined as one with a steady presence of the arc column captured with high-speed photography, or, observing the arc voltage, one for which the voltage is stable with no significant fluctuations. When considering current collection at pantograph in dynamic conditions (train speed and tolerances to accommodate the movement of the pantograph and a range of mechanical dynamic responses from the catenary system), the inter-electrode gap is set by design at significant values of about $5-10$ cm, much longer than several inter-electrode distance values considered in studies for magnetic launchers and welding applications.

The lateral movement between the catenary and the pantograph is due both to the staggering of catenary (that runs in zig-zag fashion with respect to the track axis) and lateral train movement, for example at curves and passing over switches. From the viewpoint of the direct effect on arc length, not too a fast lateral movement may be tolerated, increasing slightly the arc length; how the relative movement of anode and cathode influences arc stability has not been however satisfactorily investigated.

7.1.1.2 Average Arc Lifetime (AAL) and arc stability

The average arc lifetime is defined as the time interval from arc initiation until self-extinction; the process is in reality more complex if the attention is on the characterization of arc duration for radiofrequency emissions, because current collection occurs on a multitude of smaller arcs sharing the overall collected current and settling onto values that are a tradeoff for stability; smaller arcs will last much less and will have a lesser contribution to electromagnetic emissions in terms of radiated power, possibly occupying the highest part of the spectrum of emissions.

Klapas, Hackam and Benson [82] report investigations where ac arcs were able of sustained operation up to speeds of 460 km/h ensured when a large current intensity of 2 kA is used; the inter-electrode gap for these tests was $1.4-2$ cm. On the contrary, dc arcs are much less stable and tests performed with a similar inter-electrode gap (2.5 cm), but with lower current (250 A), indicate an AAL of few ms. They underline that arc stability is much improved by an order of magnitude if the gap is protected by a few mm of insulating material that confines the arc on the inner surfaces of electrodes: this is of course not possible for the usual current collection schemes using pantographs, whereas it is what happens on third rail systems. Again other experiments done with smaller inter-electrode gaps of few mm indicate dc stable arcs at large speeds of 150 km/h with 300 A of collected current.

As already pointed out, the arc length for pantographs is normally between a few and ten cm: a tighter gap would be difficult to manage because of the mechanical dynamic response of the catenary system and the complex control system necessary for the pantograph; a longer length breaks the arc very easily, increasing instability and wearing out electrode surfaces.

AAL depends on the arc current and inter-electrode gap: observing Fig. 3 in [82], at low current (50 A) the AAL decreases rapidly with increasing gap (90 ms at 1 cm gap, 9 ms at 4 cm and 0.5 ms at 10 cm), whereas at larger current (200 A) the arc is more stable (between 2 and 3 ms) with no apparent dependency on gap value and approximately equal to the AAL obtained for a gap of $5-6$ cm at 50 A.

Stability is favored by cathode geometry: pointed electrodes give much shorter AAL around a few ms, while a widened hollow-cratered electrode improves it by nearly an order of magnitude. It is observed that the geometry of the pantograph sliding contact is in this direction and that this favors arc stability, besides being required to reduce mechanical impact.

It is understood that when the arc current is significant (e.g. above 50 A), the negative effect of self-magnetic disturbance becomes relevant, acting to decrease arc lifetime: at very large current intensity the control of the magnetic field distribution may be quite useful in stabilizing arc, wearing and electromagnetic emissions.

7.1.1.3   Total radiated power

As an introduction to the next section 7.1.2, the overall radiated power and electromagnetic field intensity is considered as a function of the just reviewed arc characteristics.

The measured radiated power is inversely proportional to gap and arc length [91]: arcs of some mm radiate ten to a hundred times more power than longer arcs of about 10 mm. The reason is that when the arc gets longer, its impedance between anodic and cathodic roots is getting larger and consequently the flowing arc current is smaller. Following the Hertzian dipole radiation model in sec. 1.3.2 the electric field intensity is linearly proportional to the current intensity, so that it is expected that the radiated power is proportional to its square. In [91] the experimental results indicate a more drastic dependency with a reduction of two orders of magnitude doubling the gap length, that the authors do not explain. From personal experience and observing the results of other references cited in the following, a less drastic dependency is expected observing that a longer plasma column radiates more efficiently if compared to the wavelength and because in the Hertzian dipole expressions the field intensity is directly proportional to the element length. Additionally, it is not clear if the power measured in [91] captures equally well terms with arbitrary polarization and when the arc length increases in dynamic conditions (i.e. following a moving cathode) it stretches tilting onto the horizontal plane from the straight vertical position of the static condition.

Also train speed increases the arc length and thus in turn reduces radiated power: the results in [91] show a drastic reduction passing from standstill conditions to moderate speed of 14 to 24 km/h; however, a further speed increase to 30 km/h causes a significant radiated power increase that the authors cannot explain.

Electric arc radiative characteristics are not only determined by the geometry of the plasma column, but a more accurate model of arc dynamics is also needed, to better define ignition and extinction times (possibly including their distribution and not only the AAL value) and the number of individual arcs implementing the current collection, that in turn determine the arc conductance function, thus characterizing the current spectral components flowing in the plasma column. A comprehensive model is presented in [161], where the Mayr's model is accurately reviewed, giving insight into model equations. Further details appear in [24, 89, 133].

7.1.1.4 Surface wear

Another factor is introduced, that is the surface conditions of the sliding electrodes and the reciprocal influence with the phenomenon of arc discharge.

It is intuitive that the wear rate of the contact strip has a proportional relationship to the amount of arc discharge occurring during sliding, as clearly stated in [86] for a sintered metal alloy strip; the authors demonstrate that the determinant factor is the total energy involved in the discharge process. Some other results are referred to metalized carbon contact strip and copper contact wire combination [82].

In [86] a copper rotating disk simulates the sliding contact against a metalized carbon strip for a moving train: it was treated with abrasive materials to create an initial uniformly distributed roughness of $60 - 120\,\mu$m and then run for an equivalent distance of 50 km, measuring the profile with a laser scanner every 8.3 km (five minutes). At first large amounts of energy were delivered (50 J) corresponding to the largest particles (spherical metal particles and carbon fragments), decreasing to 10 J at the end of the test; it is underlined that with the progressing wearing of disk surface, being no adjustment of distance between the two sliding surfaces, the contact force was decreasing from the initial 15 N to about 4.

Arc power and repetition intervals were measured in [86] over one revolution of the wheel that simulated the sliding catenary: the speed at the wheel periphery is 27.8 m/s (or 100 km/h), the wheel diameter is 1 m, so that the wheel does a complete revolution in 113 ms; the values are thus to be multiplied by about 10 (more precisely 8.85) to obtain per-second estimates. Results for electric arc power and total time of discharge per revolution are reported in graphical form in [86]: the power level was constant at about 1600 W with the number of revolutions that stand for the run distance; the time duration of the arc discharge was longer at the beginning (about 30 ns) reducing progressively to less than 10 ns before 5000 cycles, that is at about 15 km of simulated run.

As for the dependency of arc energy on the flowing current, tests were made at 50, 100, 150 and 200 A: the arc energy is larger for larger current but the proportionality is not evident because of the larger variability of recorded energy for increasing current; it may be said that above 100 A the average values of arc energy are almost similar, with different behavior in terms of dispersion and periodicity, with always an initial five minute interval (corresponding to about 15 km of run) of larger energy and more intense wearing (about 50 to 100 J maximum), followed by the "steady wear" region, where the energy is between 10 and 50 J).

7.1.1.5   Electrode materials

In railway applications the choice of materials is limited to copper and copper alloys for the catenary, but a steel layer is used on third rails, and there is a wider range of choice for the sliding contact material.

Observing the results in [46] the best combination to keep arcing low is copper-copper, with a very similar behavior when copper is replaced by a graphite-copper composite identified as CGCM. On the contrary, even small amounts of iron and ferrous materials promotes arcing and due to the lower thermal and electrical conductivity, higher temperatures and more wearing.

An extended analysis of carbon-copper or graphite composite sliding contacts is reported in [86].

7.1.1.6   Effect of asperities and iced catenary

So far the phenomenon of electric arcing has been considered for smooth surfaces with possibly increasing wearing. It is underlined that due to exceptional events and unavoidable lack of homogeneity and uniformity, hot spots may occur at some points along the catenary (e.g. catenary junctions and neutral sections), as well as at some points where an initial flash or discharge occurred for unknown circumstances (e.g. raising of a pantograph, accidental deposition of melted material and debris, clash with a worn out or broken pantograph).

Asperities and ice increase the intensity and rate of occurrence of arcing, but the lack of quantitative assessment by measurements to author's knowledge does not allow to confirm the impression that the spectrum extension is wider for smaller faster arcs.

7.1.2   Electric arc radiated emissions

At the sliding contact arcs are characterized by a length, are assumed of negligible thickness and ignite and extinguish with a chaotic behavior. In general, they are characterized by:

- arc length L, which is related to the inter-electrode gap length and electrodes relative speed;

- arc current I, which is defined by the traction current and by the impedance of the supply line and of the locomotive equipment at high frequency;

- electrode construction and geometry: flat cathodes reduce arc stability; two dimensions arc stabilization is achieved when the distribution of arc current on the cathode surface is symmetrical around the arc root, and this is obtained with a hollow cathode; the shape of anode surface has little effect on arc stability;

- RF emissions are related to average arc lifetime (AAL), which is a direct measure of arc stability, the lifetime of an individual arc being defined as the time interval from arc ignition until arc self-extinction.

7.1.2.1 Simplified model

A simplified arc model for RF emissions corresponds to a short current element (Hertzian element) of length L and current I, the latter interpreted as the RF current, in terms of frequency domain spectrum. It may also be assumed that that the current is nearly constant $(= I_0)$ for a broad frequency range (from dc to some MHz), as confirmed by values of similar dc electrified urban rail transit systems in the literature and direct comparison of model predicted values of electric field at 10 m distance with laboratory measurements performed on free-burning arc in air. Above some MHz the current must fall with $1/f^{1.6-2.0}$ law, to have the electric field in agreement with measurements and data in the literature. Hence the expression for current I

$$I = \begin{cases} I_0 & f < f_0 \\ \\ I_0(f/f_0)^{1.8} & f > f_0 \end{cases} \qquad (7.1.1)$$

where the exponent was approximated with 1.8 and the corner frequency f_0 ranges approximately between 1 and 10 MHz.

Under far field assumption the electric and magnetic fields are expressed by (1.3.28) and (1.3.29). So, electric field strength can be expressed simply as

$$E = \frac{kIL}{4\pi\varepsilon rc} \qquad (7.1.2)$$

with a straightforward relationship between the current intensity as a function of frequency and the electric field intensity.

A current value of $I_0 = 10$ mA at frequency $f_0 = 1$ MHz is obtained to correctly interpolate the measured values shown in 7.1.2 by means of (7.1.1) and (7.1.2) calculated at a distance $r = 3$ m and for an arc length $L = 10$ mm. Of course, the measured data points are few and the estimate of I_0 and f_0 cannot accurate.

7.1.2.2 Experimental results

Previous measurements [107] indicate that dc arcs are quite stable up to 150 km/h for 300 A arc current and 3 mm inter-electrode gap; high speed and small current make the arc less stable and especially at high speed the arc is remarkably skewed towards the horizontal plane and tend to a more erratic movement of the plasma column.

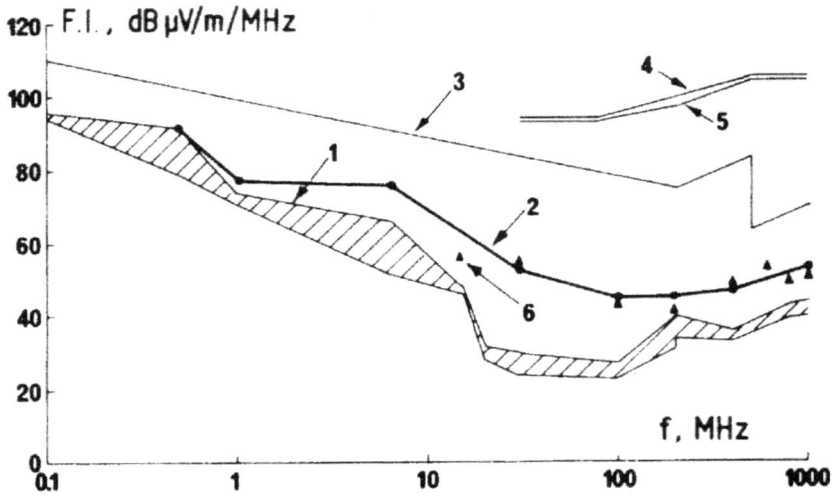

Figure 7.1.2 – Electric field emissions versus frequency measured at 3 m distance from a free-burning arc (10 mm long at 100 A arc current), curve 2; background noise is the shaded polygonal area 1; curves 3, 4 and 5 were limits considered for discussion in [83].

Electric field profiles A sample electric field profile of a free burning arc is taken from the literature [83] and shown in Figure 7.1.2.

Observing the values of electric field intensity at 3 m distance measured and reported in [83], a slightly larger level of emissions is experienced for smaller arc current (see Figure 7.1.4), as was already anticipated in sec. 7.1.1.2 considering the influence on arc stability. Emission levels are slightly lower for longer inter-electrode gap, but not significantly.

Regarding the behavior with respect to arc current and as a function of frequency, Figure 7.1.3 reports a substantial insensitivity and a peculiar behavior, for which the 30 and 100 MHz bands have the lowest emissions (as it may be clearly seen also in Figure 7.1.2) and the 100 MHz measurements show also a remarkable dependency on arc current. It is not believed that this behavior may be general, especially observing that also another curious fact is reported: the 1000 MHz band has a remarkably larger level of emissions, with the preceding 400 MHz band already increasing with respect to the minimum reached between 30 and 100 MHz; it is believed that this may be due to an abnormal response, such as a resonance, of the test setup, since the background noise is nearly 10 dB lower, even if shows the same increase of intensity of recorded emissions.

Decaying slope A confirmation of the expected decaying behavior is given in [160], Fig. 11 and 17, where the spectra of the electric field emissions for ESD experiments are reported: despite the smaller energy of an electrostatic discharge and arc current values not larger than a tens of A, the reported spectrum density in V/m/Hz matches well the expectations. Two experiments are reported: an air discharge in Fig. 11 that is faster with a more rich frequency content, beginning to decay at about 400 MHz with a −40 dB/decade slope; a contact discharge against a plate in Fig 17, slower that the air discharge, with nearly ten times the field intensity, and beginning the decaying slope of $-40\,\mathrm{dB/decade}$ at about $100\,\mathrm{MHz}$.

Tanaguchi et al.[1] report a similar frequency dependency (slope obtained by linear regression of measured spectra is about $1.8-1.9$) for electric arcs from sliding contacts inside contactors, so for a different application with different sliding speeds and materials.

The results of the measurements performed with an electric probe next to the pantograph on the train roof and inside the train at a relevant distance several coaches away such to ensure a significant attenuation are shown in [92], Figure 2: the peak intensity is $12\,\mathrm{kV/m}$ and $0.3\,\mathrm{V/m}$, respectively, and the frequency occupation extends up to $1\,\mathrm{GHz}$, for outside measurements and reduces to $100\,\mathrm{MHz}$ for the inside location.

In [91] the broadband characteristics of the electric arc phenomenon and the best RBW setting for its measurements are considered by comparing measurements made inside a reverberation chamber with increasing RBW values from $100\,\mathrm{kHz}$ to $5\,\mathrm{MHz}$: the relationship is nearly linear as expected for the stationary situation where anode and cathode are not sliding; another test performed at $30\,\mathrm{km/h}$ sees deviations from linearity up to about $7\,\mathrm{dBm}$, but distributed randomly with plus and minus sign, so to be interpreted as due to experimental variability, for which however the authors give no assessment nor indication.

Shinkansen impulsive noise Measurements of impulsive emissions from Shinkansen trains were reported in [117] in terms of crossing rate distribution (CRD) and amplitude probability distribution (APD), besides other characteristic values, such as rms and average. In this case measurements cover emissions from electric arc itself and the rest of the train. Measurements were taken using two dipoles, tuned to 50 and $100\,\mathrm{MHz}$ and alternatively oriented in the two polarizations, at a measurement location $40\,\mathrm{m}$ from the line; the height above ground was set to $4\,\mathrm{m}$. The adopted resolution bandwidth was quite small ($30\,\mathrm{kHz}$) reducing the contribution of background noise. Trains going upward (eastbound) and downward (westbound) were monitored during coasting/cruising, acceleration and braking, with the latter operation resulting in the largest emissions: for quick interpretation of results upward direction corresponds to deceleration approaching the station at $3\,\mathrm{km}$ distance, viceversa downward direction sees the trains accelerating leaving the same station; the condition of coasting/cruising occurs when the train did not stop at the station. The measurements were collecting the entire train emissions, without the possibility of distinguishing between electric arc and on-board apparatus. The traction system is $25\,\mathrm{kV}$ $60\,\mathrm{Hz}$, so that electric discharges at phase reversals are expected. The intensity of emissions increased when the train was approaching the measurement site and decreasing after its passage, lasting for about $12-14\,\mathrm{s}$, demonstrating that the direct radiation was measured, as it is expected for such a high frequency of 50 and $100\,\mathrm{MHz}$.

Horizontal polarization (indicated by "H" in the paper) resulted with the largest values in some cases of evident emissions above background noise, indicating the contribu-

---

[1] M. Taniguchi, T. Inoue and K. Mano, "The Frequency Spectrum of Electric & Sliding Contact Noise and Its Waveform Model," *IEEE Transactions on Components, Hybrids and Manufacturing Technology*, Vol. 8, no. 3, Sept. 1985, pp. 366-371.

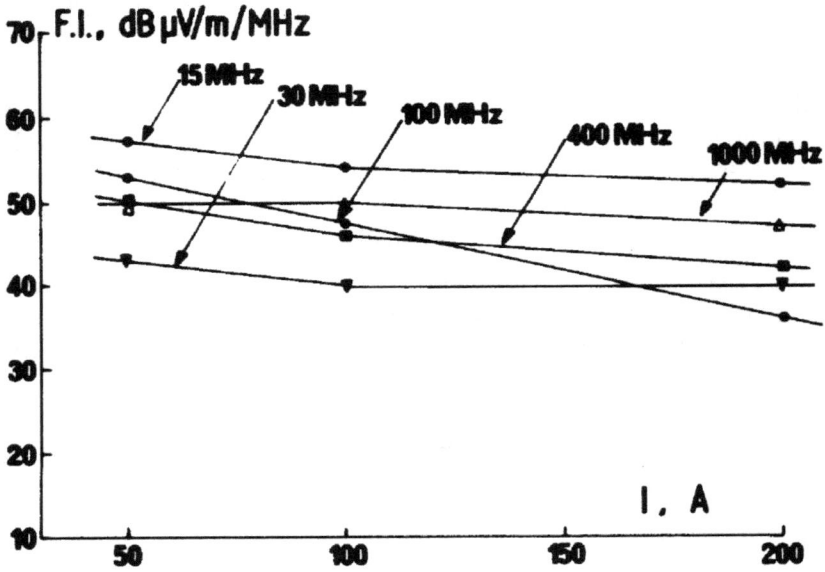

Figure 7.1.3 – Electric field emissions from a free-burning arc (10 mm long) measured at 3 m distance versus arc current and for five frequencies between 30 and 1000 MHz [83].

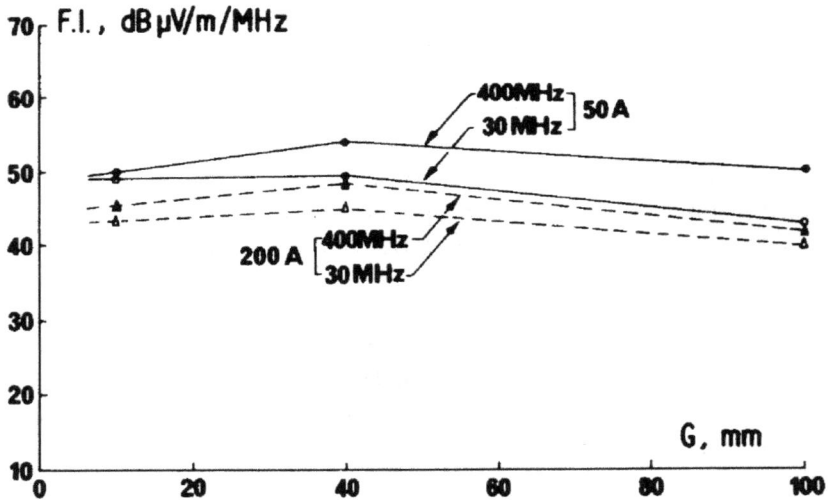

Figure 7.1.4 – Electric field emissions from a free-burning arc versus interelectrode gap measured at 3 m distance for two frequencies and two arc current values [83].

tion of the catenary system; however, in several cases vertical polarization gives values that are comparable to those of the horizontal one, indicating thus the presence of background noise independent on train emissions or directly from train pantographs.

Sample results are shown in Figure 7.1.5. The general prevalence of 50 MHz emissions in Figure 7.1.5(a) indicates a downward spectrum envelope: the difference between 50 MHz and 100 MHz is about 10 dB at the largest levels above about 50 dBμV/m; this implies a downward slope of about 33 dB/decade. Conversely, in Figure 7.1.5(b) profiles for 50 MHz and 100 MHz are not clearly separated, alternating for varying field strength and polarization.

7.1.3 Overhead contact wire radiated emissions

More significant emissions are caused by the electric arc current flowing in the catenary system, rather than directly from the electric arc itself, that is a comparatively smaller and less efficient radiating element, even including part of the pantograph in the radiating circuit. For the high frequency components that characterize electric arc emissions (e.g. between a MHz and one or few GHz), the catenary excited by the electric arc current may be seen as a long thin dipole, with waves originating at the current collection point (i.e. at the sliding contact where the electric arc is located) and propagating in both directions. The thin linear antenna assumption holds (in place of a more precise model based on cylindrical antenna equations), because the introduced approximation errors are low for large length/diameter ratios and in any case smaller than those related to the knowledge of other system characteristics (e.g. the impedance of the electric arc).

The far-field radiation pattern was defined in (1.3.30): focusing on the maximum of the E-field intensity, it can be noted that it occurs for different values of θ and always for $\theta = \pi/2$; for simplicity only the latter solution is considered, which corresponds to the smallest distance from the source, leading to

$$|E|_{\max} = \frac{60I}{r} \left[1 - \cos\left(kl/2\right)\right] \tag{7.1.3}$$

The function $\cos\left(kl/2\right)$ with argument much larger than π oscillates rapidly between -1 and $+1$; it may be said that

$$|E|_{\max} \leq \frac{120I}{r} \tag{7.1.4}$$

Following (7.1.1) and using estimated values for I_0 and f_0, it is possible to derive simple estimates of the expected electric field intensity as a function of distance from the catenary; the expression is clearly derived assuming far field and should be used at a high enough frequency, approximately above 30 MHz.

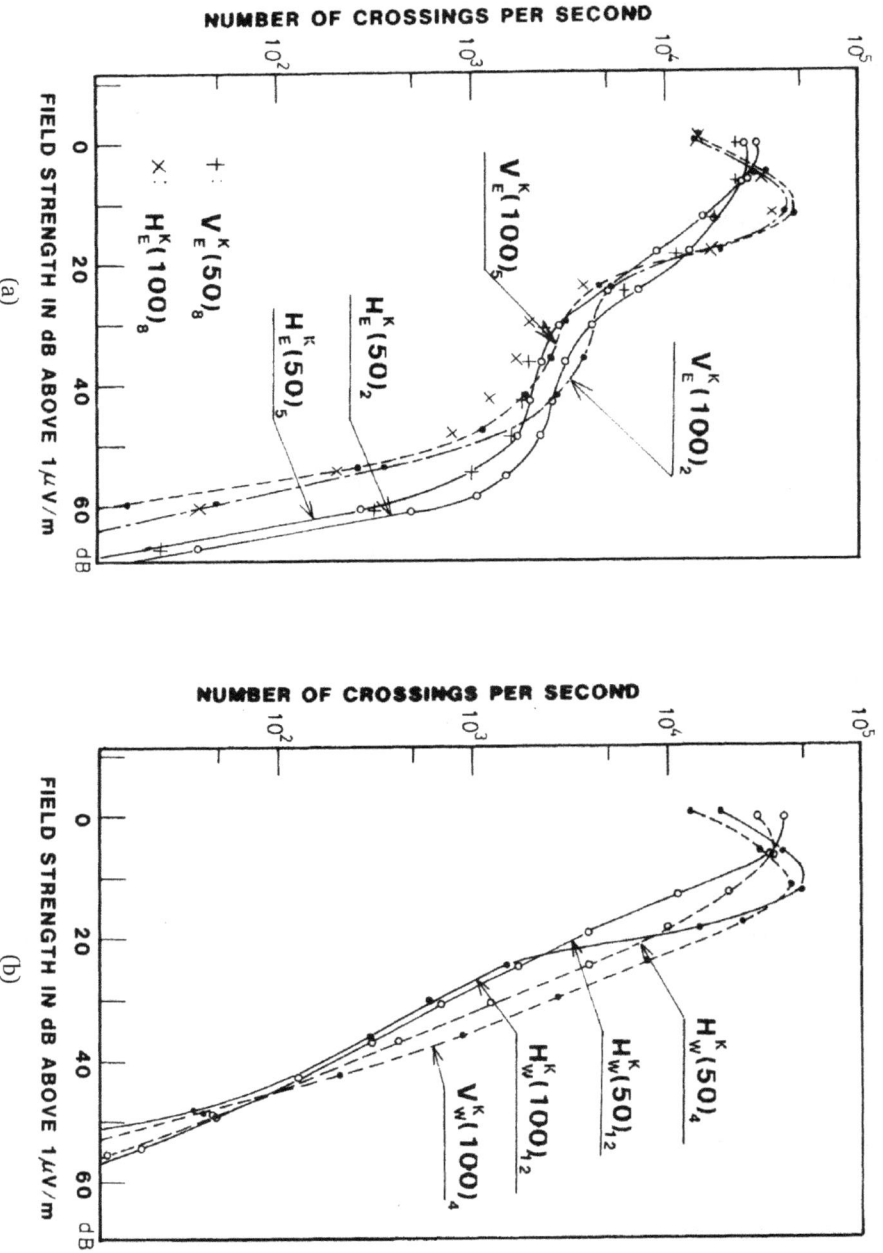

Figure 7.1.5 – Measurements of impulsive noise from Shinkansen trains displayed as crossing rate distributions: (a) eastbound direction, braking mode, Kodama type trains, and (b) westbound direction, acceleration mode, Kodama type trains. The letter "V" and "H" indicates vertical and horizontal polarization, the number "50" and "100" indicates the frequency, the small subscript refers to the set of results [117].

7.2 Interference to radio systems

High frequency emissions from railway systems and rolling stock may be intermittent and have a broad spectrum occupation. Emissions from the current collection mechanism have just been reviewed; in addition to these, switching transients inside converters may also be included as a source of high frequency emissions able to impact on the operation and performance of some radio systems.

Radio systems may be considered in different ways regarding their performance: besides radiation efficiency, radiation pattern and electric field intensity that regard their own emissions in the transmission band, spectral purity, spurious emissions and adjacent channel power are also relevant elements, that characterize the mutual effect between concomitant channels and the compliance to the assigned bandwidth, and thus potentially the threat onto adjacent radio services. The effect of the interference, not only for mutual effects between channels, but also due to external disturbance, is first of all a reduction of the signal-to-noise ratio, but most important the increase of the number of errors in the received symbols and, as a consequence, of the number of discarded messages and repeated transmissions.

Radio system performance in relationship to external interference is usually measured in terms of number of correctly received bits, or packets, or frames, depending on protocol, or more directly as receiver interference and RSSI (Receiver Signal Strength Indicator), that is a measure of receiver signal available with a software call in many systems. These parameters are usually collected under the term "Quality of Service (QoS) indicators" and are available in simulators, sniffers, etc.

Whereas the signal to noise ratio may be determined independently from the protocol by means of a double measurement or estimate (noise power measured in the assigned channel due to external disturbance and signal power measured or calculated, based on available transmitted power or minimum received power, i.e. sensitivity or a similar threshold), QoS indicators shall be determined knowing the protocol and the implemented decision criteria at the receiver, any correction algorithm (i.e. the so called forward error correcting codes, FEC), and tracing the passing data packets, including both payload bits and control bits. It is thus evident that the evaluation and analysis of QoS performance shall be tailored onto the specific protocol.

A brief overview is given of the most relevant protocols in relationship to the analysis of possible interference with respect to railway system applications:

- TETRA (Trans European Trunked Radio), operating mostly in the 380 to 420 MHz band, but possibly also in the 860 to 880 MHz;

- GSM-R (Global System for Mobile - Railway signaling system), operating in the two reserved bands for uplink ($876 - 880$ MHz) and downlink ($921 - 925$ MHz);

- GSM for passengers and public use; the most common bands are $890 - 915$ MHz for the uplink (starting from 880 MHz for the extended GSM) and $935 - 960$ MHz for the downlink (starting from 925 MHz for the extended GSM); other bands are the 1800 MHz one ($1710 - 1785$ MHz for uplink and $1805 - 1880$ MHz for downlink), and the American equivalent bands at 850 MHz (824 − 849 MHz for uplink and 869 − 894 MHz for downlink) and 1900 MHz (1850 − 1910 MHz for uplink and 1930 − 1990 MHz for downlink);

- UMTS has been added mostly in the 1700 MHz band (1710 − 1755 MHz for uplink and 1845 − 1880 MHz for downlink) and 2100 MHz band (1920 − 1980 MHz for uplink and 2110 − 2170 MHz for downlink); also 3.5 GHz services have been recently enabled in some countries to implement LTE and WiMax;

- WiFi (Wireless Fidelity) in the commercial 2.4 GHz and 5 GHz bands and sometimes on adjacent reserved bands to avoid interference and tampering from commercial devices, e.g. for train-to-wayside communication.

Stenumgaard [147] reports a thorough overview of digital systems and their basic characteristics as a starting point for analysis of interference, similar to that carried out in sec. 7.3 below for GSM-R.

Not only radio broadcasting and mobile communication systems are possible victims of electric arc disturbance: other systems used for specific communications between remote points and with particularly low reception levels may be examples of further potential victims. Aircraft communication and landing systems are a typical example, with safety and mission critical functions, such as the ILS (Instrument Landing System), operating at 110 MHz with a two-tone amplitude modulation at 90 and 150 Hz. This frequency band is not only still close to the frequency interval where rolling stock and catenary system emissions are recognizable, but occurs in the full power bandwidth of the spectrum of electric arc emissions, extending right up to a hundred MHz. In any case strong, yet out-of-band, pulses may overload the ILS input stage and cause intermodulation products to appear. In the example considered in [54] a catenary disjuncture at an insulated section exacerbates the phenomenon, increasing the amplitude of measured transients: at 100 m distance a peak of 0.8 V was entering into the ILS receiver; in terms of electric field the sensitivity specified by the International Civil Aviation Organization is 40 dBμV/m and the ILS sensitivity measured in [54] was about −100 dBm. However, for the case considered in [54] the interference onto the ILS was negligible, as directly tested with measured and synthetically reproduced signals; the reason may be a combination of the effectiveness of the ILS input filter cutting a narrow band in the order of some hundreds Hz around the said 110 MHz carrier and a particularly intense sparking at the neutral section, but not as fast and with extended spectrum, as repetitive sparks at the collection system of a high speed line may be.

7.2.1 Evaluation of interference

Modern digital communication systems, as anticipated above, feature a wide range of modulations and data rates, and spectrum occupation changes accordingly. Generally speaking, this has two consequences: the measurement resolution bandwidth used to evaluate disturbance in the radio channel may be smaller or larger than the channel bandwidth, requiring correct interpretation and use of results in either case: as pointed out simply by Stenumgaard [147], "if the measurement bandwidth is considerably smaller than the system bandwidth and the bandwidth of the interference is larger than the measurement bandwidth, then the interference power perceived in the radio system will be much higher than is measured." To cope with this problem in his paper two RBW values are proposed, one complying with emission standards (that is $200\,\mathrm{kHz}$, similar to the prescribed $120\,\mathrm{kHz}$, below $1\,\mathrm{GHz}$ and $1\,\mathrm{MHz}$ between 1 and $6\,\mathrm{GHz}$) and called *narrowband*, and another one from five to twenty times larger, called *broadband*.

The approach that links signal-to-noise ratio SNR (or signal-to-interference ratio, SIR) to the bit error rate, BER (or bit error probability, BEP) is commonly accepted, provided that comparisons are based on the same interpretation of BER (e.g. taken at the same position of the decoding and demodulation chain at receivers), that characteristics of the interfering signal are well analyzed and monitored (e.g. amount of transient peak with respect to rms value, pulse repetition rate, etc.), and that the effect of error correcting codes, if any, is duly considered.

Given a certain BEP P_0, the effect of the pulse repetition rate R_P of the interfering signal shows a saturation effect: the intensity of the pulsed interfering signal V_{rms}^{dB} as measured by the rms detector that causes the same level of BEP reduces from low repetition rates until a threshold value is reached equal to the symbol rate R_S. The reduction is linear on a log scale and was found nearly the same for several digital systems in [146], starting from the Binary PSK and including then a MSK and 64-QAM; the slope is $7.5\,\mathrm{dB}$ for each decade, so that the correcting factor for the rms value of the pulsed interfering signal is $7.5\,\log_{10}(R_P/R_S)$. It is underlined that when talking of the rms intensity V_{rms}^{dB} we are not indicating the height of the applied pulsed interfering signal, nor the width of the pulse, but its rms value as measured by the rms detector set on the radio system bandwidth and the pulse repetition rate: the demonstration is reported in [146] and is based on the assumption that the time duration of the single interfering pulse is short (electric arcs are quite fast to extinguish in our case) and that the rms detector is capturing the power of the transients falling inside the digital radio bandwidth.

The repeated pulse pattern simulates quite well the phenomenon of electric arc induced transients, whereas however the repetition interval is not constant (see for example Figure 7.3.2(d)) and the shape of the single transient impulse is not rectangular, and may have superimposed ringing (as discussed in sec. 7.3.2.1).

Another class of disturbance that is not originating from the current collection mechanism and will not be further considered here is that of emissions from co-located digital systems, such as clocks of microprocessor boards and network interfaces: in [145] it is clearly shown that modulated clocks have a more detrimental effect on BER, that increases quite fast after a threshold on the disturbance power level is reached. This threshold is remarkably lower (from 10 to $20\,\mathrm{dBm}$, that is one or two orders of magnitude) than that for unmodulated ones, that for many types of victim systems increases also more slowly.

7.2.2 Effect of error correcting codes

The digital systems considered so far were uncoded and do not make use of error correcting codes (ECC), that are on the contrary quite widespread at least for the detection of transmission/reception errors and for the correction of the most common errors. In this regard it is underlined that the effect of ECC may be quite complex and variable: ECC are specially designed to handle particularly difficult problems that cause transmission degradation and are not properly addressed by other techniques such as channel equalization, diversity, etc. However, when testing a radio system for the effects of interference, measuring the BER after error correction has several drawbacks: it is in general unclear how to interpret the results, in particular if it is not known how much of the code capacity has been used, so that the margin on the maximum acceptable interference would be unknown; additionally, reducing the BER and frame error rate as the result of applying ECC, increases testing times before a sufficient number of errors is collected in order to reach a reliable statistical estimate [48]. However, against the mere finding that ECC may be taken into account afterwards, by applying the measured BER, there is the fact that ECC performance depends on how errors due to interference are distributed: there are codes able to correct more efficiently errors occurring as bursts of corrupted bits and a smaller amount of isolated bits; additionally, not all bits in a packet or frame have the same importance and are protected in the same way, as it happens for instance for the GSM-R protocol (see sec. 7.3.1.3 below). Plainly we may say that the ECC knowledge needed to consider ECC effects a posteriori is also sufficient to interpret the results for disturbance applied including ECC; by the way for a complete and finished product such as a GSM-R or TETRA terminal installed on-board there is no way of excluding ECC or affecting its operation during tests without the presence of the designer.

In [146] a qualitative estimate is shown of the effect of ECC, protecting the integrity of messages up to very large disturbance amplitudes, with a steep slope downward to the limit $R_P = R_S$ at which the ECC protected protocol has in any case a small advantage called "coding gain" with respect to the uncoded one.

7.2.3 Spectrum distribution of interference

An interesting comparison of electric arc impulsive noise falling into the TETRA and GSM bandwidths has appeared in [126, 127]. The step change in measured noise between the default background white noise and that produced by an artificial ESD source is considered: a 25 dBm increase was observed for TETRA frequencies, whereas only 5 dBm were observed for GSM. The APD profiles of TETRA and GSM are shown in Figure 7.2.1: the offset change of 5 dBm and 25 dBm is clearly visible; the curves refer to a "WGN" situation (white Gaussian noise) and to two similar measurements of ESD disturbance performed in time domain ("OSC") and frequency domain ("EMI"), that result, as expected, in very similar, not to say overlapping, curves.

Further comparison appears in [127], where a synthetic impulsive noise generated according to EN 61000-4-4 is used superposed to AWGN. The obtained APD (shown in Figure 7.2.2) shows again the agreement between time domain and frequency domain measurements and distinguish the regions of the APD curve where the AWGN and

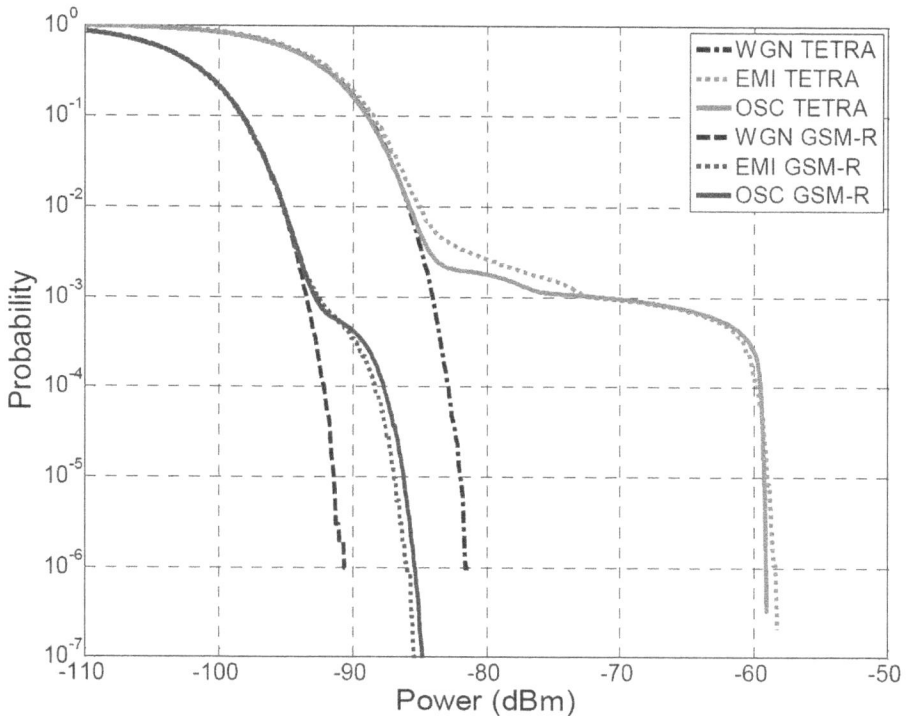

Figure 7.2.1 – APD vs. noise power for TETRA and GSM frequency bands resulting from the measurement of the artificial ESD source using a 30 kHz bandwidth [127].

the impulsive noise exert their influence: seldom large-amplitude noise is due to the impulses, reaching about -42 dBm; the AWGN populates the lower power levels and is characterized by a much larger probability.

The APD curves are useful to define optimized (or minimum) signal power levels for an assigned error probability. When designing a radio system installation, for the determination of the number of transmitters, their power level, location, coordination, which antenna or leaky cable, etc. the design team writes down a power link budget with which assumptions are made of the existing noise level, of the farthest distance from the nearest transmitter, the minimum detectable signal, the effect of obstruction of train body, ending with the so called signal coverage study. Then, during the deployment and subsequent test and commissioning phase the correct operation of the installed system is verified by mapping the measured signal level inside stations and other buildings, and along the right-of-way, the latter moving along the track with one of the first trains, thus implicitly including the effect of the disturbance coming from the current collection system.

In general, it is sufficient to ensure that train and line conditions are such to cause a large enough disturbance to take as a worst case and to give demonstration of the so-created test conditions by measuring the noise power in the channel bandwidth. Of course, when testing the radio system the channel is "occupied" by the transmitted signal, but there are always free channels at some distance from the used one: usually leaving one unused channel as additional slack besides the guard band between

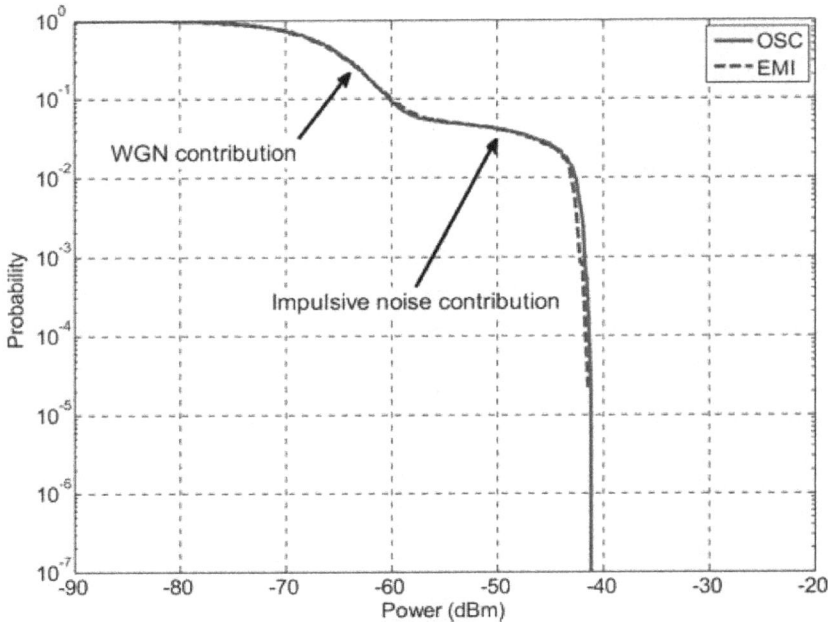

Figure 7.2.2 – APD vs. noise power for TETRA and GSM frequency bands resulting from the measurement of the artificial ESD source using a 30 kHz bandwidth [127].

channels, to reduce further any leakage into adjacent channels, is enough to avoid significant influence.

Examples of in-band noise for GSM-R are reported in Figure 7.2.3 for electric traction and diesel traction systems. Two facts are evident and confirm that electric arc is a relevant intermittent source of disturbance:

- the power level of the noise in the GSM-R channel is lower by about 15 dB for the diesel electric locomotive; the -100 dBm of the latter are mostly due to the intrinsic noise of the GSM-R receiver, being very close to the nominal sensitivity of -96 dBm;

- the power level statistics are completely different, with the 25 kV train showing a standard deviation of 5 dB and a maximum nearly 12 dB above the average value, whereas the noise for the diesel-electric locomotive case is nearly constant, with a maximum spread of ± 2 dB,

For its safety relevance and the close similarity with the most widespread communication system, GSM-R is considered in more detail in the next section.

7.3   Interference to GSM-R system

There have been some significant contributions to this topic in the last ten years, focusing on modeling of interference [20, 128], its quantification by measurement [20, 91, 128], the statistical characterization and the joint time-frequency representa-

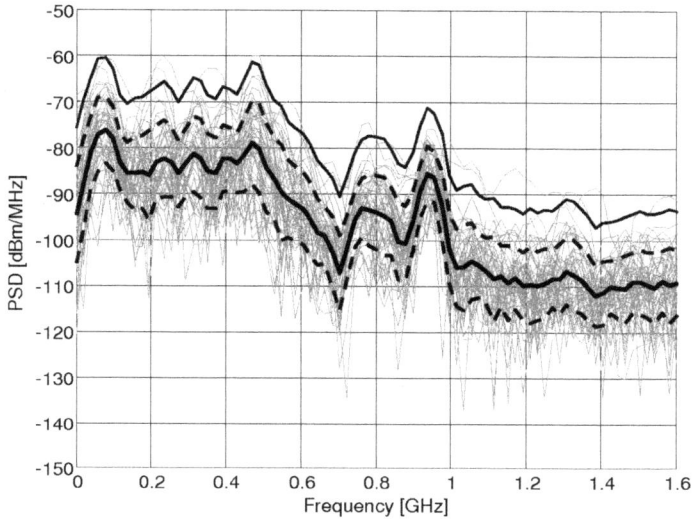

(a)

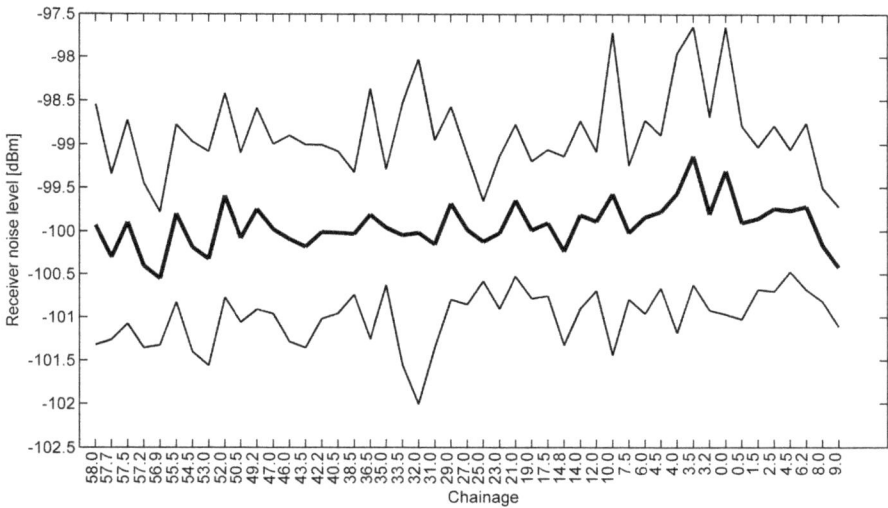

(b)

Figure 7.2.3 – GSM-R channel noise: (a) 25 kV ac high-speed train, (b) diesel electric train.

tion of disturbance [20], the prediction of the interference level and the recognition of disturbance patterns [49] and the relationship between signal-to-noise ratio and QoS parameters, including experimental observations [48].

As all modern communication protocols, GSM (and GSM-R) uses a combination of techniques:

- pulse shaping for spectrum optimization, maximizing channel exploitation: the modulation for the channel symbols is a GMSK, that is a minimum frequency shift keying, MSK, with Gaussian shaping [131];

- use of a mixed TDMA/FDMA (Time Division Multiple Access/Frequency Division Multiple Access), that, besides the subdivision in physical channel, each with its own center frequency and bandwidth, distributes cyclically in time the access to the physical channel among the connected subscribers up to the supported maximum; with burst-like disturbance, lasting longer than a single transient, the TDMA scheme distributes its effect among the subscribers, rather than heavily affecting just one;

- within the payload for each data segment, bits have different levels of protection, based on control bits and forms of encoding, such as cyclic redundant codes, able to correct bit corruption due to noise both in sparse patterns and concentrated in a single larger chunk (e.g. one major noise transient with longer duration).

7.3.1 GSM protocol

The GSM-R system employs two frequency bands of the GSM frequency interval: [921 MHz, 925 MHz] for the downlink, from the RBC to the train, and [876 MHz, 880 MHz] for the uplink, from the train back to the RBC; each band is subdivided into 20 frequency channels of 200 kHz each. The nominal GSM raw data rate is 270 kbit/s, for which a bandwidth efficiency of 1.35 bit/s/Hz may be derived; for GSM-R the data rate over one timeslot (assigned to a single train) is 9600 bit/s, so a total of 76.8 kbit/s.

7.3.1.1 Channel coding

GSM signal is the result of a Gaussian filtering applied to a Minimum Shift Keying (MSK) signal, derived from an original bit sequence of raw bit symbols before channel coding. MSK modulation is a continuous-phase Frequency Shift Keying (CP-FSK); use of a continuous phase modulation ensures that transitions between successive symbols are removed and bandwidth occupation is optimized. The MSK uses two orthogonal signals of sinusoidal shape; the frequency change between the two symbols is half of the bit rate. The use of Gaussian filtering applied to the MSK sequence allows to reduce spectral sidebands, thus reducing further band occupation: reductions of 10 to 20 dB of signal amplitude are possible for frequency offsets of 60 % to 150 % of the bit rate; the counterpart is an increased memory of the modulation mechanism and a possibly larger inter-symbol interference.

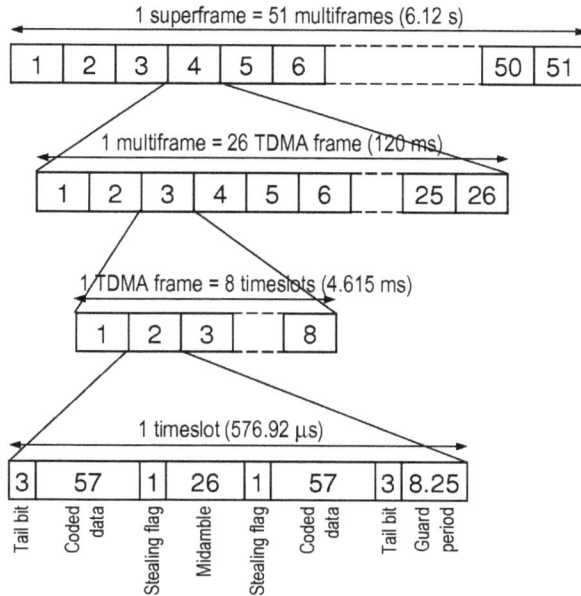

Figure 7.3.1 – TDMA structure of the GSM physical channel.

The expression of the GMSK modulation is non-linear and thus in many studies various forms of linearization are used with a certain amount of approximation of results. Another factor that is hard to include in a simple analysis framework is the effect of channel equalization, that is quite effective in reducing inter-symbol interference: results in terms of bit error rate after channel equalization show an improvement of about a factor of two for various signal-to-noise ratios. For our analysis the signal-to-noise ratio is used directly in the bit error probability expression, with a practical corrective factor for the estimated difference between MSK and GMSK (see below sec. 7.3.1.4).

7.3.1.2 TDMA/FDMA scheme

The channel-frame-slot structure is the same as GSM: after the subdivision in physical frequency channels (FDMA, Frequency Division Multiple Access), onto which a train has its on-board GSM-R equipment tuned, the communication occurs with a TDMA (Time Division Multiple Access) approach. For each frequency channel data are organized as a periodic TDMA frame, with a period of $4615\,\mu s$; each TDMA frame is divided into 8 time intervals of $576.92\,\mu s$ called "time slots" and each user (train) occupies a frequency channel only one eighth of the time with the assigned cyclical time interval of one slot duration; each "time slot" includes 156 bits, so the transmission time of one bit, or "bit time", BT, is $3.692\,\mu s$ [50, 131, 144]. This structure is shown in Figure 7.3.1.

The 26 bits in the center identified as "midamble" are a training sequence aimed at channel estimation.

By inspection of the time-slot structure it is immediate to recognize that a transient, and the consequent bit error, can occur at all positions within it, for both data bits (in total of 114 bits out of 156) and the other bits, used for control, synchronization, etc. From this the fact that bits can be protected by error control codes at different degrees depending on their function and importance.

7.3.1.3 Error control and correcting coding

When using the GSM channel for data transmission (as for GSM-R), if the full rate channel speed of 9.6 kbit/s is considered, the following logical coding mechanism is used. An input block of 240 bits is encoded for data protection with a convolutional code with half rate, so becoming a 480 bit block, brought to 456 bit by removing 1 bit every 15 (called "1 out of 15 puncturing"), so to fit 4 slots with 114 data bits each. Even with slower data rates and shorter input block lengths, the rates of the convolutional code and of puncturing action are adapted for a final output block of 456 bits. The output block is subdivided into a variable number of burst portions, depending on data rate and distinguishing speech and data traffic, in a process that is called "interleaving", aiming at increasing robustness to interference [56].

It is thus evident that a complete modeling of the data protection algorithm with respect to interference and final frame error rate values is as difficult as useless, if a reverse approach may be followed: the efficiency of data protection can be estimated by repeated tests under controlled conditions in terms of SNR and location of interfering transients and building up look-up tables that relate these parameters, the raw BER and the FER. This is a commonly followed approach that simplifies analysis and experimental characterization of channel noise [128]. Any sharp increase of the latter for small changes of the former will indicate a critical threshold for the data protection and error recovery algorithms; noise power may be evaluated including its statistical distribution by means of APD (Amplitude Probability Distribution) [105].

7.3.1.4 SNR and BER

The evaluation of the signal-to-noise ratio requires not only that noise is correctly measured and represented (e.g. broadband white assumption, correlation introduced by the measurement chain, observation interval, consistency of statistical estimates), but also that the correct signal level is defined. When evaluating GSM-R signal coverage by simulation or measurements, an estimate of the effectively available signal power at receiver may be derived; however, a more conservative approach may be followed in which the minimum detectable signal is chosen, not to impact on GSM-R performance.

GSM sensitivity (the minimum detectable signal at receiver) is very low and around −85 to −95 dBm, depending on the type of service. For a railway environment the minimum signal strength (giving sufficient coverage) has been made correspond to −90 dBm for 95% of time, but a more conservative value of −80 dBm may be adopted to account also for burst noise occurrence and to drastically reduce BER also in worst-case circumstances (e.g. partial obstruction of signal propagation, bad weather conditions, etc.). In the determination of the minimum signal strength other phenomena come into play, such as the signal decay with distance and the correct hand-over between two Radio Base Stations (or Base Terminal Stations, BTS), that at the maximum design speed of $500 \, \mathrm{km/h}$ is not a trivial aspect. A higher signal strength or sensitivity value, of course, comes at the price of a larger transmitted power and closer separation of BTSs.

Considering a stationary disturbance with Gaussian distribution in the channel bandwidth, the expression of the raw Bit Error Rate (BER) is [131], pp. 241-257.

$$P_{be} = \mathrm{BER} = Q \left(k \sqrt{\mathrm{SNR}_b} \right) \tag{7.3.1}$$

where SNR_b is the signal-to-noise ratio per bit, $Q(\cdot)$ indicates the area under the tail of the Gaussian distribution (equivalent to $1 - \mathrm{erf}(\cdot)$, with $\mathrm{erf}(\cdot)$ indicating the "error function", that is the Cumulative Distribution Function of the Gaussian PDF), and k identifies a coefficient that takes into account the type of modulation, and in particular how the information is conveyed into channel symbols and the shape of the used symbol waveform. For MSK this coefficient is $k = 2$ (since for an optimum receiver MSK symbols are considered antipodal signals, [131], pp. 271) and for GMSK this coefficient is generally assigned the value 1.8, based on experimental observations, since the theoretical analysis is not straightforward. The result of 1.8 is a compromise between the inter-symbol interference introduced by the Gaussian filter of the GMSK (that in turn compacts the occupied band) and the resulting error performance degradation with respect to MSK modulation.

7.3.2 In-band interference and noise transients characterization

GSM, and GSM-R, is quite robust to out-of-band noise thanks to the hardware filters cutting the overall frequency interval for up-link and down-link and the digital filters that define each physical channel, limiting adjacent channel power and preventing the effects of spurious emissions and channel-to-channel interference.

7.3.2.1 Noise transients: waveforms and statistics

Transient noise caused by the sliding contact at pantograph may vary depending on several system characteristics and conditions, as reviewed before in sec. 7.1. Transients captured by the GSM-R receiver through the roof top antenna have in any case some common features that may be quickly synthesized as: steep front caused by the electric arc ignition possibly slowed down by the antenna transient response; longer decay also due to the slower extinction of the individual arc; repeated oscillations due to the self resonance of the antenna triggered by the electric arc transients. The antenna is a band-pass system whose self resonance is centered in the operating band; for multi-band antennas (as for the roof top antenna considered in [20]) oscillations occur at different resonant frequencies, creating a more complex response pattern; it is possible that the transient response of the antenna influences also the time decay of the transient with additional time constants, but generally arc extinction should be a slower process than antenna time constants.

Recorded transients were processed and analyzed in [20] to extract their characteristics in terms of waveforms parameters and related statistical distributions; it may be objected that such measurements are affected unavoidably by the characteristics of the GSM-R antenna installed on the train used for tests, but it may also be observed that such antenna is quite standard and that different antennas for GSM-R applications have really minor differences, so that these results may be extended in general to all GSM-R applications in the presence of a sliding contact on a medium to high speed catenary system. The identified parameters characterizing the recorded transients are:

- rise time RT of the first peak, possibly, but occasionally, preceded by a "pre-peak";

- time duration (or time decay) TD, taken from the occurrence of the first peak up to the crossing of a convenient threshold, set to 30 % and 50 %;

- repetition interval RI, that characterizes the occurrence of subsequent transients on longer time windows, and that is the most relevant parameter in order to define the amount of interference to the frame bits in terms of repeated corrupted bits and their time distance;

- peak amplitude A of the first peak, distinguishing positive peaks A_+ and negative peaks A_-; the peak-to-peak amplitude $AA = 2A = A_+ - A_-$ is calculated over a movable time window;

- the frequency of ringing FR, quantified by the time distance of adjacent zero crossings in time domain or by a more thorough joint time-frequency analysis

During the analysis in [20] several recordings with non-unimodal PDFs and weird or inconsistent data were noticed, due to several reasons: superposition of two or more electric arc events, external disturbance (in particular in the same GSM band), wrong scale setting. Although a carefully prepared setup and instrument settings can avoid some of aberrations or mistakes, data pre-processing is always necessary using rules and heuristics derived from a set of expected behaviors of the electric arc emissions (minimum time domain decay, maximum number of oscillations, metrological quality of the recorded waveform, such as visible quantization artifacts and amount of broadband noise). Data passed to statistical analysis and joint time-frequency transforms are thus correct and do not bias negatively the results. The histograms of the waveform parameters (rise time RT, peak amplitude A, time decay TD and repetition interval RI) are shown in Figure 7.3.2, together with some PDFs proposed to fit the experimental histograms. The selection of the evaluated quantities is in agreement with the work by Uchino, Tagiri and Shinozuka [153], which was focused on the real-time measurement and quantification of pulsed RF signals, identifying: Average Crossing Rate (ACR) or Crossing Rate Distribution (CRD), that for impulsive noise is the number of pulses with amplitudes greater than or equal to the threshold within one second; Pulse Duration Distribution (PDD) is the fraction of pulses with duration larger or equal to some time τ compared to the total number of pulses; Pulse Spacing Distribution (PSD) or Interpulse Spacing Distribution (ISD) is the fraction of pulses with space duration that are larger or equal to some time τ compared to the total number of pulse spacings.

Although different systems may feature different distributions, results reported in the following figures and tables were confirmed with two measurement campaigns.

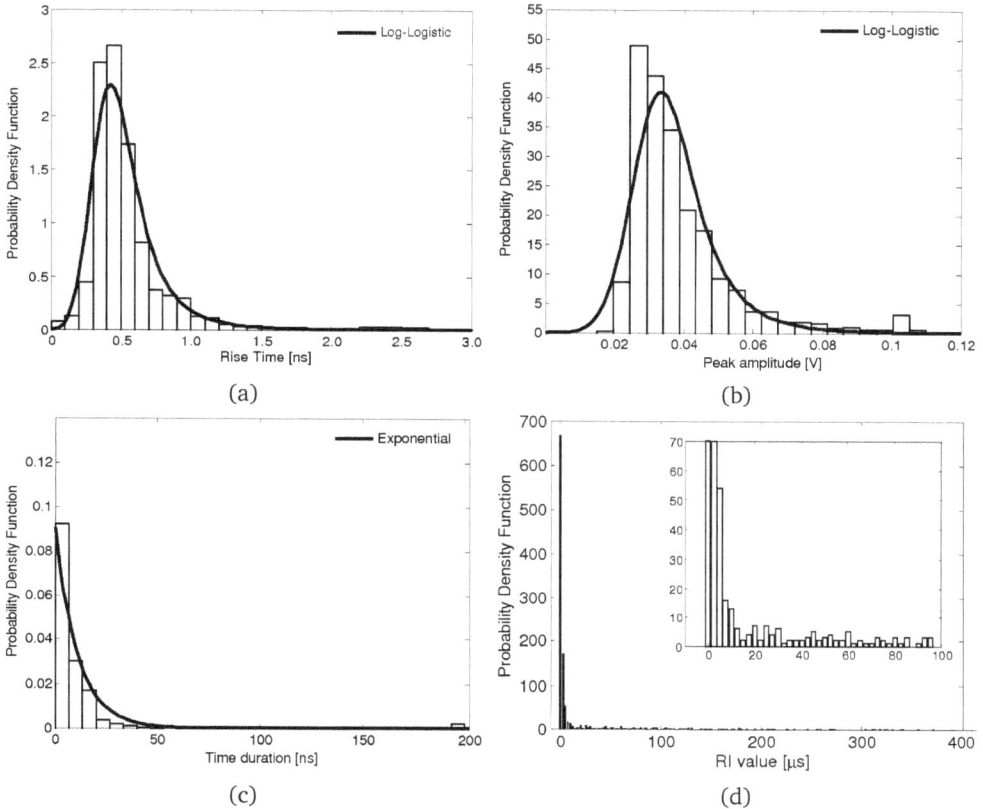

Figure 7.3.2 – Statistical distributions from experimental data for (a) rise time RT, (b) peak amplitude A, (c) time decay at 50% of peak amplitude TD50%, (d) repetition interval RI [20].

As shown before in sec. 7.2.1, when considering the effect of the pulse repetition rate and reported also in [20], the repetition interval RI is a particularly relevant parameter to establish the impact of repetitive transient disturbance on the GSM-R transmission and QoS. Recognizing if a transient peak repeats or simply rebounds due to oscillations and self resonance of one element of the measurement chain is a matter of applying criteria such as an amplitude threshold and time gating. As a consequence, results and in particular RI values are influenced by the wise choice of these analysis parameters; the criterion followed in [20] was to apply smoothly changing thresholds and observe the consequential variation of results, establishing an optimum point, where results are no longer affected significantly by moderate changes of thresholds. The results are shown in Figure 7.3.3 for an amplitude threshold varying between the 25% and 55% of the peak amplitude. More accurate estimates are summarized in Table 7.3.1.

7.3.2.2 In-band interference, APD, SNR and BER

The impact of Normally distributed noise on the bit error rate of coherent receivers was already introduced in sec. 7.3.1.4 [131]. The signal-to-noise ratio concept may

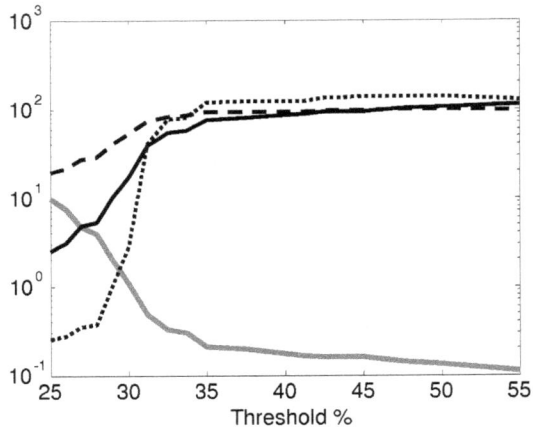

Figure 7.3.3 – Evaluation of repetition interval RI; statistical parameters as a function of threshold level in % of peak value: number of RI samples x 1000 (gray thick curve), sample mean (black curve) in μs, standard deviation (dotted curve) in μs, inter-quartile range (75 %-25 %) (dashed curve) in μs [20].

	Amplitude threshold in % of largest peak				
	30 %	**35 %**	**40 %**	**50 %**	**55 %**
RI_av [μs]	0.431	0.953	1.60	7.13	11.6
RI_sd [μs]	8.87	14.4	19.9	46.4	60.1
RI_iqr [μs]	0.0822	0.116	0.113	0.260	0.251
RI samples	1162989	520850	307706	67926	41592

Table 7.3.1 – Statistics of the repetition interval RI for 1200 recordings [20].

be adopted in order to obtain reliable estimate of BER: antipodal signals implements the easiest receiving scheme, made possible by the coherent detection also for GMSK frequency shift modulation of GSM and GSM-R; the error probability (that for antipodal binary signals corresponds to the BER) was estimated using the $Q(\cdot)$ function; the slight correlation and overlapping of successive signals sent onto the channel due to the Gaussian shaping filter was accounted for using the 1.8 coefficient.

At a first approximation this approach holds also for transient disturbance, but a certain amount of variability is expected for variable time relationship with the bit sequence clock, with transients possibly affecting parts at the extremes of the eye pattern that do not contribute much to detection. The input filter with a bandwidth that is roughly inversely proportional to the data rate in any case prolongs the time duration of each transient to about one bit time T_b; additionally, the transient waveform is expected to be somewhat different because of persistent oscillations due to the antenna transient response, also prolonging the duration of the transient itself.

A time domain approach to the evaluation of the impact of transient disturbance is proposed in [128], removing the uncertainty due to frequency domain measurements of time domain transient phenomena: the amplitude probability distribution (APD) of the transient disturbance is used to characterize its impact on GSM transmission;

for a complex modulation the APD is calculated for the modulus of the demodulated in-phase and quadrature components, as shown in [128]. The justification of the APD characterization of incoming disturbance to assess the amount of interference and the BER of digital transmissions may be found in [159], with the correction appearing in [105]. For the classification of noise and definition of APD and methods for its calculation a fundamental work is that by Middleton [109]. He proposed a classification of non-Gaussian noise that has been extensively used since then: Class A covers narrowband noise, narrower than the receiver bandwidth; Class B, conversely, covers broadband noise, larger than the receiver bandwidth; Class C is the superposition of Class A and B noise terms. The proposed models contain parameters to describe the rate of occurrence of impulsive disturbance (called "impulsive index"), the intensity of the impulsive noise components and the ratio between Gaussian and non-Gaussian noise components. Descriptive statistics are derived using PDF (probability distribution function) and APD, that, applied to both phase and amplitude, takes the name of "a posteriori distribution function". Several practical cases are considered and evaluated: emissions from ore-crushing machinery, fluorescent lights, overhead power line, automobile ignition system.

When considering interference to wireless systems, a more correct evaluation of BER and noise impact shall in general take into account also the multi-path nature of propagation and the different time of arrival of multiple copies of the transmitted signal with unavoidable overlapping and a certain amount of inter-symbol interference. This is normally addressed considering a fading channel model that for multi-path conditions is the so-called Rayleigh fading channel, whereas for substantially one ray (or line of sight) propagation is a Rician fading channel [36]. A line-of-sight model may be adopted for GSM-R propagation along a railway line that is substantially straight, with BTS at some height above ground, so that transmission is fairly unobstructed; for more complex geometries indoor or in tunnels, such as for TETRA or CBTC applications, the multi-path fading channel is almost a necessity. Simply speaking it is expected that the Rician model with a significant LOS component performs better than the Rayleigh model, where multi-path is prevailing; both will perform worse than a model where fading is not accounted for, such as a simply additive white Gaussian noise model (AWGN). This is shown in [128] for a 4DPSK modulation (the closer to the present case of two symbol FSK modulation of GSM). At a BER of 10^{-3} the required SNR values for a four symbol ($M = 4$) modulation are 10, 14 and 12 for the AWGN, Rayleigh and Rician models [128]; increasing the number of symbols the difference is even larger, with the Rayleigh multi-path channel performing increasingly worse. It is perfectly understood that in complex environments a two-symbol modulation, such as the one chosen for GSM-R, is very robust with minimum effect of multi-path.

A complete model is presented in [142], where three components are considered: the useful signal $s(t)$, the interfering signal $i(t)$ (e.g. caused by the electric arc emissions) and AWGN noise $n(t)^2$; these three signals are summed together with two weighting coefficients μ_s and μ_i to take into account the propagation through the respective

---

2 The Gaussian noise and the impulse-like interfering signal may be lead back to the classification of Class A and Class B Middleton.

channels and assigning an arbitrary phase to the interference signal to take into account the lack of synchronization and possibly the statistical nature of the interference.

$$r(t) = \mu_s s(t) + \mu_i(t) i(t) + n(t) \tag{7.3.2}$$

The GMSK signal $s(t)$ is made of two Non-Return to Zero sequences $D_I(t)$ and $D_Q(t)$ for in-phase and quadrature components, corresponding to odd and even bits of the sequence, each with a symbol duration of $2T_b$, being T_b the duration of one bit.

$$\begin{aligned} s(t) &= \sqrt{2E_s} \left[D_I(t) \cos\left(\pi t/2T_b\right) \cos(2\pi f_0 t) + D_Q(t) \sin\left(\pi t/2T_b\right) \sin(2\pi f_0 t) \right] \\ &= \sqrt{2E_s} \left[D_I(t) R_I(t) + D_Q(t) R_Q(t) \right] \end{aligned} \tag{7.3.3}$$

The in-phase and quadrature demodulation is explicitly calculated in [142]: for each of them the noise is separated by the mix of signal and interference, giving a formulation like the following one for the in-phase channel.

$$\begin{aligned} X_I &= \frac{2}{T_b} \int_{-T_b}^{+T_b} R(t) \sqrt{2E_s} R_I(t) \, \mathrm{d}t \\ &= 2\mu_s E_s D_I(t) + \frac{2\mu_i \sqrt{2E_s}}{T_b} \int_{-T_b}^{+T_b} s_i(t,\varphi) R_I(t) \, \mathrm{d}t + \frac{2\sqrt{2E_s}}{T_b} \int_{-T_b}^{+T_b} n(t) R_I(t) \, \mathrm{d}t \\ &= -A + y \end{aligned} \tag{7.3.4}$$

where A is the part that includes signal and interference and will take the role of reference threshold with respect to which the probability is evaluated for y, that is a Gaussian random variable with zero mean and variance $\sigma^2 = \dfrac{2E_s N_0}{T_b}$, with N_0 the spectral noise density.

The error probabilities that determine the bit error rate may be estimated as the probability that the Gaussian variable y is below or above the threshold A for bits with $D_I(t)$ equal to -1 and $+1$, respectively. After having expressed the interference as in-phase and quadrature components and having performed some simplifications [142], the probability of error for $+1$ and -1 are symmetrically calculated as:

$$\Pr\left\{+1 \mid y < A\right\}_I = \frac{1}{4\pi} \int_0^{2\pi} \mathrm{erfc}\left(\gamma + 2\gamma_i g_I \cos\theta\right) \mathrm{d}\theta \tag{7.3.5}$$

$$\Pr\left\{-1 \mid y > A\right\}_I = \frac{1}{4\pi} \int_0^{2\pi} \mathrm{erfc}\left(\gamma - 2\gamma_i g_I \cos\theta\right) \mathrm{d}\theta \tag{7.3.6}$$

having indicated with γ_s^2 the signal-to-noise power ratio (SNR), $\gamma_s^2 = \mu_s^2 \dfrac{E_s T_b}{N_0}$, and with γ_i^2 the input interference-to-noise ratio (INR), $\gamma_i^2 = \mu_i^2 \dfrac{E_i T_b}{N_0}$.

The g_I function projects the interfering signal onto the in-phase channel:

$$g_I^2 = \frac{1}{4 E_s E_i T_b^2} \left\{ \left[\int_{-T_b}^{+T_b} s_i(t) \sqrt{2 E_s} R_I(t)\, dt \right]^2 + \left[\int_{-T_b}^{+T_b} \widetilde{s}_i(t) \sqrt{2 E_s} R_I(t)\, dt \right]^2 \right\} \quad (7.3.7)$$

having indicated with $\widetilde{s}_i(t)$ the Hilbert transform of $s_i(t)$.

For the quadrature channel g_Q (calculated with respect to $R_Q(t)$ instead of $R_I(t)$) takes the place of g_I and sine function is used in place of cosine in the two integrals above for the calculations of probability.

Instead of passing through the integrals for the in-phase and quadrature components (that are mostly useful to implement a simulation model), Stenumgaard proposes similar expressions based directly on the theory of optimum receiver and decision thresholds in the presence of additive white Gaussian noise [146, 147, 148].

8

Low-frequency magnetic-field emissions

8.1 Sources and propagation

The interaction between the traction supply system, substations and trains causes fluctuations of system current that populate the low-frequency portion of the frequency interval. The phenomenon is particularly relevant for electric transportation systems with significant headway and supplied at dc, featuring thus many non-characteristic and transient spectral components located towards dc and extending up to a tens of Hz; ac systems are normally operated at higher voltage, lower current and lower headway and may be considered less relevant. Considering the large power absorption and dynamics of transit systems such as metro and light railways, the intensity of low frequency components may be a significant fraction of the nominal dc current.

The traction line excited by the source current I emits magnetic field that propagates at very long distance, being its attenuation of the $1/r$ type. It is thus apparent that the analysis of the impact of a transportation system shall include far locations with victim equipment, such as medical or scientific equipment, particularly susceptible for their own principle of operation an characteristics.

8.1.1 Traction line current low-frequency spectrum

Focusing on the traction line current of dc systems, we may assume that low frequency components are located at various frequency intervals that depend on different mechanisms:

- traction supply characteristic harmonics due to the rectifying process are located at multiples of the supply fundamental (normally $6k$ multiples, k integer, for three-phase rectifiers), often complemented by some low order even harmonics and the fundamental itself due to asymmetry;

- some components located at frequencies of hundreds of Hz are due to on-board converters, both choppers and inverters;

- towards low frequency all changes of operating conditions and load modulation cause transients and fluctuations, with more or less significant spectral components: in the range between a fraction of Hz and a few Hz the spectrum of single accelerations and decelerations determines components with an approximate $1/f^n$ slope, easily obtained by Fourier transformation of an approximately triangularly shaped current profile;

- at even lower frequencies the combination of more than one acceleration and deceleration from different vehicles may form repetitive patterns where the typical headway is recognized as the fundamental period; this is a sort of pulsation of the magnetic field emissions due to the typical system operating pace, as was noticed for BART in [14, 80] (see sec. 8.2.2).

This information may be recovered in principle with the Fourier analysis of long enough records of traction line current: from the fastest to the slowest components the required observation time increases from a fraction of a second to several minutes; the limit for such an investigation is imposed by the longest time interval over which an undisturbed operation of the victim equipment shall be performed (see sec. 8.2).

8.1.2 Magnetic field emission from the traction line

From a general viewpoint the magnetic field created by the flow of current along a set of conductors may be calculated using the Biot-Savart law applied to each infinitesimal current element, thus encompassing arbitrary geometries for the conductors.

8.1.2.1 Modeling

Modeling is achieved by a more or less complex set of equations that relate the current flowing into the conductors, conductors geometry and the position of the measuring point (or the volume of space where the estimation is required). To this aim a 2-D or 3-D vector formulation may be followed for a complete calculation of the magnetic field components (vector format): the well known formulation of the Biot-Savart law is based on the vector product between $\vec{\mathrm{d}l}$ taken on the wire carrying the current I and the vector $\vec{r}$ joining the $\vec{\mathrm{d}l}$ current element and the measuring point P (the point P may be a single point or a set of points covering an assigned volume of space following an assigned grid):

$$\mathrm{d}\vec{H}(\mathrm{P}) = \frac{I}{4\pi}\frac{\vec{\mathrm{d}l} \times \vec{r}}{r^3} \qquad (8.1.1)$$

The traction line geometry may be simplified to straight parallel conductors for which the Ampere law may be applied, with the assumption that the conductor length is nearly infinite (much longer than the observation distance r between the conductor and the measurement point P. Of course at the large distance required to evaluate the impact on susceptible equipment for low frequency emissions (e.g. up to $1\,\mathrm{km}$) this assumption is no longer valid, but the resulting field intensity obtained by straightforward application of Ampère law is always larger than that calculated by Biot-Savart expression, so that it is a convenient conservative approximation.

$$H(\mathrm{P}) = \frac{I}{2\pi r} \qquad (8.1.2)$$

All the expressions have been written for the magnetic field H; multiplication for the magnetic permeability of the medium, that in our case may be assumed that of free space μ_0, gives the magnetic induction field. Assuming the medium isotropic (as it is in reality), the direction also of the magnetic field vector $\vec{H}$ is preserved and $\vec{H}$ and $\vec{B}$ are co-linear.

Even when using a simplified model taking into account the current distribution in a transverse cross-section (2-D formulation), the estimation of the amount of current flowing in the conductors influences directly the estimated intensity and related accuracy: current leaving the traction return circuit and leaking into the soil, thus increasing the equivalent effective area of the traction circuit loop is an example of quite a common configuration to consider; also the use of just one rail for the return current (the other one being used for signaling purposes) is a relevant configuration, even if the asymmetry is not so relevant for magnetic field distribution; the use of compensating conductors for the overall reduction of magnetic field emissions is another relevant configuration that has a significant influence on the magnetic field distribution in close proximity to the traction system (to exemplify, in the first $20-30\,\mathrm{m}$) and then on the magnetic field intensity.

8.1.2.2 Current leaking into soil

Current in the running rails can leak into the railway infrastructure and soil because of conductance to earth: rail-to-earth conductance values are limited by design using insulating pads beneath rails, insulated fastening systems, or isolated supporting mats in order to reduce stray current impact (for dc systems in particular) and to comply with signaling requirements (when using track circuits); however, for worst-case analysis or to include the effect of overvoltage limiting devices, running rails may be sometimes assumed somehow connected to earth, increasing thus the percentage of return current leaving the rails up to some tens of % (see Figure 8.1.1). A practical, but over-estimating, approach is that of considering the current leaving the rails as lost current, thus reclosing at infinite, rather than trying to evaluate at which depth this fraction of the return current is really flowing (for dc current the distance from the running rails may be quite large making the assumption of lost current going to infinite a realistic one).

It is advisable to include calculations of the magnetic field distribution for a given amount of lost current, such as 20 %. These results are however relevant for worst-case analysis and cannot be made correspond to measurements on real systems, for which leaking current will be in reality much less (provided that the system was designed and built for protection against stray current), unless short-circuit connections to earth (such as overvoltage limiting devices) are installed.

8.1.2.3 Large distance emissions and conductor arrangements

At a first approximation, magnetic field amplitude reduces linearly with increasing distance starting from a lateral position r_0 that is larger than conductors separation (i.e. the distance between running rails and catenary); at closer distance $r < r_0$, the dependency on distance is a more complex function of higher order. More complex conductor systems, e.g. including additional earthing conductors, compensating conductors, or for ac system negative feeder, may be reduced to an equivalent system with two or three (in the case of 2x25 kV systems) equivalent conductors at different potential: magnetic field emissions sufficiently farther away cannot be distinguished for the complete and simplified system.

For ac systems induction and magnetic field emissions are reduced by the use of an overhead earthed conductor and the use of autotransformers or booster transformers, that reduce the current flowing into the running rail and bring it closer to the catenary flowing into the negative feeder or overhead earthed conductors. Compensating conductors are particularly useful to reduce magnetic field emissions in dc systems: inductively coupled circuit, autotransformers or booster transformers are of no use at dc for this purpose. The magnetic moment, and thus the magnetic field amplitude, is proportional to the area of the loop formed by the catenary and the running rails: the traction current is thus brought closer to the rails or the traction return current is brought up close to the catenary.

This observation lends itself not only to plan measurements with respect to possible critical locations and exposed victim equipment, but also to implement close-up measurements with respect to farther away locations, at which the required sensitivity of the measurement instrument would be critical and expensive (see sec. 8.4.1).

An example of magnetic field distribution farther away from a dc traction system is shown in Figure 8.1.2, where the regularity of the isolevel curves can be observed at distances from the track axis of about $50\,\mathrm{m}$; this is confirmed also by Figure 8.1.1.

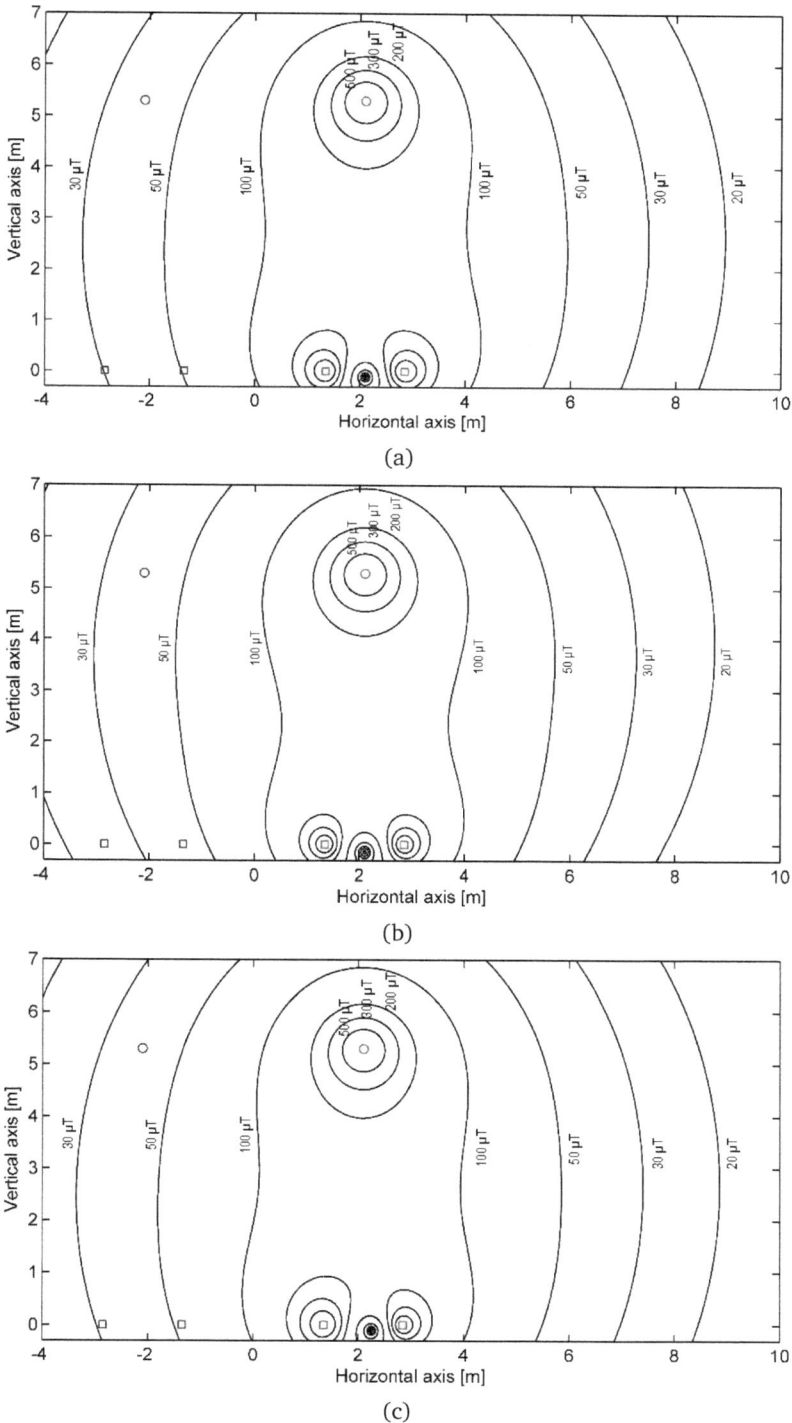

Figure 8.1.1 – Low frequency magnetic field emissions versus distance from the traction line: (a) ideal configuration $I_{R1} = I_{R2} = 50\,\%$, $I_{gnd} = 0\,\%$; (b) symmetric configuration with 20 % soil leakage $I_{R1} = I_{R2} = 40\,\%$, $I_{gnd} = 20\,\%$; (c) asymmetric configuration with 20 % soil leakage $I_{R1} = 30\,\%$, $I_{R2} = 50\,\%$, $I_{gnd} = 20\,\%$.

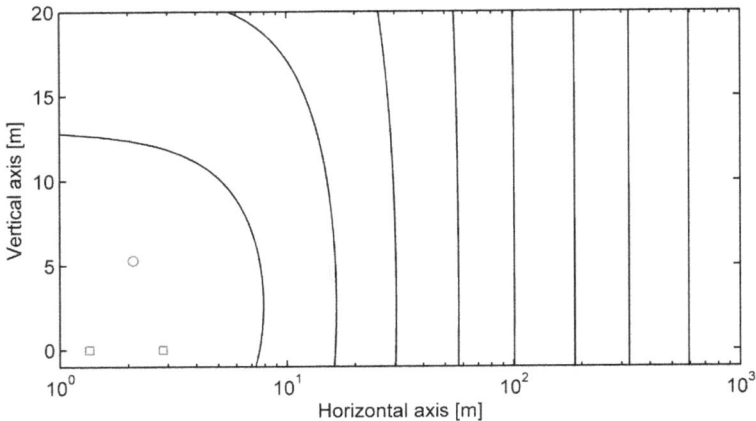

Figure 8.1.2 – Distribution of magnetic field induction at long distance from the traction system: isolevel lines beyond $50\,\mathrm{m}$ are straight and evenly spaced (they are calculated for 3 and 10, so not exactly the $10\,\mathrm{dB}$ that would be for 3.2 and 10).

8.1.3 Influence other than emissions

The movement of trains in the earth's magnetic field may affect magnetic field distribution, having an impact on sensitive instrumentation similar to low frequency magnetic field emissions. This occurs because trains made of steel distort the geomagnetic field; for modern ones the extensive use of aluminum mitigates the effect. The perturbation may be very small relative to the value of the earth's field; in absolute terms, however, the perturbation may be comparable with those caused by electrical systems emissions.

Fluctuations of geomagnetic field (that at rest has a nominal value of $580\,\mathrm{mG}$) were measured at a distance of about $25\,\mathrm{m}$ from a LRT system in Minnesota [108]: while two- and three-car trains were moved with other-than-electric means, fluctuations between 1 and $2\,\mathrm{mG}$ were measured. Similarly, in [116], some reasoning and site measurements led to estimate the influence of heavy passing traffic as $0.1\,\mathrm{mG/axle}$ for heavy trucks at $25\,\mathrm{ft}$ distance and decay with distance between square and cubic power.

8.2 Victims, reference standards and limits

Railway standards do not stipulate limits for dc and low-frequency emissions, but a few reference values are reported in EN 50121-2.

From the standpoint of no impact on existing systems, a modern transportation system shall comply with the limits established by the susceptibility of nearby systems. Whereas electric systems and equipment are generally immune to low-frequency magnetic field and withstand as per EMC standards a test for immunity to magnetic field at supply frequency (namely 50 or 60 Hz for industrial, office-like and residential applications and environments), a few systems may be identified that deserve special care and attention: medical systems, and in particular magnetic resonance equipment, as well as any scientific equipment, such as electron microscopes, that is particularly suscep-tible or necessitate to perform accurate measurements exactly in the same frequency interval, resulting evidently in significant in-band susceptibility.

Of course, limits valid for the low frequency interval are already established for human exposure and health of workers and passengers (see Chapter 9), but such levels are much larger and of no real usefulness to define suitable limits for susceptibility.

8.2.1   Medical diagnostic equipment

8.2.1.1   Magnetic Resonance

The relevant field component for interference to NMR (or MRI) machines is that paral-lel to the machine axis, that is horizontal or vertical with respect to the soil depending on the type of machine. The body of the machine shields the external field with an attenuation that may vary around an order of magnitude (between 15 and 27 dB, as reported in [125]); what is peculiar is a delay of several seconds for the external fluc-tuation to perturb the internal field.

The sensitivity of NMR probes in terms of amplitude and frequency range represents the intrinsic limit: more classical NMR atomic magnetometers eliminate the need for cryogenic cooling and push the sensitivity to lower frequency, thus more exposed to emissions from transportation systems; sensitivity was reported in the range of $1\,\mathrm{fT}/\sqrt{\mathrm{Hz}}$ for both pick-up coils and alkali RF magnetometers, while SQUID devices on particular frequency ranges well above low frequency (in the tens or hundred kHz) allow a reduction by about a factor of 30 [139].

Low frequency operation is typical of zero-field devices, e.g. superconducting quantum interference devices (SQUIDs) or atomic magnetometers that enable the use of weak magnetic field, proximal to the earth's static field, and thus exposing experiments to a wide range of external disturbance [88]. The control of the applied magnetic field from zero to small levels allows a better identification of chemical species and has attracted much attention. Again, this poses strict requirements on the tolerated magnetic field disturbance: in [88] results are shown for control levels between 44 and 264 nT for the axial component of the magnetic field; in [157] the measurement field is declared to be 6.3 µT.

On SQUID sensitivity, low-level NMR and necessary shielding, there are some literature references indicating quite low reference levels of magnetic field intensity: $5\,\mathrm{fT}/\sqrt{\mathrm{Hz}}$ are indicated in [61] as intrinsic device noise with additional unavoidable fluctuations of the internal Helmholtz-coil generated field of $3\,\mathrm{fT}/\sqrt{\mathrm{Hz}}$ along the sensitive device axis; in [88] the atomic magnetometer has an intrinsic noise of $40-50\,\mathrm{fT}/\sqrt{\mathrm{Hz}}$; in [157] the used SQUID has a noise floor of $8\,\mathrm{fT}/\sqrt{\mathrm{Hz}}$.

Other specifications for the maximum acceptable external interference may be derived by the internal sources of interference or instability, that in some references are enough detailed; units operating at weak field mentioned above are the most exposed to ex-ternal magnetic field fluctuations and disturbance: with detection field in the range of 50 µT obtained with an Helmholtz coil pair, the stability of the feeding current source of 25 ppm establishes a corresponding fluctuation of the detection field intensity of

1.3 nT [61]; even fluctuations of the earth's magnetic field on very long observation times (used to exploit averaging and improve the signal-to-noise ratio of the spectra of recorded signals) may be relevant because in the same order of magnitude.

Given the very demanding requirements for external magnetic field disturbance in the broad frequency interval of small fraction of Hz ("fluctuations") to tens of Hz, and up to a few hundreds Hz ("noise"), performance is achieved using shielded rooms, built with a mix of eddy current shielding (e.g. aluminum) and field shunting by permeability difference (e.g. mu metal and high permeability materials). A magnetically shielded room (BMSR-2) is provided of 2.8 m side and with walls protected by 7 layers of Mu-metal, one layer of aluminum, and a surrounding RF shield, with a shielding factor of $7\,10^4$ at 0.01 Hz [61]; additional active shielding may be used with triaxial Helmholtz coil controlled by magnetometers, imposing a reference stable field, reaching a shielding factor of $6\,10^6$ at 0.01 Hz. External field sources are thus largely attenuated; what is relevant in addition is the resulting field gradient due to change of permeability of medium using the designed cubic shape with edges and angles at corners: the residual magnetic field gradient is 6 pT/cm, but internal sources, such as feeding cables for the various coils, increase this gradient to 100 pT/cm.

A 2.7x2.5x2.5 m room is used in [88] where aluminum and Mu-metal layers ensure a shielding factor of about 40 at 1 Hz, about 200 at 60 Hz, with a residual DC magnetic field below 0.1 μT, that is thus far larger than the previous one. Being the polarization fields of the same order of magnitude in the two cases (but larger for [88] up to a factor of five), the discrepancy may be explained simply by observing that at the same field level of $50-60$ μT the quality of results in [61] is much better and peaks much more recognizable, implying a better design and an underlying stricter requisite regarding signal quality and reduction of external sources.

JEOL specifies for its 64x0 MP machines a 1 mG peak-to-peak limit for ac fields and a non-better defined dc field change of 0.5 mG at the maximum acceleration voltage of 30 kV, and proportionally lower limits for lower acceleration voltages [76].

8.2.1.2   Scanning electron microscope

Only B-field components orthogonal to the SEM beam are relevant for distortion; the effect of an external field superimposed to the scanning coils that center the beam is to cause beam displacement.

Less than 0.2 mG rms are needed for clean STEM images at 0.2 nm[115]; such resolution is now commonplace for Scanning Transmission Electron Microscopy, but image quality may be deteriorated by external interference both from nearby sources (even inside the equipment) and from external farther sources, especially at power supply frequency: power supply frequency magnetic field is quite a common problem, propagates at long distance and is scarcely attenuated by walls and buildings. Magnetic field emissions from railway systems are very similar, covering the wide frequency range of a small fraction of Hz up to the supply frequency of 50 or 60 Hz and a few harmonics. The microscope shall be shielded against external fields, but magnetic field is not the only external source of interference, where airflow and air pressure may also play a significant and more relevant role, so that the decision on the maximum acceptable magnetic field intensity from external sources shall be balanced also with other non-electric sources of interference.

As reported in [120] induction field levels of less than 1 down to $0.1\,\mathrm{mG}$ (thus between 10 and $100\,\mathrm{nT}$) at 60 Hz are considered compatible with the operation of various electron microscopes; similarly, in other publications values between 80 and $200\,\mathrm{nT_{pp}}$ are given, and that confirm and nearly overlap to the previous ones; also in [140] a threshold of $0.1-0.4\,\mu\mathrm{T}$ is given for a generic "electron microscope".

The sensitivity of high-performance electron microscopes was identified as $0.05\,\mathrm{nm/mG}$ for the scan subsystem and about $1\,\mathrm{eV/mG}$ for post-column spectrometers [116].

Magnetoencephalography (MEG) requirement is reported as $<1\,\mathrm{nT}$ [140].

8.2.2 Geomagnetic instrumentation

The study reported in [80] reports curves of ULF geomagnetic activity of $1\,\mathrm{pT}/\sqrt{\mathrm{Hz}}$ increasing by one decade for each decade of reduction in frequency, and observed levels for seismic phenomena about one order of magnitude larger. In the same work they point the finger at BART (Bay Area Rapid Transit) system and the generated magnetic disturbance by its operation: noise levels larger than the background noise by about a factor of two to five were observed, namely about $10-3\,\mathrm{nT}/\sqrt{\mathrm{Hz}}$ at 0.1 Hz, reducing to about $0.3\,\mathrm{nT}/\sqrt{\mathrm{Hz}}$ at 1 Hz and then not distinguishable from the background noise profile above a few Hz. These measurements were performed at $30\,\mathrm{km}$ from the BART line and showed the typical periodicity between 5 and $20\,\mathrm{s}$ typical of the traffic pattern. Ambient background noise was measured during the BART off periods between 2:00 am and 4:00 am.

Similar results are confirmed in [14] within a factor of two in amplitude; including measurements at other locations farther away from BART area of influence ($40\,\mathrm{km}$), the measured noise keeps almost constant during the day. The two publications agree on the frequency band occupation with maxima in $0.1-0.2$ Hz frequency interval, that may be led back to the typical duration of acceleration phases, where the largest change of current in the traction circuit is expected. As confirmation, the amplitude of that frequency interval is strongly correlated with the overall system load estimated by the authors proportional to the number of running BART cars.

8.3 Used instrumentation

Low frequency magnetic field including dc components shall be measured with magnetic field probes able to extend down to very low frequency and dc, for which drift, wander and flicker noise are all relevant elements to include in evaluation of uncertainty: very low frequency intervals are characterized by erratic behavior, slow moving drift and chaotic phenomena of the flicker type, which are not so often detailed in instruments datasheets and technical literature. If for human exposure the limits are so large that they have no relevant impact on required instrumentation sensitivity, when evaluating possible interference to susceptible equipment, reference levels are correspondingly lower, posing serious problems to identify suitable instruments, at the point that the same victim equipment (or similar, based on the same physical principles and construction) might be used as probing element. Of course, in this case, the victim equipment shall be characterized as a measurement instrument itself, in particular for repeatability and reproducibility.

Magnetic field measurements need a long observation time in order to estimate the lowest frequency components that are required for the evaluation of disturbance to some medical and scientific equipment reviewed in sec. 8.2. This implies that their long-term noise in various forms shall be known or need to be characterized beforehand: low frequency noise, flicker components, wandering and short/medium term drift (characterization of noise sources may be found in [99] and references). Flicker noise is normally caused by the instrument electronics (semiconductors feature such a noise profile with an approximate $1/f$ slope); wandering and drift are forms of very low frequency instability, due to a combination of sources and mechanisms: thermal effects and internal temperature variations, chaotic processes, etc. Averaging, that was used to reduce noise effects in terms of variance or average value, is quite ineffective with these effects, due to the intrinsic non-stationarity and the ever increasing spread of recorded values for increasing recording times. Definitely, when acquisitions last for several minutes or hours these factors shall all be carefully considered, performing an accurate and thorough characterization of the instrument, because very often datasheets and manuals are weak and lack complete information.

8.4   Measurement procedures and techniques

The magnetic field distribution around the most relevant sources outside rolling stock is quite easy to follow based on physical principles and knowledge of the role and operation of equipment and elements (cables, contactors, etc.). When circuits and geometry are complex and the distance between sources and measurement points is short, measurement procedures shall be more detailed and articulated, taking into account less intuitive distributions and significant gradient; this is however already considered for human exposure in Chapter 9.

On the contrary, when measuring low-frequency magnetic field emissions at very long distance the most challenging aspects are the noise sources inside the instrumentation and in the surrounding environment, the influence of conductive and magnetic objects nearby and that of movement (e.g. due to mechanical instability, wind, vibrations, etc.) or other varying conditions during measurements.

8.4.1   Sensing and acquisition

Magnetic field probes were reviewed in sec. 3.2.3, identifying their characteristics, principles of operation, and an estimate of performance and uncertainty. The best types of sensor for the weakest field intensities are coil and GMR probes: coils shall be built with an extremely large number of turns thus limiting the high frequency response because of the unavoidable parasitic capacitance, the ultimate limiting factor being the noise of the amplification stage and the required amplification to feed the signal to the acquisition system [97]; GMR is very sensitive and feature a moderate internal resistance between hundreds Ω and few kΩ, not adding much noise.

Acquisition may be performed in time or frequency domain: in both cases over long observation times the stability of the equipment and of its characteristics, including in particular internal noise sources, is mostly important; since the mission time for a diagnostic operation e.g. using NMR is several minutes, up to about one hour, the observation time will be trimmed accordingly and the lowest observed frequency will be in the range of $0.1 - 1$ mHz, at which flicker noise and wandering are quite relevant and badly documented. It is thus extremely important that the measuring equipment is characterized beforehand, using e.g. tests and processing techniques such as those described in Chapter 2 and 3, and more extensively in [99].

Since probe sensitivity may be an issue for very low target intensity (as seen in sec. 8.2 for the susceptibility of many scientific and medical equipment), measuring field intensity closer to the railway system ruling out many sources of uncertainty and then extrapolating to the victim location is a viable technique: in sec. 8.1.2.1 modeling was considered not only to take in the due account the various elements of the traction circuit that are relevant to magnetic field emissions, but also to define a usable decay profile for magnetic field intensity versus distance. It is evident that at a significant distance from the traction line the magnetic field has the same distribution as for an equivalent conductor featuring an equivalent current and that extrapolation to other distances may be achieved by using Ampere's law (so with a $1/r$ law of variation); by inspection of Figure 8.1.2, the "significant distance" may be conservatively 50 m, where the field distribution is already quite regular.

8.4.2 Intrinsic instrumentation noise and sensitivity

We saw that the expected relevant noise caused by external sources may vary between a few $nT/\sqrt{Hz}$ down to $fT/\sqrt{Hz}$ for magnetic resonance imaging; the measuring system shall have similar performance in terms of internal noise and sensitivity. This figure of performance shall include thermal fluctuations and other electrical instabilities, as was underlined in [97] during the characterization of a low-noise general purpose magnetic field probe: the internal noise was measured as $9\,nV/\sqrt{Hz}$ at 1 Hz and $30\,nV/\sqrt{Hz}$ at 0.1 Hz, to be multiplied by the gain of the connected probe. Lower noise is attainable by using even less noisy operational amplifiers (a reduction by a factor of two is possible) and paralleling several of them possibly reaching a further reduction by a factor of three with nine devices in parallel. Then, beneath noise voltage density values of about $1\,nV/\sqrt{Hz}$, further increase of sensitivity can only be achieved by increasing the area and the number of turns of coil probes, or using more sensitive probes of the Giant Magneto-Resistive (GMR) type or those sensors (atomic magnetometer and SQUID) already reviewed as sensing elements of victim medical and scientific equipment above.

Because of internal noise fluctuations due to thermal gradients, when measuring weak field levels, not only warm-up prescriptions shall be complied to, but also direct sunlight and changes of environmental conditions shall be avoided and annotated.

8.4.3 Conductive and magnetic objects

As said in sec. 8.1.3, moving conductive objects may be a source of magnetic field fluctuation and during measurements this shall be avoided evaluating the effect of typical object and their distance beforehand, in order to determine a minimum safe distance and instructions for participants. The identification of relevant disturbing objects shall be made in advance to prepare adequate test procedures. Understanding the influence on magnetic field distribution is part of both preliminary measurement preparation and evaluation of effects other than emissions (see sec. 8.1.3): the Earth's magnetic field and its spatial distribution variations due to metallic masses and buildings is a relevant factor for dc measurements.

8.4.4 Probe movement and vibrations

Undesirable movement of the measuring probe may be the consequence of many factors and interactions: bumps, instability of tripod and holding system, wind, vibrations induced by walking and, most of all, traffic. All these phenomena are extremely relevant when measuring magnetic field components of very small amplitude with frequencies that require observation times of fraction of a second to several seconds, thus perfectly corresponding to slow fluctuations and changes, that may sometimes go unnoticed. Although induction effects due to movement through magnetic field lines may be less relevant if magnetic field intensity is weak over the entire frequency range, variations of reading due to magnetic field gradient are also relevant, whatever the magnetic field intensity. The latter usually affects reproducibility, because even small positioning errors may compromise the resulting uncertainty with respect to instrumentation uncertainty alone or repeatability.

8.4.5 Measurement in off conditions

When measuring weak magnetic field components, it is always advisable to perform off readings with the source of emissions switched off or not operating. This is the equivalent of measuring background noise for RF phenomena; the aim is the determination of instrumentation internal noise and fluctuations, as well as external sources.

With railway transportation systems already in commercial service this requires to perform measurements during off times, that unavoidably occur during night, usually for a few hours only.

Similarly, the contribution of external sources that are out of control of the operator, such as overhead and underground power lines, affects the accuracy of the determination of off conditions, due to the unavoidable contribution at the supply frequency and very low frequency following network fluctuations. It is thus absolutely necessary to map all these external sources, selecting far away measurement locations.

9

Human exposure to electromagnetic field

9.1 Problem description

The problem of human exposure to electromagnetic fields has become more and more relevant for several applications and environments, including railway applications and in particular on-board rolling stock. The assessment of exposure is disciplined by the CENELEC standard EN 50500 [33], based on the one hand on the limits set forth by the 1998 ICNIRP publication [64] and on the other hand on a set of methods and considerations analyzed in the literature [11], regarding spectrum calculation, frequency leakage and improvement of spectrum accuracy, and spectral analysis. The EN 50550 covers up to $20\,\mathrm{kHz}$: we will comment on the possible extension up to a hundred kHz, or even to the MHz range, using the same measurement methods. However, for modern railway applications other kind of exposures to radio sources shall be also evaluated against several CENELEC standards (EN 50383, EN 50400, EN 50420 and EN 50492) and specific contractual requirements or national regulations, being outside the scope of application of the EN 50500.

The exposure may be considered particularly significant on-board for several reasons: reduced distance between sources and victims, passengers both seating and standing (in many commuter trains practically filling the entire available space), sources characterized by switching waveforms and time-varying sources, for which the methods developed in [11] may be applied.

In particular, many spectral components populate the spectrum of emissions and may also overlap: the fundamental, e.g. dc or ac, whether the line frequency or the motor frequency, and the components deriving from the switching process, usually located between some hundreds Hz and several kHz, thus located in the frequency range where exposure limits are at the minimum value (see end of sec. 9.3). In this case much care shall be given during design to cable routing for those cables where the largest switching components are flowing, as well as filter inductors and transformers: the former especially are a significant source of stray magnetic field due to their open, or quasi-open construction, necessary to avoid saturation; the gap in the core necessary to avoid saturation and the gaps in the enclosure to allow ventilation are the reason for a relevant amount of stray magnetic field.

For many high-speed and heavy trains, bulky converters and "magnetics" are located under-frame, so closer than roof-top apparatus to the portion of space where passengers are seated.

9.2 Theoretical background

Whereas wayside and at platforms the geometry of exposure is clear and simple for a person standing next to the track and exposed to the magnetic field from the traction circuit, when considering on-board exposure there are many equipment and circuits all located in a narrow space, so that the local magnetic field intensity may be considerably large.

For railway applications the standard EN 50500 [33] reduces the range of assessment to 20 kHz, claiming that there are no relevant emissions beyond that for rolling stock and railway equipment: besides the presence of radio systems used for train-to-wayside communication, signaling, support to police, fire fighting teams, etc., that need a separate assessment using standards for radiofrequency sources, it is also well established that on-board apparatus is able of emissions in the hundreds kHz range, still of significant amplitude if compared with ICNIRP limits and considering their contribution to the overall ICNIRP index.

On-board apparatus includes traction converters and other large power converters in the main power circuit (e.g. front-end choppers, four-quadrant converters, etc.), all interconnected by cables, that prevalently run longitudinally along the vehicle or train, using the underfloor or the space in the roof: in both cases cables are distributed sources of emissions that affect several locations inside the vehicle or unit, whereas the converters are lumped sources for which protection by shielding is a little bit easier. Other equipment, such as air conditioning, electric braking circuit and similar, may be additional sources of emissions that affect a significant portion of space. There are conversely spot sources that are relevant only in a limited volume of space and that may be difficult to tackle: examples based on experience are some power supplies for emergency lights, small motors for fans and windshield wipers, etc.

It is thus evident that the geometry of exposure is complex and that the selection of the measurement points is not trivial without a preliminary analysis for the most complex cases (such as modern high-speed trains or multiple traction voltage units).

For the presence of metal and conductive parts everywhere, it may be assumed that the electric field intensity is not relevant for the entire frequency range except at very high frequency in the presence of intentional transmitters: not only transmitters for radio communication systems, such as GSM-R, VHF radios, WiFi, etc., but also signaling systems that use radio energy and a relevant intensity of the electromagnetic field, such as the balise antenna, radio links for traffic lights management, etc. (even if they are normally directed outside the vehicle and towards portions of space, beneath and ahead, not occupied by personnel and passengers).

Models for analysis and prediction may focus thus on magnetic field emissions, assuming the typical circuit geometries: loops and straight wires. These two may be quite helpful for a reasonable modeling effort regarding emissions around a power converter and along the connecting cables:

- loops may represent emissions from internal circuits where current with significant amplitude and spectral components flows, such as snubber circuits around semiconductors, internal bars and heavy cabling interconnecting semiconductors, transformers, filters, etc;

- cables may be simulated by straight wires, using either real currents or equivalent current, taking into account the deal of cancellation of differential components.

These models keep valid up to the largest frequency that is of interest: it was said that significant emissions are expected also above $20\,\mathrm{kHz}$; although the spectrum of emissions is highly dependent on the conversion principle and architecture, and its implementation, based on experience and some general knowledge of converters for medium power applications, it is possible to establish an upper frequency limit in the range of the hundreds kHz, maybe conservatively saying "up to $1\,\mathrm{MHz}$". For this frequency range all emitting elements are still lumped, being smaller than the shortest wavelength.

For straight conductors the models shown in sec. 8.1.2.1 may be used as well, provided that corrective coefficients are introduced for the presence of metal around [23] and for the non-negligible diameter of cables with respect to their separation (reduction of differential mode emissions and proximity effect). Loops may undergo similar modeling using closed-form equations for loops or using Biot-Savart formulation on elementary segments, integrated along a circular path, that shall be closed, but of arbitrary shape [26, 44].

9.3   Reference standards and limits

From a normative viewpoint the EN 50500 is the mandatory reference for Europe and a useful reference also for other countries, being the only international standard on the subject to author's knowledge. It is grounded on the ICNIRP publication issued in

1998 [64], regarding frequency domain evaluation of time-varying electric and magnetic fields. The limits are given for the measurable quantities, that is electric and magnetic field intensity, that are related back to the dosimetric internal quantities by means of reliable models and assumptions. As an example, the effect of an external magnetic field in the frequency range of interest for railway applications causes an induced current flowing inside the body of the victim and interfering with endogenous regulation, cardiac activity, etc., for which conservative limits are set and equivalent limits for the external magnetic field are estimated based on the model of induction, the known electric characteristics of tissues, etc.

Depending on several characteristics regarding also the position, the awareness and the medical surveillance, and the duration of exposure, different limits are derived for residential and occupational exposures, that is those occurring to the general public and to workers, respectively: in the first case a margin of about five is applied, by correspondingly reducing the limit values; in both cases the ICNIRP publication states that safety margins are intrinsic to the limit values because of the underlying assumptions. It is often questioned how fulfillment of limits shall be considered and if additional margins shall be considered to further increase the reliability and confidence of exposure limitation; when elaborating models and assumptions and then the limits themselves (first, as dosimetric quantities and basic restrictions, then for external measurable electromagnetic quantities as reference levels), ICNIRP declares an intrinsic safety margin associated with the limits due to overestimating assumptions on modeling and tissue electric and electromagnetic properties, regarding the geometry of exposure (orientation, posture, etc.), and the inclusion of particularly weak categories of victims, such as children, women in wait, elders.

The limits (reference levels) for residential and occupational exposure still in use are those published by ICNIRP in 1998 [64][1].

Other more recent publications are not referenced in standards nor regulations, so that they lack of statutory authoritativeness. The 2003 publication [66] deals with a more accurate time-domain evaluation, suitable for largely varying intensity of field components, such as for impulse-like waveforms. The method is quite straightforward, taking the frequency domain limits and translating them in the frequency response of an equivalent weighting filter (the inverse of the limit curve is used), that processes in real time the incoming time-domain signal. This publications was motivated chiefly by the widespread use of anti-theft systems and the awareness of the field intensity, not only for customers passing through the gates, but also for the continuous exposure of personnel stationing nearby (e.g. at supermarkets and airports). Correspondingly, CENELEC released the standard EN 50357 [31].

The frequency-domain limits were reviewed by ICNIRP increasing slightly the values in the low frequency range and changing the cutoff frequencies of the curves, that formerly were around $800\,\mathrm{Hz}$ and are now located at $300\,\mathrm{Hz}$ and $400\,\mathrm{Hz}$. The new limits for the frequency interval up to $100\,\mathrm{kHz}$ were published in 2010 [65]. As said, up to now no regulation is invoking explicitly the new ICNIRP statement and implementing these limits.

---

[1]   ICNIRP publications are available for download at www.icnirp.de.

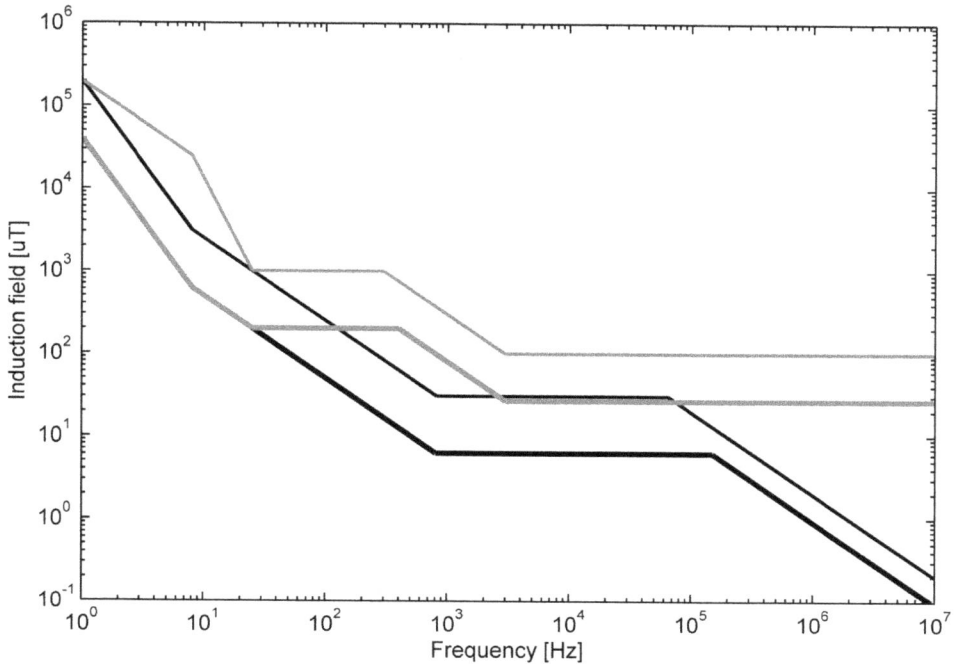

Figure 9.3.1 – Graphical comparison of limits for residential (thick) and occupations (thin) exposure in the 1998 (black) and 2010 (gray) editions of ICNIRP guidelines.

A graphical comparison of the ICNIRP limits for residential and occupational exposure for 1998 and 2010 editions of ICNIRP guidelines is shown in Figure 9.3.1: it is easy to see that all limits have been increased for both residential and occupational exposures. Although the graphical representation of induction field limits extends up to 10 MHz, above 100 kHz ICNIRP recommends that in addition radiofrequency basic restrictions need to be taken into account. The criteria are considered in more detail in sec. 9.4.

9.4 Verification of compliance

Verification of compliance, given the limits issued by ICNIRP in 1998 and recalled in regulations and standards (seen in sec. 9.3), is performed in frequency domain: the verification of compliance is performed, first, by comparing the amplitude of spectral components with the corresponding limit value at corresponding frequency and verifying that the intensity of all components is lower than the limit, and, second, by calculating the combined effect of all relevant components in the so called ICNIRP index, which weights the spectrum components by the respective limit values and that shall be lower than unity. The word "relevant" was used because in the calculation of the ICNIRP index the background noise and the effects of spectral leakage have an important role: the former increases the overall index by summing a huge number of small terms, if they are not properly pruned (see sec. 9.4.3), the latter lowers the height of the real component and creates coherent side components (see sec. 9.4.3 and sec. 2.3.3 for an introduction to spectral leakage and windowing).

9.4.1 Comparison with frequency-domain limits

ICNIRP limits are defined in the frequency domain assuming a single-tone exposure or multiple tones and the possibility of a frequency domain representation (either by DFT of a time-domain acquired signal or by direct frequency-domain measurement). Of course this excludes in principle the possibility of evaluating exposure for transients, for which a specific guideline was prepared [66]; in reality, only large transients should be excluded, as in real systems (and in particular in railway applications) several small transients occur superimposed to the main traction waveforms and they cannot be separated, nor excluded; in general they do not preclude the possibility of a frequency domain analysis as it is analyzed in [11]. Examples of these small transients are switching transients inside converters, step regulation, etc.

9.4.1.1 Stationarity of signals

Frequency domain analysis and assessment of compliance can be done by means of a frequency domain scan (e.g. by FFT analyzer or spectrum analyzer) or a time domain acquisition followed by DFT [2], provided that the signal can be Fourier transformed, or, in other words, that is ergodic or stationary at least in the considered interval (see sec. 2.5.2). In [11] this was identified as "interval of stationarity" and was estimated by examining several recordings, analyzing among others their correlation function: by inspection of Figure 6 and 7 in [11], it is possible to estimate such interval as ranging between $0.25\,\mathrm{s}$ and $0.5\,\mathrm{s}$ for the examined cases, where the persistence was observed of some low frequency components, mainly related to motor currents and possibly input filter transients; high-pass filtering the data with a cutoff frequency of $20-50\,\mathrm{Hz}$ reduced this interval to about $0.15-0.25\,\mathrm{s}$.

A different and more pragmatic approach was then used to obtain a confirmation of the first estimate: signal windows progressively reducing from about $2\,\mathrm{s}$ to less than $0.1\,\mathrm{s}$ were used to calculate the ICNIRP index and its behavior was considered, identifying undue increase as related to "inadequate representation of non-stationary components"; the minimum of the ICNIRP index was found for 256 samples, corresponding to $0.125\,\mathrm{s}$, whereas the optimum value of the energy descriptor associated to the ICNIRP index around that interval was for 512 samples, that is $0.25\,\mathrm{s}$.

Last, an adaptive algorithm for the definition of the optimal time window length, while moving along the signal and calculating Short Time Fourier Transforms, gave values of $0.3\,\mathrm{s}$ or shorter for more than 80 % of processed windows, including all transients.

The relevance of the fixed and adaptive choice of the stationarity interval was evaluated on real examples, one of which was reported in Figure 13 of [11], shown below in Figure 9.4.1. The largest difference between the two approaches occurs for transient operating conditions, when e.g. sudden acceleration is applied; the nearly zero value occurring between 80 and $105\,\mathrm{s}$ corresponds to coasting.

---

[2] The term STFT (Short Time Fourier Transform) is used in [11], including thus the quantity "time".

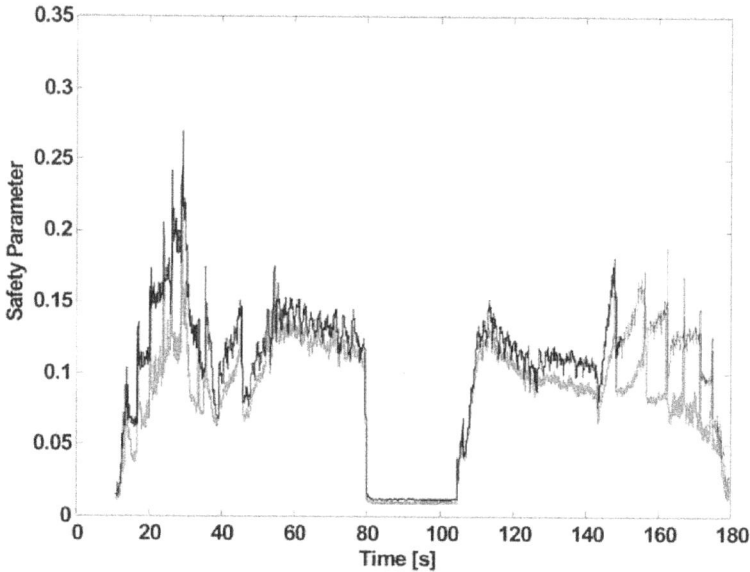

Figure 9.4.1 – ICNIRP index value (called "safety index" [11]) calculated for the [5 − 800 Hz] frequency interval and plotted for nearly 3 min. of test run: black and gray curves refer to the fixed-length (0.25 s) and adaptive-length STFT procedure.

9.4.1.2 Application of DFT or settings of frequency domain analysis

The application of the STFT (Short Time Fourier Transform) to the acquired signal, or alternatively the sequence of spectra acquired using frequency-domain equipment, is then processed before comparison with limits.

Once the interval of stationarity is identified (it was said $T_{w0} = 0.25$ s), a decision shall be taken for the frequency resolution that cannot be smaller than the reciprocal of T_{w0}, in this case 4 Hz. Similarly, for frequency domain equipment the resolution settings shall be similar, to imply a time window of the same length of T_{w0}.

A slightly coarser frequency resolution may be used (e.g. 10 Hz) with a correspondingly larger number of transformation windows fitting into T_{w0}, possibly overlapping, increasing the number of windows available for calculations (with 50% overlap and 10 Hz resolution a total of four windows fit into T_{w0}). Using coarser frequency resolution has the benefit of a smoother amplitude spectrum profile but a higher noise floor, due to the noise power for each frequency bin proportional to the frequency resolution and thus 2.5 times larger.

The spectra of subsequent windows may feature some amount of instability of component intensity, due to captured minor transients caused by the several sources onboard: averaging the spectra collected over the interval of stationarity T_{w0} preserves the original time resolution (one spectrum is output every T_{w0} seconds) with a slightly worse frequency resolution (that is however non critical), but featuring improvement of spectrum profile and removal of short artifacts. The worsening of the signal-to-noise ration due to the larger frequency resolution is only apparent: first, coherent averaging can correct the additional noise bringing the average noise floor back to the previous value (see sec. 2.6.2.2), and, second, noise pruning removes any noise component below an assigned threshold whatever its height, provided it is not excessively large.

What averaging cannot reduce is coherent noise, and in particular that caused by spectral leakage, that is considered later on in sec. 9.4.3.2.

The sampling frequency when using time sampled data shall of course comply to the minimum requirements of the Sampling Theorem (larger than twice the maximum frequency in the input signal spectrum), but also necessitates a certain amount of oversampling, whose benefits from different viewpoints are: better signal-to-noise ratio when considering quantization noise (whose power is spread over a large frequency interval up to half the new higher sampling frequency), better definition of slopes of transients and better phase resolution (that, however, it's useless for the most direct approach to compliance evaluation based on amplitude only).

9.4.1.3 Comparison criteria

The comparison with the ICNIRP limits shall be done using the rms values resulting from the DFT or frequency-domain measurements: for the former, when using the instantaneous values of the sampled signal, a reduction by $\sqrt{2}$ shall be applied; the latter is usually already expressed in terms of rms value.

In general none of the standards or guidelines considered so far specifies any margin with respect to limit values nor how equipment accuracy and uncertainty reflect into the assessment of compliance. So, rather than a statistical assessment based on confidence interval and confidence level, the comparison is strictly numeric with a "less than" condition fulfilling the human exposure limits. The same applies to the ICNIRP index in the next section.

The approach may be justified by the fact that human safety is concerned and that no statement of "probably less than" is accepted, but it's a fact also that equipment uncertainty is in any case a probabilistic quantity and that measurement errors are random variables (see sec. 2.7).

9.4.2 Calculation of ICNIRP index

The ICNIRP guideline [64] establishes different rules for the calculation of the ICNIRP index, depending on the relevant phenomena and thus the frequency range. The same subdivision is also established at normative level by the standard EN 50392 [32].

For induced current density and electrical stimulation effects, relevant up to 10 MHz, it is required to fulfill both of the following expressions:

$$I_E^{(i)} = \sum_{m=1}^{M_1} \frac{E_m}{E_{L,m}} + \sum_{n=M_1+1}^{M_2} \frac{E_n}{E_{L,n}} \leq 1 \qquad (9.4.1)$$

$$I_H^{(i)} = \sum_{m=1}^{N_1} \frac{H_m}{H_{L,m}} + \sum_{n=N_1+1}^{N_2} \frac{H_n}{H_{L,n}} \leq 1 \qquad (9.4.2)$$

where the indexes m and n shall be intended as integer indexes along E or H vectors resulting from measurements pointing at the first M_1 (or N_1) significant components in the first frequency interval and at the M_2 (or N_2) significant components in the second frequency interval; the two intervals are explicitly indicated in the notation adopted in the ICNIRP guideline and their upper bounds correspond to 65 kHz for the E field (M_1), 1 MHz for the H field (N_1) and 10 MHz for both E and H field (M_2 or N_2), with $m = 1$ starting from the minimum frequency (1 Hz). The quantities with subscript "L" refer to the limit curves: the ICNIRP guideline replaces $E_{L,n}$ and $H_{L,n}$ with two constants because in the two frequency intervals [65 kHz, 10 MHz] and [1 MHz, 10 MHz] the limit curves are horizontal and the values constant and equal to 610 V/m and 87 V/m for the E field, occupational and general public exposure, respectively, and 24.4 A/m and 5 A/m for the H field, occupational and general public exposure again, respectively.

For thermal considerations, relevant above 100 kHz , the following two requirements are applied:

$$I_E^{(t)} = \sum_{m=1}^{P_1} \frac{E_m}{E_{L,m}} + \sum_{n=P_1+1}^{P_2} \frac{E_n}{E_{L,n}} \leq 1 \qquad (9.4.3)$$

$$I_H^{(t)} = \sum_{m=1}^{Q_1} \frac{H_m}{H_{L,m}} + \sum_{n=Q_1+1}^{Q_2} \frac{H_n}{H_{L,n}} \leq 1 \qquad (9.4.4)$$

where the indexes m and n shall be intended as integer indexes along E or H vectors resulting from measurements pointing at the first P_1 (or Q_1) significant components in the first frequency interval and at the P_2 (or Q_2) significant components in the second frequency interval; the two intervals are explicitly indicated in the notation adopted in the ICNIRP guideline and correspond to 1 MHz (P_1 or Q_1) and 300 GHz (M_2 or P_2) for both E and H field. The quantities with subscript "L" refer to the limit curves: the ICNIRP guideline replaces $E_{L,m}$ and $H_{L,m}$ with two letters suggesting a constant value, that is not as reported also in the same page, so that a more general notation is preferred.

It may be stated that for railway applications, as covered by the EN 50500, the first ICNIRP indexes related to induction effect are the most relevant (and in particular that for the magnetic field), while thermal effects may be observed for RF exposure due to radio transmitters (e.g. GSM-R or TETRA antennas, VHF systems, RF tag readers). While generally in these cases the exposure can be assessed by comparing a single measured value for the channel in use and the occupied bandwidth, in case of broadband systems with multiple carriers or in case of exposure to concomitant sources, the use of the thermal ICNIRP indexes (9.4.3) and (9.4.4) is necessary.

9.4.3 Considerations on noise, spectral leakage and uncertainty

9.4.3.1 Background noise

Noise is present in the spectrum of the signal both as result of data acquisition and DFT calculation, and as preexisting background noise entering the measuring instrument through the input connector. Noise is surely relevant as contributing to the uncertainty of components estimate, but also heavily influences the evaluation of the ICNIRP index; this was underlined in [11] with the necessity of trimming the noise floor (nicknamed "grass"), e.g. by simply applying a hard threshold to component amplitude (noise pruning).

The correct threshold value may be assigned observing on the one hand what is the maximum background noise in terms of confidence interval and expected probability (see sec. 2.2.1 for Gaussian noise) and on the other hand that such a threshold is a low percentage of the limit with which spectrum components are compared (e.g. 1 % might be a good compromise, also considering the expected instrumental uncertainty).

9.4.3.2 Spectral leakage

Spectral leakage is comprehensively considered in sec. 2.3.3, where distinction is made between short-range and long-range spectral leakage: the former describes the fact that the real underlying signal component may lie between two discrete points of the chosen DFT representation, thus worsening or biasing the frequency estimate; the latter is the real spectral leakage, that causes the spread of components to the adjacent bins. Additionally, when considering countermeasures, the two phenomena are addressed differently: besides correct synchronization of the observation period with the periodicity of the signal (if possible), smoothing (or tapering) windows are used for the long-range leakage, whereas interpolation is used for the short-range leakage.

Spectral leakage is particularly relevant when evaluating compliance to limits: spectral leakage lowers the amplitude of spectrum components thus facilitating the compliance by comparison with limits; additionally, the coherent lateral components of lower amplitude created by leakage all clutter the noise floor and add all in phase to the calculation of the ICNIRP index.

9.4.3.3 Uncertainty

As for all measurements involving rolling stock, the driving style and the response of on-board converters and traction chain represent significant elements in the uncertainty budget. However, for human exposure the approach is characterized by a certain amount of overestimation at several points of the procedure, such as the use of max hold during post-processing of data and the determination of limits themselves, that may hide somehow variability in the final results.

Human exposure measurements are quite expensive because they require complete runs for each measurement and measurement point, so that it is extremely difficult that time and money constraints allow to repeat measurements in order to collect a statistically significant set of results, enabling the analysis of repeatability and Type A uncertainty.

The evaluation of instrumental uncertainty, Type B approach [15, 42, 43], is the approach indicated by EN 50500 [33]: "the uncertainty of the complete measurement chain from the field probes to the final display unit shall be not more than 20 %", that for low frequency equipment is quite a large and permissive value, but takes into account the prevailing uncertainty of probes in terms of flatness, isotropicity, etc. (see sec. 3.2). Needless to say that such statement of uncertainty is not complete because it is not accompanied by the assigned level of confidence.

With a Type B approach applied to the available information for the elements of the measurement chain, the obtained distribution when several sources of uncertainty of approximately the same order of magnitude are identified, is approximately Gaussian and indicating a 1.96σ confidence interval corresponds to a 95 % confidence level (see sec. 2.7.4).

9.5   Used instrumentation

As long as the required post-processing is the comparison of amplitude of components with limits and the calculation of the ICNIRP index based on amplitude only, frequency domain equipment (such as spectrum analyzer) may be used; considering the very low minimum frequency (5 Hz as per EN 50500 and 1 Hz as per the new ICNIRP statement), an FFT analyzer is a better choice, since heterodyne spectrum analyzers cannot measure profitably and effectively such a low frequency (the minimum frequency is often around $100\,\mathrm{kHz}$, sometimes at $10\,\mathrm{kHz}$ or slightly behind, but rarely down to the tens of Hz; an local oscillator feed-through is always negatively affecting the lowest portion of the frequency range, as shown in sec. 3.1.5). As long as measurements are performed for compliance to the EN 50500 and are limited to a frequency range of a hundred kHz or in exceptional cases a few MHz, RF frequency domain equipment is not strictly needed, except to ensure a lower noise floor and immunity to out of band signals. A sampling oscilloscope or data acquisition system may be used instead, with all the flexibility of time-domain measurements, as it was done in [11].

9.6   Measurement procedures and techniques

The most important aspect is to clarify the scope of the measurement, that in principle extends up to a significantly large frequency, including dc, and necessitate the evaluation of both electric and magnetic field with respect to the interaction mechanisms and effects identified by the ICNIRP. The EN 50500 standard limits considerably both frequency extension and phenomena deemed relevant for railway applications: "As the detectable emission of rolling stock, traction power supply and signaling equipment is in the frequency range from d.c. up to 20 kHz, measurements, simulation and calculation are restricted to this range. Accordingly only one summation regime[3] is applied. In this frequency range the magnetic field is dominant and the electric field can be neglected." The identification of the magnetic field as the most relevant one with electric field intensity nearly irrelevant is shared and backed by theoretical rationale and observations; the exclusion of thermal effects is again well grounded on

---

[3]   The mentioned summation regime refers to the ICNIRP index expressions reported in (9.4.1) through (9.4.4)

the assumption that, if radiofrequency emissions are observed, they are due additional equipment that is not really part of rolling stock, line and substation, and that may be separately tested. On the contrary, limiting measurements to 20 kHz is not justified: converter emissions may be relevant up to a significant harmonic order if Pulse Width Modulation (PWM) switching pattern is considered; with a switching frequency in the range of one to some kHz and above (e.g. auxiliaries, HVAC converters, etc.), harmonic groups are expected up to several tens of kHz; additionally, observations confirm that in many cases spectral components appear well above the noise floor up to some hundred kHz, due to steep switching transients and some internal ringing.

Regarding the selection of measurement height and displacement with respect to rolling stock elements, the EN 50500 standard [33] gives different prescriptions for workers and passengers: for the former two measurements performed for a person standing at the height of 0.9 m and 1.5 m, for the latter including also a third point lower towards the floor at 0.3 m. The distance from rolling stock walls shall be minimum 0.3 m, or a larger one compatible with the space that workers or passengers may occupy.

Passengers, for example, occupy all seats and also the space available along the entire vehicle when standing: this would require a number of points that is extremely large and a correspondingly large number of test runs. Corridors and decks shall be covered in the same way and for the driver's cabin one or two seats shall be included. The standard does not mention the situation for a sit person: based on the ICNIRP intentions the measurement should cover the most important parts of the body, that is pelvis, torso and head, so approximately 20 cm and 80 cm above the seat. Since this part is not covered by the EN 50500 nor any other international standard, it shall be in general justified by simple observation and preferably documented with pictures.

Outside rolling stock (e.g. at platform) the distance from the rolling stock external surface shall be again 0.3 m; measurements hall be taken at three different heights, 0.5 m, 1.5 m and 2.5 m, measured from the top of the running rails. If this choice may be at first not convenient when the interest is in the exposure of a person standing at platform, with the feet at the height of the platform floor, these height values cover in any case the space occupied by the person using a reference point that is quite commonly used and clarified in almost all drawings of the track, the top of rail (TOR). Considering a common situation of a platform of a modern metro or light railway whose floor is about $0.7 - 0.8$ m above TOR, the first point shall be discarded and the remaining two are covering two positions at about $0.7 - 0.8$ m and $1.7 - 1.8$ m, located approximately in the upper part of the legs and the head of a tall person.

The selection of the measurement points on-board and outside, but near the rolling stock, shall be driven by a preliminary knowledge of sources (e.g. apparatus and cables) and of the most critical areas in terms of occupation by personnel and passengers and time of staying.

Observing that in many cases the magnetic field sourced by windings and cables has a $1/r^2$ dependency on distance, moving a point away from a near source triplicating the distance (passing e.g. from 10 cm to 30 cm, or from 30 cm to 1 m) reduces the intensity by nearly an order of magnitude. Thus to pick up all and only the locations that are relevant for exposure, sensing large volumes of space from a convenient distance doesn't work: the field intensity decays rapidly and, overlapping all the active sources, the picture or signature of the magnetic field changes remarkably, so that it's extremely difficult to extrapolate from one position to the other. A complete mapping, sufficiently accurate in terms of spatial resolution, would require a large number of points.

In reality, a preliminary analysis can identify the most significant points with a satisfactory trade-off between complete and exhaustive mapping and limitation of time duration of tests:

- sources shall be identified on a drawing using past experience and knowledge of magnetic field propagation, spotting out the relevant ones;

- as anticipated in sec. 9.1, on-board sources may be very well exemplified by converters, transformers, filter inductors and in general cables; distinction shall be made based on rated current and frequency spectrum, remembering that the ICNIRP limits monotonically decrease for increasing frequency, until the plateau is reached that in 1998 version began at about $800\,Hz$ and that now has moved back to $300\,Hz$ and $400\,Hz$.

Depending on the construction criteria, converters and filters may be found on the roof or underfloor, possibly located only inside motorized units of metro and LRT trains. Cables as well may be routed along the roof, underfloor or vertically at coaches and unit edges; relevance of cables depends on many factors, such as flowing current, high-frequency components (e.g. motor cables, braking resistors cables, etc.) and size and relative positioning, influencing the amount of magnetic compensation achieved for polarities or phases.

Two examples of typical locations of measurement points are shown in Figure 9.6.1 for a light tramway vehicle taken as working example.

Some assumptions have been done on the location of typical traction equipment including traction inverters (although dc motors are installed on old tramway vehicles), as described in the figure caption. In all vehicles traction inverters are located either on the roof or under-frame close to the bogies:

- in case they are positioned on the roof, long cables connect to the motors in the bogies with problems not only of human exposure, but in general of electromagnetic emissions; the positive aspect is that the line voltage cables are very short and remain on the roof and this is particularly advantageous when the vehicle is supplied at high voltage because the required safety distances, air clearance and cable insulation are much larger than for motor cables; for heavier vehicles designed for higher speeds bulky converters cannot be located on the roof for a simple mass distribution problem;

- if inverters are located under-frame close to the bogie, motor cables are short and line voltage cables follow a similar path; differently, when there are multiple units some motorized and some trailed, but with on-board large power auxiliaries supplied from the line voltage as well, electric distribution occurs through the line voltage cables.

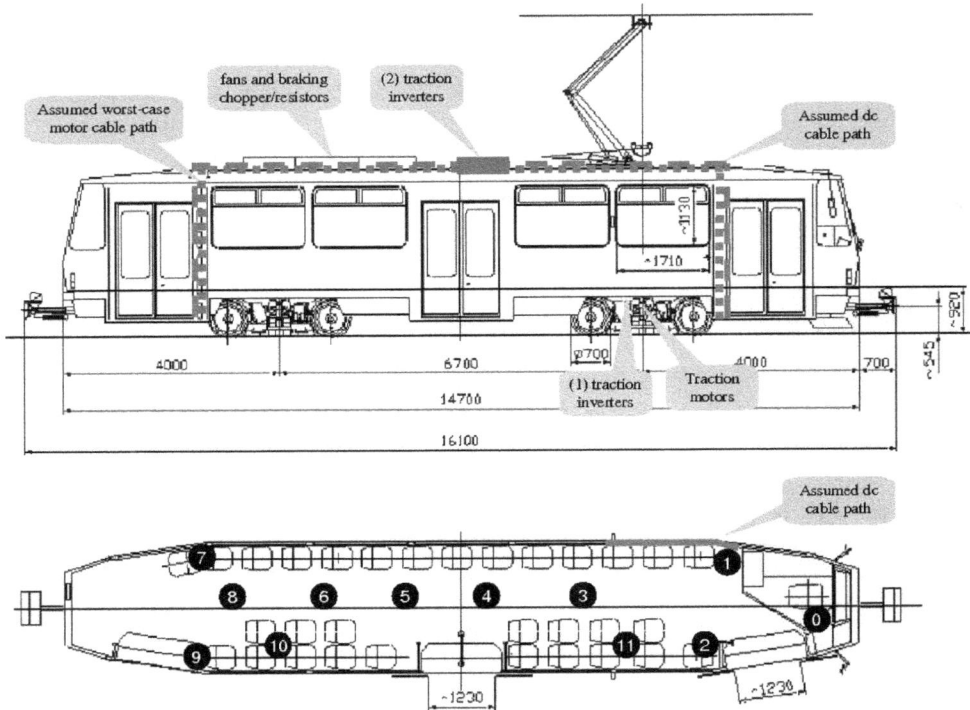

Figure 9.6.1 – Example of measurement point locations on a tramway vehicle taken as example: two assumptions are made on where the traction inverter may be located ((1) near bogies or (2) on the roof) and cables are consequently routed (gray dotted); the line voltage dc cable is routed along the roof and then vertically down to the boogie if assumption (1) holds (gray dashed).

The braking resistors and braking chopper are another source of magnetic field: resistor elements have poor magnetic compensation between positive and negative polarity because of the necessary spacing between conductors to ensure ventilation and heat exchange; the braking chopper may be also located under-frame with much longer cables going to the resistor bank on the roof; as for traction inverters, the braking chopper is a static converter polluting with its own switching frequency and harmonics.

Another source of magnetic field common to many vehicles is the low voltage bus (e.g. 220 V or 380 V) connected to all low-power loads and that at curves and crossing may deviate from a neat three-wire or four-wire symmetric arrangement, thus losing magnetic field compensation between phases and creating hot spots.

The measurement locations in the example have been thus chosen with the following rationale:

- the driver is always a relevant stakeholder and shall be always included (0);

- line voltage cables (and in many case low voltage supply bus, or motor cables if inverters are on the roof) are routed along the edge behind the driver's cabin (1);

- similarly, if using the right side of the vehicle for some of the cables, point (2) shall be included, and also to measure emissions from the motor bogie beneath;

- along the vehicle axis people standing is exposed to emissions from the roof (3, 4, 5, 6, 8), but also for the lower part of the body to possible emissions from under-frame (cables and converters); if both on the roof and in the under-frame cables are distributed also near vehicle sides, then additional measurement points shall be foreseen on the passengers seats, not only centrally in the corridor;

- symmetrically, line voltage cables and/or motor cables are routed along edges in the rear portion of the vehicle to connect to the other bogie (7, 9);

- two points (10, 11) are added to monitor passengers' seats closer to the motor bogies underneath.

Bibliography

[1] Agilent, "Spectrum Analysis Basics," *Application Note 150*, 1971 (revised in 2014).

[2] Agilent, "Spectrum Analyzer – Measurements and Noise," *Application Note 1303*, code 5966-4008E, Apr. 2, 2008.

[3] D. Agrez, "Weighted Multipoint Interpolated DFT to Improve Amplitude Estimation of Multifrequency Signal," *IEEE Transactions on Instrumentation and Measurements*, Vol. 51, no. 2, Apr. 2002, pp. 287-292. doi: 10.1109/19.997826.

[4] B. Andò, S. Baglio, A.R. Bulsara, and V. Sacco, "RTD Fluxgate: A Low-Power Nonlinear Device to Sense Weak Magnetic Fields," *IEEE Instrumentation & Measurement Magazine*, Oct. 2005, pp. 64-73. doi: 10.1109/MIM.2005.1518626.

[5] F. Auger, P. Flandrin, P. Gonçalvès and O. Lemoine, Time-Frequency Toolbox, [online] tftb.nongnu.org/tutorial.pdf (last access March 2016).

[6] B.A. Austin and A.P.C. Fourie, "Characteristics of the Wire Biconical Antenna Used for EMC Measurements," *IEEE Transactions on Electromagnetic Compatibility*, Vol. 33, no. 3, pp. 179-187, Aug. 1991. doi: 10.1109/15.85131.

[7] M. Babinet, "Memoires d'optique metrologique," *Comptes Rendus de l'Academie de Science*, Vol. 4, pp. 638 - 648, 1837.

[8] C.A. Balanis, *Antenna Theory – Analysis and Design*, 2nd ed., John Wiley & Sons, 1997.

[9] S.A. Barengolts, G.A. Mesyats and A.G. Chentsov, "Spontaneous Extinguishing of a Vacuum Arc in Terms of the Ecton Model," *IEEE Transactions on Plasma Science*, Vol. 27, no. 4, Aug. 1999, pp. 817-820. doi: 10.1109/27.782244.

[10] Bartington Instruments, *Mag-03 Three-Axis Magnetic Field Sensors*, [online] www.bartington.com/Literaturepdf/Datasheets/Mag-03%20DS0013.pdf (last access March 2016).

[11] D. Bellan, A. Gaggelli, F. Maradei, A. Mariscotti and S. Pignari "Time-Domain Measurement and Spectral Analysis of Non-Stationary Low-Frequency Magnetic Field Emissions on Board of Rolling Stock," *IEEE Transactions on Electromagnetic Compatibility*, Vol. 46, no. 1, Feb. 2004, pp. 12-23. doi: 10.1109/TEMC.2004.823607.

[12] J.S. Bendat and A.G. Piersol, *Engineering Applications of Correlation and Spectral Analysis*, John Wiley & Sons, Canada, 1980.

[13] J.S. Bendat and A.G. Piersol, *Random Data - Analysis and Measurement procedures*, 2nd ed., John Wiley & Sons, Canada, 1986.

[14] A. Bernardi, C. Fraser-Smith, and G. Villard, Jr., "Measurements of BART Magnetic Fields with an Automatic Geomagnetic Pulsation Index Generator," *IEEE Transactions on Electromagnetic Compatibility*, Vol. 31, no. 4, Nov. 1989.

[15] BIPM, *Evaluation of measurement data — Guide to the expression of uncertainty in measurement*, JCGM 100:2008.

[16] M. Bittera V. Smiesko K. Kovac J. Hallon, "Directional properties of the Bilog antenna as a source of radiated electromagnetic interference measurement uncertainty," *IET Microwaves, Antennas & Propagation*, 2010, Vol. 4, no. 10, pp. 1469-1474. doi: 10.1049/iet-map.2009.0187.

[17] M. Bittera V. Smiesko K. Kovac J. Hallon, "Influence of Directivity Pattern of Bilog Antenna to Radiated EMI Measurement Uncertainty," 14th IEEE Conference on Microwave Techniques, COMITE 2008, Prague, Czech Republic, Apr. 23-24, 2008, pp. 1 - 4. doi: 10.1109/COMITE.2008.4569924.

[18] L.I. Bluestein, "A linear filtering approach to the computation of the discrete Fourier transform," Northeast Electronics Research and Engineering Meeting Record, Vol. 10, 1968, pp. 218-219. (later published as "A linear filtering approach to the computation of the discrete Fourier transform,", *IEEE Transactions on Audio Electroacoustics*, Vol. 18, no. 4, pp. 451-455, 1970. doi: 10.1109/TAU.1970.1162132)

[19] J.T. Bolljahn, "Antennas near conducting sheets of finite size", University of California, 1950.

[20] G. Boschetti, A. Mariscotti and V. Deniau, "Assessment of the GSM-R susceptibility to repetitive transient disturbance", *Measurement*, Elsevier, vol. 45, May 2012, pp. 2226-2236. doi: 10.1016/j.measurement.2012.04.004.

[21] R.N. Bracewell, *The Fourier Transform and Its Applications*, 2nd Ed., McGraw-Hill, New York, 1986.

[22] S. Braun, T. Donauer and P. Russer, "A real-time time-domain EMI measurement system for full-compliance measurements according to CISPR 16-1-1," *IEEE Transactions on Electromagnetic Compatibility*, Vol. 50, no. 2, Apr. 2008, pp. 259-267. doi: 10.1109/TEMC.2008.918980.

[23] R. Briante and P. Imbesi, "Train carbody EMC shielding measurement," 20th IMEKO TC4 International Symposium, Benevento, Italy, 2014, Sept. 15-17, 2014, pp. 1064-1068.

[24] G. Bucca and A. Collina, "A procedure for the wear prediction of collector strip and contact wire in pantograph–catenary system," *Wear*, Elsevier, Vol. 266, 2009, pp. 46-59. doi: 10.1016/j.wear.2008.05.006.

[25] B. Carlson, *Communication Systems – An Introduction to Signals and Noise in Electrical Communication*, 3rd ed., New York, McGraw Hill, 1988.

[26] C. Carobbi, S. Lazzerini, L. Millanta, "High frequency operation of the coils for standard magnetic field generation using the bulk-current technique," *IEEE Transactions on Instrumentation and Measurements*, Vol. 54, no. 4, Aug. 2005, pp. 1427-1432.

[27] CENELEC EN 50121-2 (IEC 62236-2), *Railway applications - Electromagnetic compatibility - Part 2: Emission of the whole railway system to the outside world*, 2006-07.

[28] CENELEC EN 50121-3-1 (IEC 62236-3-1), *Railway applications - Electromagnetic compatibility - Rolling stock - Part 3.1: Train and complete vehicle*, 2006-07.

[29] CENELEC EN 50121-2 (IEC 62236-2), *Railway applications - Electromagnetic compatibility - Part 2: Emission of the whole railway system to the outside world*, 2015-03.

[30] CENELEC EN 50121-3-1 (IEC 62236-3-1), *Railway applications - Electromagnetic compatibility - Rolling stock - Part 3.1: Train and complete vehicle*, 2015-03.

[31] CENELEC EN 50357, *Evaluation of human exposure to electromagnetic fields from devices used in Electronic Article Surveillance (EAS), Radio Frequency Identification (RFID) and similar applications*, 2001-10.

[32] CENELEC EN 50392, *Generic standard to demonstrate the compliance of electronic and electrical apparatus with the basic restrictions related to human exposure to electromagnetic fields (0 GHz – 300 GHz)*, 2008-02.

[33] CENELEC Std. EN 50500, *Measurement procedures of magnetic field levels generated by electronic and electrical apparatus in the railway environment with respect to human exposure*, 2008-07.

[34] Liu Chang, Min Wang, Lei Liu, Siwei Luo and Pan Xiao, "A brief introduction to giant magnetoresistance," arXiv preprint arXiv:1412.7691. 2014 Dec 21.

[35] R.A. Chipman, *Schaum's outline of Theory and Problems of Transmission Lines*, McGraw Hill, 1968.

[36] D.K. Chy and Md. Khaliluzzaman, "Evaluation of SNR for AWGN, Rayleigh and Rician Fading Channels Under DPSK Modulation Scheme with Constant BER," *International Journal of Wireless Communications and Mobile Computing*, 2015, Vol. 3, no. 1, pp. 7-12. doi: 10.11648/j.wcmc.20150301.12.

[37] CISPR 12, *Vehicles, boats and internal combustion engines – Radio disturbance characteristics – Limits and methods of measurement for the protection of off-board receivers*, 2009-03.

[38] CISPR 16-1-1, *Specification for radio disturbance and immunity measuring apparatus and methods – Part 1-1: Radio disturbance and immunity measuring apparatus - Measuring apparatus*, 2015-09.

[39] CISPR 16-1-4, *Specification for radio disturbance and immunity measuring apparatus and methods – Part 1-4: Radio disturbance and immunity measuring apparatus - Antennas and test sites for radiated disturbance measurements*, 2012-07.

[40] CISPR 16-2-1, *Specification for radio disturbance and immunity measuring apparatus and methods – Part 2-1: Methods of measurement of disturbances and immunity - Conducted disturbance measurements*, 2014-02.

[41] CISPR 16-2-3, *Specification for radio disturbance and immunity measuring apparatus and methods – Part 2-3: Methods of measurement of disturbances and immunity - Radiated disturbance measurements*, 2014-03.

[42] CISPR 16-4-1, *Specification for radio disturbance and immunity measuring apparatus and methods – Part 4-2: Uncertainties, statistics and limit modelling - Uncertainty in standardized EMC tests*, 2009-02.

[43] CISPR 16-4-2, *Specification for radio disturbance and immunity measuring apparatus and methods – Part 4-2: Uncertainties, statistics and limit modelling - Uncertainty in EMC measurements*, 2003-11.

[44] S. Ciurlo, A. Mariscotti and A. Viacava, "A Helmholtz coil for high frequency high field intensity applications", *Metrology and Measurement Systems*, Vol. XVI, n. 1, Jan. 2009, pp. 117-125.

[45] R.E. Collin, *Foundations for Microwave Engineering*, 2nd ed., IEEE Press, New York, 2001.

[46] Da Hai He, "Test Method of Arcing Behaviour for Railway Current Collection System," World Congress on Railway Research, WCRR 2001, Koln, Germany, Nov. 25-29, 2001.

[47] A.W. Drake, *Fundamentals of Applied Probability Theory*, McGraw-Hill, New York, 1967.

[48] S. Dudoyer, V. Deniau, S. Ambellouis, M. Heddebaut, A. Mariscotti and N. Pasquino, "Test bench for the evaluation of GSM-R operation in the presence of electric arc interference," ESARS 2012, Bologna, IT, Oct. 17-19, 2012, doi: 10.1109/ESARS.2012.6387486.

[49] S. Dudoyer, V. Deniau, S. Ambellouis, M. Heddebaut and A. Mariscotti, "Classification of Transient EM Noise Depending on their Effect on the Quality of GSM-R Reception", *IEEE Transactions on Electromagnetic Compatibility*, Vol. 55, no. 5, Oct. 2013, pp. 867-874, doi: 10.1109/TEMC.2013.2239998.

[50] ETSI Std. ETS 300 577, *Digital cellular telecommunications system (Phase 2); Radio transmission and reception*, Mar. 1997.

[51] European Directive 2004/104 adapting to technical progress Council Directive 72/245/EEC relating to the radio interference (electromagnetic compatibility) of vehicles and amending Directive 70/156/EEC on the approximation of the laws of the Member States relating to the type-approval of motor vehicles and their trailers, Oct. 14, 2004.

[52] D.A. Frickey, "Conversions between S, Z, Y, H, ABCD, and T parameters which are valid for complex source and load impedances," *IEEE Transactions on Microwave Theory and Techniques*, Vol. 42, no. 2, Feb. 1994, pp. 205-211. doi: 10.1109/22.275248

[53] D.A. Frickey, "Using the Inverse Chirp-Z Transform for Time-Domain Analysis of Simulated Radar Signals," International Conference on Signal Processing Applications and Technology (ICSPAT), Oct. 18-21, 1994.

[54] R. Geise, O. Kerfin, B. Neubauer, G. Zimmer and A. Enders, "EMC Analysis Including Receiver Characteristics - Pantograph Arcing and the Instrument Landing System," IEEE International Symposium on Electromagnetic Compatibility, Dresden, Germany, Aug. 16-22, 2015, pp. 1213-1217. doi: 10.1109/ISEMC.2015.7256342.

[55] T. Grandke, "Interpolation Algorithms for Discrete Fourier Transforms of Weighted Signals," *IEEE Transactions on Instrumentation and Measurements*, Vol. 32, no. 2, June 1983, pp. 350-355. doi: 10.1109/TIM.1983.4315077.

[56] GSM, Recommendation 05.03, *Channel Coding*, ver. 3.6.1., Oct. 1994.

[57] A.E. Guile and S.F. Mehta, "Arc Movement Due to the Magnetic Field of Current Flowing in the Electrodes," *Proceedings of IEE – Part A: Power Engineering*, Vol. 1, 1958, pp. 533-540. doi: 10.1049/pi-a.1957.0126.

[58] E. Hallén, "On antenna impedances", *Transactions of the Royal Institute of Technology*, Stockholm, n. 13, 1947.

[59] R.C. Hansen, "Fundamental limitations in antennas," *Proceedings of the IEEE*, Vol. 69, no. 2, pp. 170-182, Feb. 1981.

[60] F.J. Harris, "On the Use of Windows for Harmonic Analysis with the Discrete Fourier Transform," *Proceedings of IEEE*, Vol. 66, no. 1, Jan. 1978, pp. 51-83. doi: 10.1109/PROC.1978.10837

[61] S. Hartwig, H.H. Albrecht, H.J. Scheer, M. Burghoff and L. Trahms, "A Superconducting Quantum Interference Device Measurement System for Ultra Low-Field Nuclear Magnetic Resonance," *Applied Magnetic Resonance*, Vol. 44, 2013, pp. 9–22.

[62] P. Horowitz and W. Hill, *The art of electronics*, 2nd ed., Cambridge University Press, 1989.

[63] R.M. Howard, *Principles of Random Signal Analysis and Low Noise Design*, John Wiley & Sons, New York, 2002.

[64] ICNIRP, "ICNIRP Guidelines for limiting exposure to time-varying electric, magnetic, and electromagnetic fields (up to 300 GHz)," *Health Physics*, Vol. 74, no. 4, pp. 494-522, Apr. 1998, [online] www.icnirp.org/cms/upload/publications/ICNIRPemfgdl.pdf (last access March 2016).

[65] ICNIRP, "ICNIRP Statement – Guidelines for limiting exposure to time-varying electric and magnetic fields (1 Hz to 100 kHz)," *Health Physics*, Vol. 99, no. 6, pp. 818-836, Dec. 2010. doi: 10.1097/HP.0b013e3181f06c86, [online] http://www.icnirp.org/cms/upload/publications/ICNIRPLFgdl.pdf (last access March 2016).

[66] ICNIRP, "Guidance on determining compliance of exposure to pulsed and complex non-sinusoidal waveforms below 100 kHz with ICNIRP guidelines", *Health Physics*, Vol. 84, pp. 383-387, 2003, [online] http://www.icnirp.org/cms/upload/publications/ICNIRPpulsed.pdf (last access March 2016).

[67] ICNIRP, "Guidelines on limits of exposure to static magnetic fields," *Health Physics*, Vol. 96, no. 4, pp. 505-514, Apr. 2009, [online] http://www.icnirp.org/cms/upload/publications/ICNIRPstatgdl.pdf (last access March 2016).

[68] IEC 60096-1, *Radio-frequency cables – Part1: General requirements and measuring methods*, 1986-01.

[69] ANSI/IEEE Std. 376, *IEEE Standard for the Measurement of Impulse Strength and Impulse Bandwidth*, 1975.

[70] IEEE Std. 644, *IEEE Standard Procedures for Measurement of Power Frequency Electric and Magnetic Fields From AC Power Lines*, 1994.

[71] IEEE Std. 748, *IEEE Standard for Spectrum Analyzers*, 1979.

[72] ANSI Std. C63.2, *Electromagnetic Noise and Field Strength, 10 kHz to 40 GHz – Specifications*, 1987.

[73] ISO/IEC Guide 98-3/Suppl. 1, *Uncertainty of measurement – Part 3: Guide to the expression of uncertainty in measurement (GUM:1995) – Supplement 1: Propagation of distributions using a Monte Carlo method*, 2008.

[74] A.J. Iverson, Electro-optic pockels cell voltage sensors for accelerator diagnostics, Thesis of Masters of Science in Electrical Engineering, Montana State University, Bozeman, MT, USA, July 2004.

[75] V.K. Jain, W.L. Collins and D.C. Davis, "High-Accuracy Analog Measurements via Interpolated FFT," *IEEE Transactions on Instrumentation and Measurements*, Vol. 28, no. 2, June 1979, pp. 113-122. doi: 10.1109/TIM.1979.4314779.

[76] JEOL, *64x0 MP series environmental specification sheet*, rev. 1c, June 26, 2008.

[77] H.W. Johnson and M. Graham, *High-Speed Digital Design - A Handbook of Black Magic*, Prentice Hall, Englewood Cliffs, New Jersey, 1988.

[78] J.F. Kaiser, "Nonrecursive Digital Filter Design Using the I0-sinh Window Function," Proc. IEEE Symp. Circuits and Systems, Apr. 1974, pp. 20-23.

[79] T.-W.Kang, Y.-C.Chung, S.-H.Won and H.-T.Kim, "Interlaboratory comparison of radiated emission measurements using a spherical dipole radiator," *IEE Proceedings – Science Measurement Technology*, Vol. 148, No. 1 , Jan. 2001, pp. 35-40. doi: 10. 1049/ip20010184.

[80] D. Karakelian, S.L. Klemperer, A.C. Fraser-Smith, and G.C. Beroza, "A Transportable System for Monitoring Ultra Low Frequency Electromagnetic Signals Associated with Earthquakes," *Seismological Research Letters*, Vol. 71, no. 4, July/Aug. 2000, pp. 423-436.

[81] S.M. Kay and S.L. Marple, "Spectrum Analysis – A Modern Perspective," *Proceedings of the IEEE*, Vol. 69, no. 11, Nov. 1981, pp. 1380-1419. doi: 10.1109/PROC.1981.12184.

[82] D. Klapas, R. Hackam and F.A. Benson, "Electric arc power collection for high-speed trains," *Proceedings of the IEEE*, Vol. 64, no. 12, Dec. 1976, pp. 1699-1715. doi: 10.1109/PROC.1976.10410.

[83] D. Klapas, R.H. Apperley, R. Hackam and F.A. Benson, "Electromagnetic Interference from Electric Arcs in the Frequency Range 0.1-1000 MHz," *IEEE Transactions on Electromagnetic Compatibility*, Vol. 20, no. 1, Feb. 1978, pp. 198-202. doi: 10.1109/TEMC.1978.303647.

[84] J.D. Krause and R.J. Marhefka, *Antennas*, 3rd ed., McGraw Hill, New York, 2002.

[85] A. Kriz, "Calculation of Antenna Pattern Influence on Radiated Emission Measurement Uncertainty," IEEE International Symposium on Electromagnetic Compatibility, Detroit, USA, 18-22 Aug. 2008. doi: 10.1109/ISEMC.2008.4652052.

[86] S. Kubo and K. Kato, "Effect of arc discharge on wear rate of Cu-impregnated carbon strip in unlubricated sliding against Cu trolley under electric current," *Wear*, Elsevier, Vol. 216, 1998, pp. 172-178. doi: 10.1016/S0043-1648(97)00184-1.

[87] E.A. Laport, "Long-wire antennas" in *Antenna Engineering Handbook*, (R.C. Johnson ed.), 3rd ed., McGraw Hill, New York, 1993.

[88] M.P. Ledbetter, T. Theis, J.W. Blanchard, H. Ring, P. Ganssle, S. Appelt, B. Blümich, A. Pines and D. Budker, "Near-Zero-Field Nuclear Magnetic Resonance," *Physical Review Letters*, Vol. 107, Sept. 2011, pp. 107601-1/5. doi: 10.1103/PhysRevLett.107.107601.

[89] Yu-Jen Liu, G.W. Chang and H.M. Huang, "Mayr's Equation-Based Model for Pantograph Arc of High-Speed Railway Traction System," *IEEE Transactions on Power Delivery*, Vol. 25, no. 3, July 2010, pp. 2025-2027. doi: 10.1109/TPWRD.2009.2037521.

[90] R.G. Lyons, *Understanding Digital Signal Processing*, Prentice Hall, 2001.

[91] Lan Ma, A. Marvin, Y. Wen, E. Karadimou and R. Armstrong, "An Investigation of the Total Radiated Power of Pantograph Arcing Measured in a Reverberation Chamber," International Conference on Electromagnetics in Advanced Applications (ICEAA), Aruba, Aug. 3-8, 2014, pp. 550-553. doi: 10.1109/ICEAA.2014.6903919.

[92] Lan Ma, A. Marvin, E. Karadimou, R. Armstrong and Y. Wen, "An Experimental Programme to Determine the Feasibility of Using a Reverberation Chamber to Measure the Total Power Radiated by an Arcing Pantograph," 2014 International Symposium on Electromagnetic Compatibility (EMC Europe 2014), Gothenburg, Sweden, September 1-4, 2014, pp. 269-273.

[93] A. Mariscotti and P. Pozzobon, "Synthesis of line impedance expressions for railway traction systems," *IEEE Transactions on Vehicular Technology*, Vol. 52, no. 2, March 2003, pp. 420-430. doi: 10.1109/TVT.2003.808750.

[94] A. Mariscotti, "Evaluation and Testing of Off-the-shelf Hall Sensors for Compliant Magnetic Field", IEEE International Measurement Technical Conference IMTC 2006, Sorrento, Italy, April 20-23, 2006. doi: 10.1109/IMTC.2006.328176.

[95] A. Mariscotti, "On Time- and Frequency-domain equivalence for compliant EMI measurements", IEEE International Measurement Technical Conference IMTC 2007, Warsaw, Poland, May 2-4, 2007. doi: 10.1109/IMTC.2007.379171.

[96] A. Mariscotti, "Measurement Procedures and Uncertainty Evaluation for Electromagnetic Radiated Emissions from Large Power Electrical Machinery," *IEEE Transactions on Instrumentation and Measurement*, Vol. 56, no. 6, Dec. 2007, pp. 2452-2463. doi: 10.1109/TIM.2007.908351.

[97] A. Mariscotti, "A Magnetic Field Probe with MHz Bandwidth and 7 decades Dynamic Range," *IEEE Transactions on Instrumentation and Measurement*, Vol. 58, no. 8, Aug. 2009, pp. 2643-2652. doi: 10.1109/TIM.2009.2015693.

[98] A. Mariscotti, "Assessment of Electromagnetic Emissions from Synchronous Generators and its Metrological Characterization," *IEEE Transactions on Instrumentation and Measurement*, Vol. 59, no. 2, Feb. 2010, pp. 450-457. doi: 10.1109/TIM.2009.2024696.

[99] A. Mariscotti, *RF and Microwave Measurements – Device characterization, signal integrity and spectrum analysis*, ASTM, Chiasso, Switzerland, 2015. ISBN: 978-88-941091-0-8.

[100] S.L. Marple, *Digital Spectral Analysis*, Prentice Hall, Englewood Cliffs, NJ, 1987.

[101] W. Marshall Leach, Jr., *Fundamentals of Low-Noise Electronic Analysis and Design*, Kendall Hunt publishing, 2000.

[102] W. Martin and P. Flandrin, "Wigner-Ville Spectral Analysis of Nonstationary Processes," IEEE Transactions on Acoustics, Speech, and Signal Processing, Vol. 33, no. 6, Dec. 1985, pp. 1461-1470. doi: 10.1109/TASSP.1985.1164760.

[103] A.C. Marvin and J. Ahmadi, "Comparison of open-field test sites used for radiated emission measurements," *IEE Proceedings - Part A*, Vol. 140, no. 2, March 1993, pp. 161-165.

[104] S.A. Maas, *The RF and Microwave Circuit Design Cookbook*, Artech House, Boston, 1998.

[105] Y. Matsumoto, "On the relation between the amplitude probability distribution of noise and bit error probability," *IEEE Transactions on Electromagnetic Compatibility*, Vol. 49, no. 4, pp. 940-941, Nov. 2007. doi: 10.1109/TEMC.2007.908280.

[106] Y. Matsumoto and K. Gotoh, "A method of defining emission limits including the gradient of an amplitude-probability-distribution curve," International Symposium of Electromagnetic Compatibility, Gothenburg, Sweden, Sep. 1-4, 2014, pp. 895-900. doi: 10.1109/EMCEurope.2014.6931030.

[107] A.J. Mauriello and J.M. Clarke, "Measurement and analysis of radiated electromagnetic emissions from rail-transit vehicles," *IEEE Transactions on Electromagnetic Compatibility*, Vol. 25, no. 4, Nov. 1983, pp. 405-411. doi: 10.1109/TEMC.1983.304129.

[108] Metropolitan Council, "Electromagnetic Interference - Measurement and Assessment," St. Paul, MN, May 2008.

[109] D. Middleton, "Statistical-Physical Models of Electromagnetic Interference," *IEEE Transactions on Electromagnetic Compatibility*, Vol. 19, no. 3, Aug. 1977, pp. 106-127. doi: 10.1109/TEMC.1977.303527.

[110] MIL STD 461-F, *Requirements for the Control of Electromagnetic Interference Characteristics of Subsystems and Equipment*, 2007-10.

[111] S. Midya, D. Bormann, T. Schutte and R. Thottappillil, "DC Component From Pantograph Arcing in AC Traction System – Influencing Parameters, Impact, and Mitigation Techniques," *IEEE Transactions on Power Delivery*, Vol. 53, no. 1, Feb. 2011, pp. 18-27. doi: 10.1109/TEMC.2010.2045159.

[112] S. Midya, D. Bormann, T. Schutte and R. Thottappillil, "Pantograph Arcing in Electrified Railways – Mechanism and Influence of Various Parameters – Part I: With DC Traction Power Supply," *IEEE Transactions on Power Delivery*, Vol. 24, no. 4, Oct. 2009, pp. 1931-1939. doi: 10.1109/TPWRD.2009.2021035.

[113] S. Midya, D. Bormann, T. Schutte and R. Thottappillil, "Pantograph Arcing in Electrified Railways – Mechanism and Influence of Various Parameters – Part II: With AC Traction Power Supply," *IEEE Transactions on Power Delivery*, Vol. 24, no. 4, Oct. 2009, pp. 1940-1950. doi: 10.1109/TPWRD.2009.2021036.

[114] D.C. Montgomery and G.C. Runger, *Applied statistics and probability for engineers*, John Wiley & Sons, New York, 2003.

[115] D.A. Muller and J. Grazul, "Optimizing the environment for a sub-0.2 nm scanning transmission electron microscopy," *Journal of Electron Microscopy*, Vol. 50, no. 3, 2001, pp. 219-226.

[116] D.A. Muller, E.J. Kirkland, M.G. Thomas, J.L. Grazul, L. Fittinga and M. Weyland, "Room design for high-performance electron microscopy," *Ultramicroscopy*, Vol. 106, no. 11-12, Oct.-Nov. 2006, pp. 1033-1040. doi: 10.1016/j.ultramic.2006.04.017.

[117] T. Nakai and Z. Kawasaki, "On Impulsive Noise from Shinkansen," *IEEE Transactions on Electromagnetic Compatibility*, Vol. 25, no. 4, pp. 396-404, Nov. 1983. doi: 10.1109/TEMC.1983.304128.

[118] NIST/SEMATECH, *e-Handbook of Statistical Methods*, [online] http://www.itl.nist.gov/div898/handbook/ (last access Aug. 2014).

[119] A. Ogunsola and A. Mariscotti, *Electromagnetic Compatibility in Railways – Analysis and Management*, Springer, 2012.

[120] M.A. O'Keefe, J.H. Turner, J.A. Musante, J.D. Hetherington, A.G. Cullis, B. Carragher, R. Jenkins, J. Milgrim, R.A. Milligan, C.S. Potter, L.F. Allard, D.A. Blom, L. Degenhardt, and W.H. Sides, "Laboratory Design for High-Performance Electron Microscopy," *Microscopy Today*, May 2014, pp. 8-14.

[121] A.V. Oppenheim and R.W. Schafer, *Discrete-Time Signal Processing*, Prentice Hall, Englewood Cliffs, NJ, 1989.

[122] A. Papoulis, Probability, *Random variables and Stochastic processes*, 2nd ed., McGraw-Hill, 1987.

[123] C.R. Paul, *Multiconductor Transmission Lines*, 2nd. ed., John Wiley & Sons, New Jersey, 2008.

[124] S. C. Pei and J. J. Ding, "Relations between Gabor transforms and fractional Fourier transforms and their applications for signal processing," IEEE Transactions on Signal Processing, Vol. 55, no. 10, pp. 4839-4850, Oct. 2007. doi: 10.1109/TSP.2007.896271.

[125] M. Pluska, Ł. Oskwarek, R.J. Rak, and A. Czerwinski, "Measurement of Magnetic Field Distorting the Electron Beam Direction in Scanning Electron Microscope," *IEEE Transactions on Instrumentation and Measurement*, Vol. 58, no. 1, Jan. 2009, pp. 173-179. doi: 10.1109/TIM.2008.928415.

[126] M. Pous and F. Silva, "APD radiated transient measurements produced by electric sparks employing time-domain captures," International Symposium on Electromagnetic Compatibility EMC Europe 2014, Gothenburg, Sweden, Sept. 1-4, 2014, pp. 813-817. doi: 10.1109/EMCEurope.2014.6931016.

[127] M. Pous and F. Silva, "Full-spectrum APD measurement of transient interferences in time domain," *IEEE Transactions on Electromagnetic Compatibility*, Vol. 56, no. 6, pp. 1352-1360, Dec. 2014. doi: 10.1109/TEMC.2014.2352393.

[128] M. Pous, M.A. Azpúrua and F. Silva, "Measurement and Evaluation Techniques to Estimate the Degradation Produced by the Radiated Transients Interference to the GSM System," *IEEE Transactions on Electromagnetic Compatibility*, Vol. 57, no. 6, Dec. 2015, pp. 1382-1390. doi: 10.1109/TEMC.2015.2472983.

[129] D.M. Pozar, *Microwave Engineering*, John Wiley & Sons, 2nd ed., 1998.

[130] F. Primdahl, "The fluxgate mechanism, Part 1: the gating curves of parallel and orthogonal fluxgates," *IEEE Transactions on Magnetics*, Vol. 6, no. 2, pp. 376-383, 1970. doi: 10.1109/TMAG.1970.1066795.

[131] J.G. Proakis and D.G. Manolakis, *Digital Signal Processing - Principles, Algorithms, and Applications*, 3rd ed., Prentice Hall, 1996.

[132] L.R. Rabiner, R.W. Shafer and C.M. Rader, "The chirp z-transform algorithm," *IEEE Transactions on Audio Electroacoustics*, Vol. 17, no. 2, pp. 86-92, 1969. doi: 10.1109/TAU.1969.1162034.

[133] S. Ramakrishnan, A.D. Stokes and J.J. Lowke, "An approximate model for high-current free-burning arcs," *Journal of Physics – Part D: Applied Physics*, Vol. 11, 1978, pp. 2267-2280. doi: 10.1088/0022-3727/11/16/014.

[134] C. Rauscher, V. Janssen and R. Minihold, *Fundamentals of Spectrum Analysis*, Rohde & Schwarz, 2001.

[135] C. Reig, M.-D. Cubells-Beltran and D. Ramirez Muñoz, "Magnetic Field Sensors Based on Giant Magnetoresistance (GMR) Technology: Applications in Electrical Current Sensing," *Sensors*, Vol. 9, 2009, pp. 7919-7942. doi:10.3390/s91007919.

[136] P. Ripka, M. Pribil and M. Butta, "Fluxgate Offset Study," *IEEE Transactions on Magnetics*, Vol. 50, no. 11, Nov. 2014. doi: 10.1109/TMAG.2014.2329777.

[137] E. Rubiola and F. Vernotte, "The cross-spectrum experimental method," arXiv:1003.0113v1 [physics.ins-det], Feb. 27, 2010, pp. 1-39.

[138] N. Saulig, "Rényi Entropy Based Complexity Estimation of Nonstationary Signals," Ph.D. Thesis, Faculty of Electrical Engineering and Computing, University of Zagreb, 2015.

[139] I.M. Savukov, S.J. Seltzer and M.V. Romalis, "Detection of NMR signals with a radio-frequency atomic magnetometer," *Journal of Magnetic Resonance*, Vol. 185, 2007, pp. 214-220. doi: 10.1016/j.jmr.2006.12.012.

[140] E. Schmautzer, G. Propst and K. Friedl, "The Effect of Reduction Conductors on the Magnetic Field of Electrified Railway Systems near Hospitals," International Conference on Electrical Systems for Aircraft, Railway and Ship Propulsion (ESARS), Bologna, Italy, Oct. 19-21, 2010, pp. 1-5. doi: 10.1109/ESARS.2010.5665261.

[141] A.W. Scott, *Understanding Microwaves*, John Wiley & Sons, 1993.

[142] Q. Shan and Y. Wen, "Research on the BER of the GSM-R Communications Provided by the EM Transient Interferences in High-Powered Catenary System Environment," International Conference on Electromagnetics in Advanced Applications (ICEAA), Sidney, Australia, Sept. 20-24, 2010, pp. 757-760. doi: 10.1109/ICEAA.2010.5653976.

[143] Zhi Wei Sim and Jian Song, "Radiated spurious emission measurements using fast Fourier transform-based time domain scan," *IET Science, Measurement & Technology*, Vol. 9, no. 7, 2015, pp. 882-889. doi: 10.1049/iet-smt.2014.0333.

[144] N.B. Slimen, V. Deniau, J. Rioult, S. Dudoyer and S. Baranowski, "Statistical characterisation of the EM interferences acting on GSM-R antennas fixed above moving train," *The European Journal of Applied Physics*, Vol. 48, p. 21202-1–21202-7, 2009.

[145] M. Stecher, "Possible effects of spread-spectrum-clock interference on wideband radiocommunication services," International Symposium on Electromagnetic Compatibility, Chicago, IL, USA, Aug. 8-12, 2005, Vol. 1, pp. 60-63. doi: 10.1109/ISEMC.2005.1513472.

[146] P.F. Stenumgaard, "Using the root-mean-square detector for weighting of disturbances according to its effect on digital communication services," *IEEE Transactions on Electromagnetic Compatibility*, Vol. 42, no. 4, Nov. 2000, pp. 368-375. doi: 10.1109/15.902306.

[147] P.F. Stenumgaard, "On radiated emission limits for pulsed interference to protect modern digital wireless communication systems," *IEEE Transactions on Electromagnetic Compatibility*, Vol. 49, no. 4, Nov. 2007. doi: 10.1109/TEMC.2007.908284.

[148] P.F. Stenumgaard and K.C. Wiklundh, "An improved method to estimate the impact on digital radio receiver performance of radiated electromagnetic disturbances," *IEEE Transactions on Electromagnetic Compatibility*, Vol. 42, no. 2, May 2000, pp. 233-239. doi: 10.1109/15.852418.

[149] P. Stoica, and R.L. Moses, *Introduction to Spectral Analysis*, Prentice Hall, 1997.

[150] J.E. Storer, "The impedance of an antenna over a large circular screen," *Journal of Applied Physics*, Vol. 22, pp. 1058-1066, Aug. 1951.

[151] J.E. Storer, "The radiation pattern of an antenna over a circular ground screen," *Journal of Applied Physics*, Vol. 23, pp. 588-593, May 1952.

[152] T. Thayaparan and S. Kennedy, "Application of Joint Time-Frequency Representations to a Maneuvering Air Target in Sea-Clutter: Analysis Beyond FFT," Defence R&D Canada, Technical Memorandum, DRDC Ottawa, TM 2003-090, March 2003. [online] http://www.dtic.mil/dtic/tr/fulltext/u2/a416872.pdf (last access March 2016).

[153] M. Uchino, O. Tagiri and T. Shinozuka, "Real-Time Measurement of Noise Statistics," *IEEE Transactions on Electromagnetic Compatibility*, Vol. 43, no. 4, Nov. 2001, pp. 629-636. doi: 10.1109/15.974644.

[154] G.E. Valley and H. Wallman, *Vacuum Tube Amplifiers*, MIT Radiation Laboratory Series 18, McGraw-Hill, 1948.

[155] N. van Dijk, "Uncertainties in 3-m Radiated Emission Measurements Due to the Use of Different Types of Receive Antennas," *IEEE Transactions on Electromagnetic Compatibility*, Vol. 47, no. 1, Feb. 2005, pp. 77-85. doi: 10.1109/TEMC.2004.842112.

[156] M.E. van Valkenburg, *Reference Data for Engineers: Radio, Electronics, Computer and Communication*, 8th ed., Newnes, 1998.

[157] P. Volegov, A.N. Matlachov, M.A. Espy, J.S. George and R.H. Kraus Jr., "Simultaneous Magnetoencephalography and SQUID Detected Nuclear MR in Microtesla Magnetic Fields," *Magnetic Resonance in Medicine*, Vol. 52, 2004, pp. 467-470. doi: 10.1002/mrm.20193.

[158] B.C. Wadell, *Transmission Line Design Handbook*, Artech House, Norwood, MA, USA, 1991.

[159] K. Wiklundh, "Relation between the amplitude probability distribution of an interfering signal and its impact on digital radio receivers," *IEEE Transactions on Electromagnetic Compatibility*, Vol. 48, no. 3, pp. 537-544, Aug. 2006. doi: 10.1109/TEMC.2006.877782.

[160] P.F. Wilson and M.T. Ma, "Fields Radiated by Electrostatic Discharges," *IEEE Transactions on Electromagnetic Compatibility*, Vol. 33, no. 1, Feb. 1991, pp. 10-18. doi: 10.1109/15.68245.

[161] Wu Xi-xiu, Li Zhen-Biao, Tian Yun, Mao Wenjun, Xie Xun, "Investigate on the Simulation of Black-box Arc Model," 1st IEEE International Conference on Electric Power Equipment – Switching Technology, Xi'an, China, Oct. 23-27, 2011, pp. 629-636. doi: 10.1109/ICEPE-ST.2011.6123163.

Index

It is really quite impossible to say anything with absolute precision, unless that thing is so abstracted from the real world as to not represent any real thing.

Richard Feynman